Nanotechnologies and Clusters in the Spaces of Higher Dimension:

Emerging Research and Opportunities

Gennadiy Vladimirovich Zhizhin
Independent Researcher, Russia

A volume in the Advances in Chemical and Materials Engineering (ACME) Book Series

IGI Global
PUBLISHER of TIMELY KNOWLEDGE

Published in the United States of America by
IGI Global
Engineering Science Reference (an imprint of IGI Global)
701 E. Chocolate Avenue
Hershey PA, USA 17033
Tel: 717-533-8845
Fax: 717-533-8661
E-mail: cust@igi-global.com
Web site: http://www.igi-global.com

Library of Congress Cataloging-in-Publication Data

Names: Zhizhin, G. V. (Gennadiĭ Vladimirovich), author.
Title: Nanotechnologies and clusters in the spaces of higher dimension : emerging research and opportunities / by Gennadiy Vladimirovich Zhizhin.

Description: Hershey, PA : Engineering Science Reference, an imprint of IGI Global, [2021] | Includes bibliographical references and index. | Summary: "This book examines nanotechnologies and clusters in the spaces of higher dimension"-- Provided by publisher.
Identifiers: LCCN 2019059915 (print) | LCCN 2019059916 (ebook) | ISBN 9781799837848 (hardcover) | ISBN 9781799851653 (paperback) | ISBN 9781799837855 (ebook)
Subjects: LCSH: Nanotechnology--Mathematics. | Quasicrystals. | Cluster analysis.
Classification: LCC T174.7 .Z445 2020 (print) | LCC T174.7 (ebook) | DDC 620/.50151953--dc23
LC record available at https://lccn.loc.gov/2019059915
LC ebook record available at https://lccn.loc.gov/2019059916

This book is published in the IGI Global book series Advances in Chemical and Materials Engineering (ACME) (ISSN: 2327-5448; eISSN: 2327-5456)

British Cataloguing in Publication Data
A Cataloguing in Publication record for this book is available from the British Library.

Advances in Chemical and Materials Engineering (ACME) Book Series

ISSN:2327-5448
EISSN:2327-5456

MISSION

The cross disciplinary approach of chemical and materials engineering is rapidly growing as it applies to the study of educational, scientific and industrial research activities by solving complex chemical problems using computational techniques and statistical methods.

The **Advances in Chemical and Materials Engineering (ACME) Book Series** provides research on the recent advances throughout computational and statistical methods of analysis and modeling. This series brings together collaboration between chemists, engineers, statisticians, and computer scientists and offers a wealth of knowledge and useful tools to academics, practitioners, and professionals through high quality publications.

COVERAGE

- Chemical Engineering
- Coatings and surface treatments
- Fatigue and Creep
- Metallic Alloys
- Polymers Engineering
- Thermo-Chemical Treatments
- Wear of Materials
- Materials to Renewable Energies
- Heat Treatments
- Computational methods

IGI Global is currently accepting manuscripts for publication within this series. To submit a proposal for a volume in this series, please contact our Acquisition Editors at Acquisitions@igi-global.com or visit: http://www.igi-global.com/publish/.

The Advances in Chemical and Materials Engineering (ACME) Book Series (ISSN 2327-5448) is published by IGI Global, 701 E. Chocolate Avenue, Hershey, PA 17033-1240, USA, www.igi-global.com. This series is composed of titles available for purchase individually; each title is edited to be contextually exclusive from any other title within the series. For pricing and ordering information please visit http://www.igi-global.com/book-series/advances-chemical-materials-engineering/73687. Postmaster: Send all address changes to above address.

Titles in this Series

For a list of additional titles in this series, please visit:
http://www.igi-global.com/book-series/advances-chemical-materials-engineering/73687

Inspiration and Design for Bio-Inspired Surfaces in Tribology Emerging Research and Oportunities
Hisham Abdel-Aal (Drexel University, USA)
Engineering Science Reference • ©2021 • 367pp • H/C (ISBN: 9781799816478) • US $195.00

Advanced Applications of 2D Nanostructures
Subhash Singh (National Institute of Technology Jamshedpur, India) and Kartikey Verma (Indian Institute of Technology, Kanpur, India)
Engineering Science Reference • ©2020 • 300pp • H/C (ISBN: 9781799829911) • US $225.00

Innovative Applications of Layered Materials and Structures
Sarmila Sahoo (Heritage Institute of Technology, Kolkata, India)
Engineering Science Reference • ©2020 • 300pp • H/C (ISBN: 9781799835875) • US $225.00

Using Lasers as Safe Alternatives for Adhesive Bonding Emerging Research and Opportunities
Barbara Ewa Ciecińska (Rzeszow University of Technology, Poland)
Engineering Science Reference • ©2020 • 248pp • H/C (ISBN: 9781799846345) • US $175.00

New Challenges and Industrial Applications for Corrosion Prevention and Control
Younes El Kacimi (Ibn Tofail University, Morocco) Savas Kaya (Cumhuriyet University, Turkey) and Rachid Touir (Ibn Tofail University, Morocco)
Engineering Science Reference • ©2020 • 300pp • H/C (ISBN: 9781799827757) • US $235.00

For an entire list of titles in this series, please visit:
http://www.igi-global.com/book-series/advances-chemical-materials-engineering/73687

IGI Global
PUBLISHER of TIMELY KNOWLEDGE

701 East Chocolate Avenue, Hershey, PA 17033, USA
Tel: 717-533-8845 x100 • Fax: 717-533-8661
E-Mail: cust@igi-global.com • www.igi-global.com

Dedicated to Tatiana Fedorova,
who inspired me to these studies all these years!

Table of Contents

Preface

Nanometer 1 - 200 nm measurement range opens up new properties and approaches to the study of substances. Currently, a large amount of scientific literature is devoted to the study of nano -objects (Gubin, 2019; Haberland, 1994; Gusev & Rempel, 2000; Suzdalev, 2005). Nanotechnology includes both research on the nano - objects themselves (in particular, clusters), and methods for their preparation, as well as methods for producing nanomaterials consisting of a set of clusters. It is difficult to list already obtained nanoclusters and nanomaterials. Nanomaterials can have one, two, or three dimensions, which adds a variety of phenomena that can be studied in nanoscience and technology (Pan et al., 2014; The different dimensions of nanotechnology, 2009). In this paper, it will be shown that nanoclusters often have dimensions greater than three. Therefore, nanomaterials can also have a higher dimension. This opens up new prospects for creating nanomaterials with unique properties. Nanotechnology is already playing a significant role in the processes of obtaining new substances and in their practical use in various fields of human activity. One of the surprising properties of the nano - crystalline state discovered in recent decades is the discovery of the existence of quasicrystals in which there is no translational symmetry (Shechtman et al., 1984; Janssen et al., 1984; Shechtman & Blech, 1985; Pauling, 1987; Cratias & Cahn, 1986; Janssen et al., 2007). It was not possible to explain the observed differences from classical crystals with translational symmetry within the framework of three - dimensional Euclidean geometry. However, the author of this work as a result of analyzing experimental diffraction patterns of quasicrystals (intermetallic compounds) Al_6 Mn (Shechtman et al., 1984), Al_{72} Ni_{20} Co_8 (Eiji Abe et al., 2004), Al_{70} Fe_{20} W_{10} (Mukhopadhyay et al., 1993), Ti_{54} Zr_{26} Ni_{20} (Zhang & Kelton, 1993) found that the absence of translational symmetry in quasicrystals is only apparent. If we assume that the diffraction pattern of quasicrystals is a projection of a crystal lattice from a space of higher dimensionality, then a translational symmetry appears (Shevchenko, Zhizhin & Mackay, 2013; Zhizhin, 2014a, b, c; Zhizhin & Diudea, 2016).

There are abstract methods for describing clusters, strictly speaking, not related to specific chemical compounds (Diudea & Nagy, 2007; Ashrafi, Cataldo, Iraumanesh & Ori, 2013). In these works, proceeding from the well - known three - dimensional polyhedrons of Plato and Archimedes, they are transformed by various operations: construction of a polyhedron by the midpoints of edges, truncation, construction of a dual polyhedron, adding vertices of a polyhedron, etc. At the same time, such transformations are in no way connected with real chemical compounds. It is noted that adding vertices to a polyhedron can lead to an increase in its dimension. It should be noted here, and it is significant that if the original polyhedrons corresponded to some specific chemical compounds with certain chemical bonds between the atoms and the edges corresponding to them, then the transformation of these polyhedrons leads, generally speaking, to the elimination of these chemical bonds. At the same time, it remains unclear what chemical bonds exist in the transformed bodies and what chemical compounds they correspond to. In addition, consideration of the transformed bodies is carried out in a certain abstract space, as if "forgetting" about the real dimension of chemical compounds (MacMullen & Schulte, 2002). This cluster research direction is most clearly formulated in a generalizing monograph of M.V. Diudea "Multi – shell Polyhedral Clusters" (Diudea, 2018). In the preface to it, it is directly emphasized that the cluster models built in it are not associated with specific crystallographic objects: real crystals, networks, and quasicrystals.

In the monograph "The Structure of Chemical Compounds and Higher Dimensional Molecules: Emerging Research and Opportunities" (Zhizhin, 2018), it was shown that most molecules have a higher dimension if their atoms (or functional groups) are considered as the vertices of a convex polytope. Clusters, as larger chemical compounds than molecules, naturally therefore may also have a higher dimension. At present, the question of determining the structural principles of nature, in which clusters of atoms and molecules appear, is important. The literature presents known concrete clusters of chemical compounds (see, for example, Lord, Mackay & Ranganathan, 2006). However, the dimension of the clusters in accordance with the Euler – Poincaré equation (Poincaré, 1895) was not determined, but a priori was assumed to equal three.

The characteristics of higher-dimensional polytopes corresponding to these molecules have been studied in monograph *The Geometry of Higher: Dimensional Polytopes* (Zhizhin, 2019a). It first posed the question of the importance of studying the incidence of elements of different dimensions in these polytopes. The question of the formation of polytopic prismahedrons from these polytopes capable of splitting *n* - dimensional spaces without gaps is investigated. This solves 18 Hilbert's problem (Hilbert, 1901; Delone, 1969) in the space of higher dimension.

Nanotechnology issues are important in modern medicine and biology. The monograph *Attractors and Higher Dimensions in Population and Molecular Biology: Emerging Research and Opportunities* (Zhizhin, 2019b) proved the necessity to take account of higher dimensions in the analysis of biological processes in living organisms. It is shown that the transmission of inherited traits occurs in objects of very high dimension.

In this monograph, starting with the study of spaces of higher dimensionality of intermetallic compounds (Chapter 1), clusters of intermetallic compounds (ligand - free clusters) are sequentially considered (Chapter 2), then chemical compounds of the metal atom with an environment of non - metal atoms (Chapter 3). After that, linear chains of metal atoms are considered with ligands attached to them (Chapter 4). Chapter 5 is devoted to the analysis of chemical compounds of a curved chain of metal atoms with ligands, and Chapter 6 deals with clusters of metals with ligands at the polyhedral core of metal atoms. It is significant that in all cases, and this is the principal novelty of the study, convex models of clusters are built and the dimension of the corresponding polytopes is determined. Moreover, it turned out that in all cases the dimension of clusters is higher three. Thus, clusters have a higher dimension. In this regard, the question arises about the principles of constructing clusters of higher dimension. In Chapter 7, a new law is formulated and proved: the law of the preservation of incidents in higher dimension polytopes. The law states that the total sum of incidents from elements of higher dimension to elements of lower dimension is equal to the total sum of incidents from elements of higher dimension to elements of lower dimension. Moreover, with an increase in the dimension of the polytopes, this amount increases dramatically. The incidence of natural elements of different dimensions can be interpreted as the exchange of information between elements. It follows from the law, for example, that in biological objects (DNA and RNA molecules responsible for heredity) there is a very intensive exchange of information between the components, which apparently provides a hereditary process.

The last two chapters (8 and 9) of the book are devoted to the study of the mathematical construction of nanomaterials from clusters of a higher dimension. Chapter 8 discusses the methods of transition from a single cluster of a higher dimension to a larger cluster with a higher dimension, which includes the original cluster as a component. Chapter 9 addresses the issue of extended (in principle infinite) nanomaterials from clusters. This question is mathematically formulated as partitions of n - dimensional spaces with the participation of clusters a lower dimension. Thus, the problem of nanotechnology for producing clusters and nanomaterials is solved mathematically using spaces of higher dimension.

Thus, one can assume that the process of obtaining quasicrystals (intermetallic compounds), associated with high thermal stresses, leads to the creation of spaces of higher dimensionality in these materials. Along with the translational filling of the higher dimension space in the diffraction patterns of quasicrystals, one can see the hierarchical filling of the space around its arbitrary point (Zhizhin, 2012). Quasicrystals can be considered as cluster aggregates (Eiji, Yanfa & Pennycook, 2004). All of them can be the bases of polytopic prismahedrons without changing their chemical structure (Zhizhin, 2018). Thus, polytopic prismahedrons, and not stereohedrons, of which dreamed B.N. Delone (Delone, 1961; Delone & Sandakova, 1961), solve the eighteenth Hilbert problem about constructing a space from congruent figures in the case of n - dimensional space. The fact is that B.N. Delone did not succeed in citing at least one example of a specific type of stereohedron that provided the partition of n - dimensional space. The direct construction of the partition of the n - dimensional space by polytopic prismahedrons (Zhizhin, 2015, 2019a), showed the impracticability of the contact conditions of polytopes (stereohedrons) adopted in the theory of B.N. Delone. The polytopic prismahedrons in the partition of n - dimensional space can contact each other along faces of different dimensions, and not only along faces of dimension n - 1, as is customary in the theory of B.N. Delone.

Due to the need for modern analysis of the structure of materials based on the use of clusters of chemical compounds in this book:

1. A geometric analysis of known clusters of chemical compounds is carried out with the determination of their dimensions.
2. Various types of clusters are systematized in general form.
3. Polytopic prismahedrons with bases in the form of clusters of a higher dimension are constructed and investigated.
4. Partitions of n - dimensional spaces with a polytopic prismahedrons face into a face with bases in the form of clusters of a higher dimension are constructed and investigated.

Based on the research conducted, a new paradigm of the discrete world is formulated. In it take account of the discoveries of quasicrystals, scaling processes in materials (Kadanoff, 1966; Wilson, 1971a, b), second - order phase transition processes (Landau, 1937; Zhizhin, 2014e), the formation of fractals in solids (Mandelbrott, 2007; Zhizhin, 2014a, b; Zhizhin, 2018), the highest dimension of chemical compounds (Zhizhin, 2014d, 2018). These discoveries do not fit into the paradigm of the discrete world, advanced by B.N. Delone in 1937 (Delone, 1937; Galiulin, 2003), since the basic principles of this paradigm contradict the listed discoveries.

The results of the work can be used to study the geometry of dissipative structures in n - dimensional space (Zhizhin, 2010).

This monograph completes the cycle of works by the author of four monographs (Zhizhin, 2018, 2019a, b) devoted to the study of the influence of higher dimensionality on natural processes in nature. These works create a new scientific basis, which should change our understanding of nature and lead to the modernization of all natural science disciplines: mathematics, physics, chemistry and biology. A number of the results obtained in these monographs already testify to the new opportunities opening up for using the higher dimensional accounting in the near world surrounding us. The continuation of this work is waiting for its researchers.

This monograph naturally partially uses the materials of the author's previous monographs (Zhizhin, 2018, 2019a, b) in which the geometry of individual molecules of chemical compounds was mainly considered. The material used is necessary for understanding the geometry of clusters of higher dimension, which is the main goal of this study.

Gennadiy Vladimirovich Zhizhin
Independent Researcher, Russia

REFERENCES

Ashrafi, A. R., Cataldo, F., Iranmanesh, A., & Ori, O. (Eds.). (2013). *Topological Modelling of Nanostructures and Extended Systems*. Springer. doi:10.1007/978-94-007-6413-2

Delone, B. (1937). Geometry of positive quadratic forms. *Advances in Mathematical Sciences.*, *3*, 16–62.

Delone, B. (1961). Proof of the main theorem of the theory of stereohedrons. *Reports of the Academy of Sciences of the USSR*, *138*(6), 1270–1272.

Delone, B. (1969). To the eighteenth problem of Hilbert. In P. Aleksandrov (Ed.), *Problems of Hilbert* (pp. 200–203). Moscow: Science.

Delone, B., & Sandakova, N. (1961). Theory of stereohedrons. *Proceedings of the Mathematical Institute. V.A. Steklov, 64*, 28 – 51.

Diudea, M. V. (2018). *Multi – shell Polyhedral Clusters*. Springer. doi:10.1007/978-3-319-64123-2

Diudea, M. V., & Nagy, C. L. (2007). *Periodic Nanostructures*. Springer. doi:10.1007/978-1-4020-6020-5

Eiji, A., Yanfa, Y., & Pennycook, S. J. (2004). Quasicrystals as cluster aggregates. *Nature Materials*, *3*(11), 759–767. doi:10.1038/nmat1244 PMID:15516956

Galiulin, R. (2003). Systems B.N. Delone as the mathematical foundation of a discrete world. *J. Computed. Mat. and Math. Physics*, *43*(6), 790–801.

Gratias, D., & Cahn, J. W. (1986). Periodic and quasiperiodic crystals. *Scripta Metallurgica*, *20*(9), 1193–1197. doi:10.1016/0036-9748(86)90030-X

Gubin, S. P. (1987). *Chemistry clusters*. Moscow: Science.

Gusev, A. I., & Rempel, A. A. (2000). *Nanocrystalline materials*. Moscow: Fismatlit.

Haberland, H. (Ed.). (1994). *Clysters of atoms and molecules*. Berlin: Springer-Verlag.

Hilbert, D. (1901). Article. *Gesammelte Abhandlungen*, *3*(1), 44 – 63, 213 – 237.

Janssen, T., Chapuis, G., & De Boissieu, M. (2007). *Aperiodic Crystals. From Modulated Phases to Quasicrystals*. Oxford: Oxford University Press. doi:10.1093/acprof:oso/9780198567776.001.0001

Kadanoff, L. P. (1966). Scaling Laws for Ising Models Near τ_c^*. *Physics*, *2*(6), 263–272. doi:10.1103/PhysicsPhysiqueFizika.2.263

Landau, L. D. (1937). To Theory of Phase Transitions. *Journal of Experimental and Theoretical Physics*, *7*, 19–38.

Lord, E. A., Mackay, A. L., & Ranganathan, S. (2006). *New Geometry for New Materials*. Cambridge: Cambridge University Press.

Mandelbrott, B. B. (1982). *The Fractal Geometry of Nature*. New York: Freeman.

McMullen, P., & Schulte, P. (2002). *Abstract Regular Polytopes*. Cambridge Univ. Press. doi:10.1017/CBO9780511546686

Mukhopadhyay, N. K. (1993). Diffraction studies of icosahedral phases in $Al_{70}Fe_{20}W_{10}$. *Journal of Non - Crystalline Solids, 153 – 154*, 1193 – 1197.

Pan, Y. T., Yin, X., Kwok, K. S., & Yang, H. (2014). Higher - Order Nanostructures of Two-Dimensional Palladium Nanosheets for Fast Hydrogen Sensing. *Nano Letters, 14*(10), 5953–5959. doi:10.1021/nl502969g PMID:25198201

Poincare`, A. (1895). Analysis situs. *J. de é Ecole Polytechnique, 1,* 1 – 121.

Pouling, L. (1987). So - called Icosahedral Decagonal Quasicrystals Are Twins of an 820-atom cubic crystal. *Physical Review Letters, 58*(4), 365–368. doi:10.1103/PhysRevLett.58.365 PMID:10034915

Shechtman, D., & Blech, I. (1985). Article. *Trans. Ser. A., 16*, 1005–1009.

Shechtman, D., Blech, I., Gratias, D., & Cahn, J. (1984). Metallic phase with longer angle orientational order and no translational symmetry. *Physical Review Letters, 53*(20), 1951–1953. doi:10.1103/PhysRevLett.53.1951

Shevchenko, V., Zhizhin, G., & Mackay, A. (2013). On the structure of the quasicrystals in the high dimension space. In M. V. Diudea (Ed.), *Diamond and Related Nanostructures*. Dordrecht: Springer. doi:10.1007/978-94-007-6371-5_17

Suzdalev, I. P. (2005). *Nanotechnology. Physical chemistry of nanoclusters, nanostructures and nanomaterials*. Moscow: URSS.

The different dimensions of nanotechnology. (2009). *Natura Nanotechnology, 4*, 135.

Wilson, R. G. (1971b). Renormalization Group and Critical Phenomena. II. Phase – space Cell Analysis of Critical Behavior. *Phys. Rev. B., 4*(9), 3184–3205. doi:10.1103/PhysRevB.4.3184

Wilson, R. G.Renormalization Group and the Kadanoff Scaling Picture. (1971a). Renormalization Group and Critical Phenomena. I. Renormalization Group and the Kadanoff Scaling Picture. *Phys. Rev. B.*, *4*(9), 3174–3183. doi:10.1103/PhysRevB.4.3174

Zang, X., & Kelton, K. (1993). High - order crystal approximant alloys $Ti_{54}Zr_{26}Ni_{20}$. *Journal of Non-Crystalline Solids*, *153-154*, 114–118. doi:10.1016/0022-3093(93)90325-R

Zhizhin, G. V. (2010). *Geometrical bases of the dissipative structures*. St. Petersburg: Polytechnic.

Zhizhin, G. V. (2012, October). *Hierarchical filling of spaces with polytopes.* Paper presented at "St. Petersburg Scientific Forum: Science and Human Progress". 7th St.- Petersburg meeting of Nobel Prize laureates, St. Petersburg, Russia.

Zhizhin, G. V. (2014a). The fractal nature of disproportionate phases. In *10th all-Russian Scientific School "Mathematical research in the natural sciences"*. Apatity: Geological institute KSC RAS.

Zhizhin, G. V. (2014b). Incommensurable and fluctuating structures in the terrestrial space. *Biosphere*, *3*, 211–221.

Zhizhin, G. V. (2014c). *World – 4D*. St. Petersburg: Polytechnic Service.

Zhizhin, G. V. (2014d). On the higher dimension in nature. *Biosphere*, *6*(4), 313–318.

Zhizhin, G. V. (2014e). Phase transitions of the second kind with a fluctuation of the geometric structure. In *10th All-Russian Scientific School "Mathematical research in the natural sciences."* Geological institute KSC RAS.

Zhizhin, G. V. (2015, November). *Polytopic prismahedrons – fundamental regions of the n-dimension nanostructures.* Paper presented at The International conference "Nanoscience in Chemistry, Physics, Biology and Mathematics", Cluj-Napoca, Romania.

Zhizhin, G. V. (2018). *Chemical Compound Structures and the Higher Dimension of Molecules: Emerging Research and Opportunities*. Hershey, PA: IGI Global. doi:10.4018/978-1-5225-4108-0

Zhizhin, G. V. (2019a). *The Geometry of Higher – Dimensional Polytopes*. Hershey, PA: IGI Global. doi:10.4018/978-1-5225-6968-8

Zhizhin, G. V. (2019b). *Attractors and Higher Dimensions in Population and Molecular Biology: Emerging Research and Opportunities*. Hershey, PA: IGI Global. doi:10.4018/978-1-5225-9651-6

Zhizhin, G. V., & Diudea, M. V. (2016). Space of Nanoworld. In M. V. Putz & M. C. Mirica (Eds.), *Sustainable Nanosystems, Development, Properties, and Applications* (pp. 214–236). New York: IGI Global.

Chapter 1
Higher-Dimensional Space of Nanoworld

ABSTRACT

In this chapter, a geometrical model to accurately describe the distribution of light points in diffraction patterns of quasicrystals is proposed. It is shown that the proposed system of parallel lines has axes of the fifth order and periodically repeating the fundamental domain of the quasicrystals. This fundamental domain is 4D-polytope, called the golden hyper-rhombohedron. It consists of eight rhombohedrons densely filling the 4D space. Faces of the hyper-rhombohedron are connected by the golden section; they can be scaled as needed. On this universal lattice of the vertices of the golden hyper-rhombohedrons, famous crystallographic lattices—Bravais, Delone, Voronoi, etc.—can be embedded. On the lattice of the vertices of the golden hyper-rhombohedrons, projections of all regular three-dimensional convex bodies—Plato's bodies—can be constructed.

INTRODUCTION

The consideration of clusters in the space of higher dimension must begin with the presentation of general ideas about the highest dimension of the nanoworld, arising from the geometric analysis of intermetallic diffractograms.

Incommensurate (modulated) structures (Izyumov, 1984) are widespread in nature. They include liquid crystals (chiral smectics), quasi - crystals, intercalated graphite compounds (structure consisting of alternating layers

DOI: 10.4018/978-1-7998-3784-8.ch001

of carbon and layers of metal atoms), hardening alloys, etc. They may also include various types of ore bodies, consisting usually of many layers and distributed inclusions of different shape from different substances. The content of the definition "incommensurate" means that in these structures some basic structure with the translational symmetry can be distinguished and a substructure, which either has no translational symmetry, or has translational symmetry, but its period is incommensurate with the period of translating the basic structure (Gridnev, 1977). In both cases, the overall structure consisting of basic structure and substructure has no translational symmetry. Such structures are the result of different influences on the initial body with translational symmetry – thermal, mechanical, electrical, magnetic ones. The absence of translational symmetry in the incommensurate phase combines them with quasi - crystals, which got its name precisely because of the lack of translational symmetry in their diffraction patterns.

The concept of earth reality space introduced by Vernadsky (1965) and his evaluation of Pierre Curie's (Curie, 1966) principle of dissymmetry as a tool for the study of this space are of considerable interest to both the construction of the theory of rocks, and for the understanding of biological morphogenesis processes (Belousov, 2013). According to Pierre Curie's principle of dissymmetry, for the results of earthly reality phenomena some deviations from mentally attainable symmetry limit are characteristic. Rocks are formed as a result of complex physical, chemical and mechanical processes, in this regard Landau's (Landau, 1937) theory of phase transitions, linking the decrease of order of the symmetry group of a substance occurring in it under the influence of temperature and mechanical external effects is of great interest (Zhizhin, 2014e). One important consequence of this theory of Landau discovered in recent decades is the experimental proof of the existence of incommensurate phases. The decrease of the order of symmetry group is also characteristic of living organisms in the course of their development (Belousov, 2013), these changes being also associated with phase transitions as a stage in their development. Thus, it can be claimed that the decrease of order of symmetry is a general property of natural bodies being in the space of earthly reality (Zhizhin, 2014b). The most studied (both experimentally and theoretically) are the issues of crystalline solids structure changing. Therefore, one turn to studies of incommensurate crystalline solids structures taking into account the modern concepts of generalized crystallography (Lord, Mackay, & Ranganathan, 2006; Shevchenko, Zhizhin, & Mackay, 2013a, b; Zhizhin, 2014d; Zhizhin, & Diudea, 2016; Zhizhin, 2018).

Finding in 1982 of ordered structures deprived of translational symmetry (Shechtman et al., 1984), next called "quasicrystals", had marked the beginning of numerous cycles of papers and books devoted to the experimental and theoretical study of these unusual materials. The main problem was to prove that the lack of translational symmetry does not contradict the existence of the crystal, even it is mandatory in the classical crystallography. It was noted (Janssen et al., 2007) that the absence of visible strict periodicity of quasicrystals does not mean randomness (i.e. disorder) of such structures. Approximation of quasicrystals can be described in terms of almost - periodic functions (Bohr, 1952). Almost - periodic (or quasi - periodic, aperiodic) functions occur when periodic functions with incommensurable periods (expressed by irrational numbers) are involved. It was shown that, in order to obtain a clear diffraction pattern, it is not necessary to have a strict periodicity of the crystal, but rather a quasi - periodicity. Moreover, the diffraction pattern retains its form by modulation of periodic structures by irrational numbers. However, the quasi - periodic functions do not conflict with the axes of the fifth and tenth - order diffraction patterns, observed in quasicrystals, banned in the strictly periodic structure in Euclidean three – dimensional space. The presence of these axes in the diffraction patterns of quasicrystals promoted attempts to explain the structure of quasicrystals by means of convex polyhedrons with the axes of the 5-th order, i.e., icosahedron and dodecahedron (Shechtman & Blech, 1985). Many studies have been devoted to the qualitative explanation of diffraction patterns of quasicrystals by Penrose tiling (Penrose, 1979). However, icosahedrons cannot fill the three - dimensional space without cracks and gaps and therefore the description of the diffraction patterns of quasicrystals using icosahedrons impossible; also, the geometric elements of Penrose rhombic tiling in the diffraction patterns are absent. Pauling (Pauling, 1987) explained the diffraction patterns of icosahedron apparent symmetry by multiple twinning of cubic crystals. He based his arguments only on the radial intensity distribution of the spots of diffraction patterns. However, this model could not explain the high - resolution micrographs and diffraction patterns with the distribution of spots different from those proposed by Pauling (Gratias & Cahn, 1986). Application of expansions of functions in a Fourier row in classical crystallography to describe the direction of rays reflected from a periodic lattice, obtained in recent years continued to describe his spots diffraction patterns of quasicrystals (Janssen et al., 2007). It uses the idea of modulation grids terms in the Fourier row with incomparable periods. However, such modulation may not lead to an accurate strictly ordered arrangement of spots observed in the diffraction patterns.

Figure 1. Electron diffraction pattern of compound Al_6Mn

Figure 1 illustrates the electron diffraction pattern of the cluster Al_6Mn (Shechtman et al., 1984).

The same countenance is shown by the electronic diffraction patterns of many other compounds: $Al_{72}Ni_{20}Co_8$ (Eiji Abe et al., 2004) (Figure 2), $Al_{70}Ni_{20}W_{10}$ (Mukhopadhyay et al., 1993) (Figure 3), $Ti_{54}Zr_{26}Ni_{20}$ (Zhang & Kelton, 1993) (Figure 4). The brightest detailed diffraction pattern was obtained in the work of Eiji Abe et al. (Figure 2). This diffraction pattern will

Figure 2. Electron diffraction pattern of compound $Al_{72}Ni_{20}Co_8$

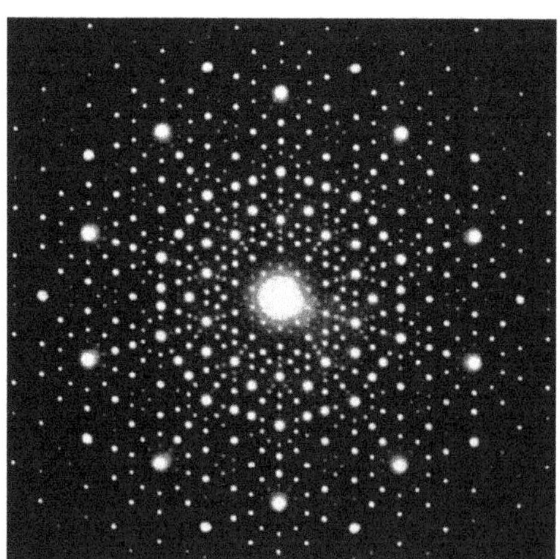

be used in the future for additional geometric constructions characterizing the structure of a quasicrystal. Therefore, this diffraction pattern is shown on a larger scale compared to other ones.

It is seen clearly that the diffraction patterns are at the center of a bright spot correspond to the crash of the electron beam on the sample. Less bright spots of various sizes form a complex geometric pattern away from the center. The main elements of this diffraction patterns are straight lines emanating from the center, forming the axis of rotation of the 10th order, passing through the center perpendicular to the plane of diffraction patterns. The structure of the pattern on the diffraction patterns of all the same. In the vicinity of the center spot, the smaller ones form regular pentagons, adjacent to each other. On external diagonal these pentagons larger pentagons rely on and on their external diagonal rely on yet larger pentagons, etc. An important feature of these structures is to increase the distance between the spots away from the center of the diffraction patterns along the lines passing through the center, as well as the formation of pentagons smaller pentagons into pentagons large size. Spot sizes and the distance between them can vary depending on the experimental conditions, the composition of matter and the used device.

THE FRACTAL NATURE OF INCOMMENURATE PHASES

The system of lines, on which the diffraction patterns spots are located (Figures 1 - 4), includes five families of parallel lines (Zhizhin, 2014a, b, c;

Figure 3. Electron diffraction pattern of compound $Al_{70}Ni_{20}W_{10}$

Figure 4. Electron diffraction pattern of compound $Ti_{54}Zr_{26}Ni_{20}$

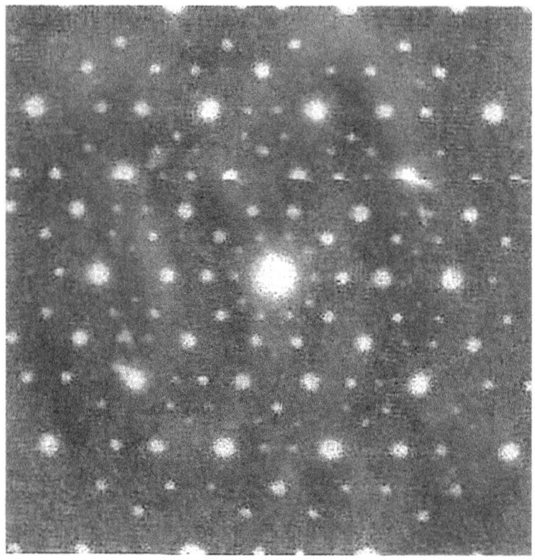

Shevchenko, Zhizhin, & Mackay, 2013a, b) having an inclination angle of $i\times36^o(i=0\div4)$, integer). In each family, lines are spaced apart by a distance *d,* constant within set lines. Let us choose an arbitrary point O on one of the lines of the family $i = 0$ (Figure 5a).

Figure 5a. Representative of families of parallel lines

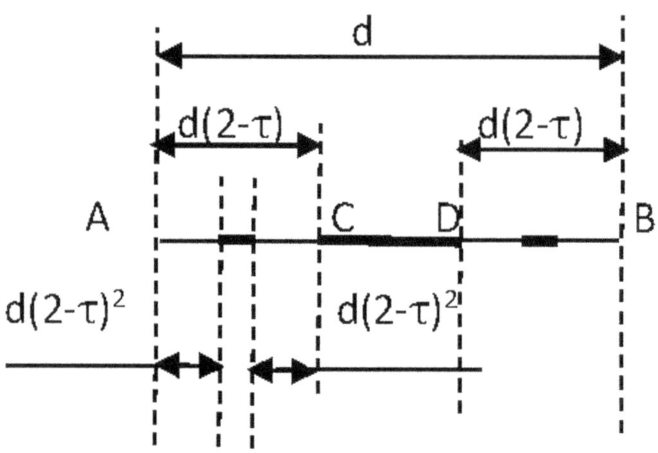

From this point, more four straight lines spread, with angles $i\times36^o(i=0\div4)$). Each of these will provide a family of parallel lines, spaced at a distance *d*.

Figure 5b. Dividing a segment d by a system of parallel lines in the golden section

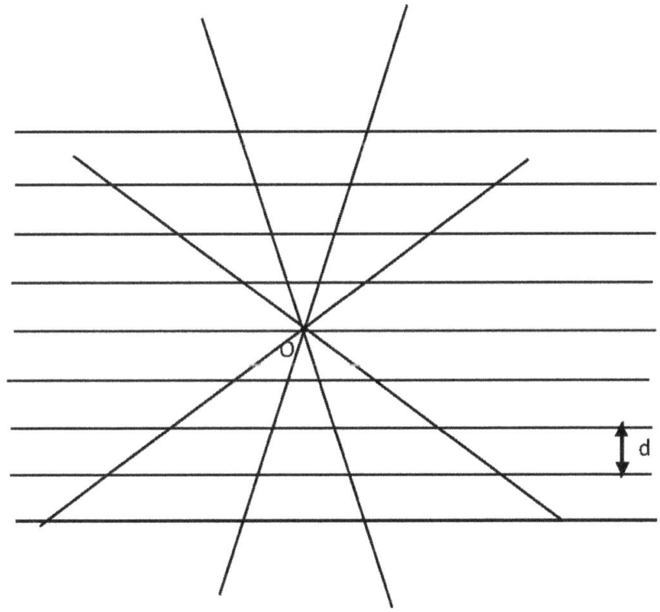

Now step $d = AB$ (Figure 5b) in all of the parallel lines and divide the segment AB into three parts, according to the golden section, by points C and D so that

$$AC = BD = d\,(2 - \tau),$$

$$CD = d\,(2\,\tau - 3),$$

$$\tau = (\sqrt{5} + 1)\,/\,2 = 1.618033988.$$

Through the points C and D on each step in the sequence of all the families of parallel lines, draw lines parallel to the lines of this family. It is obvious that the line passing through the points C, at each step, will form a plurality of parallel lines with a pitch d. Treat similarly the lines passing through the point D. Each of the intervals AC, BD and CD are divided also into three golden sections section and this process can be continued. However, the distance between lines will decrease rapidly and, accordingly, reduce (in the diffraction patterns) the size and intensity of the spots located on these newly constructed lines. The result of the mutual disposition and the intersection of all the families of parallel lines is shown in Figure 6.

Figure 6. Geometric model of electronic diffraction patterns of quasicrystals

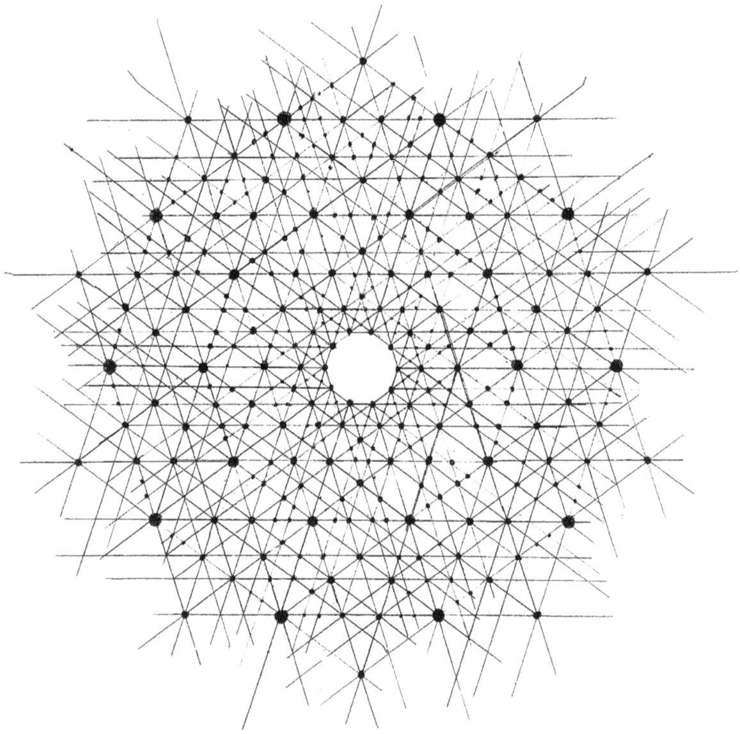

The intersection of these families create a plane system of regular pentagons, of various size, whose vertices superimpose over the spots of the electron diffraction patterns. Thus, the observed diffraction patterns of quasicrystals originate in the system of five families (and subfamilies) of parallel lines lying with respect to each other at the distance *d*, i.e. in there is a (hidden) periodicity in 5 - directions in the plane.

It should be noted that the location of the spots in Figure 6 along any straight line passing through the center of the figure coincides with the location of reflexes (upon irradiation with fast electrons of the surface layers) of In on As, In on Ge, Pb on Si (Galicin et al.,1998; Ichikawa, 1981; Weiering, 1992). Figure 7 gives an example of reflections on the surface layers of In on As (Galicin et al., 1998).

Geometric structure on Figure 6 can be viewed as a hierarchical filling of the dual plane by regular pentagons. Indeed, in the center of each diffraction pattern is clearly visible a regular decagon, which is a result of blending (with rotation) to each other of two regular pentagons. By the way, the Penrose

Figure 7. The XRD pattern of the reflexes of In on As

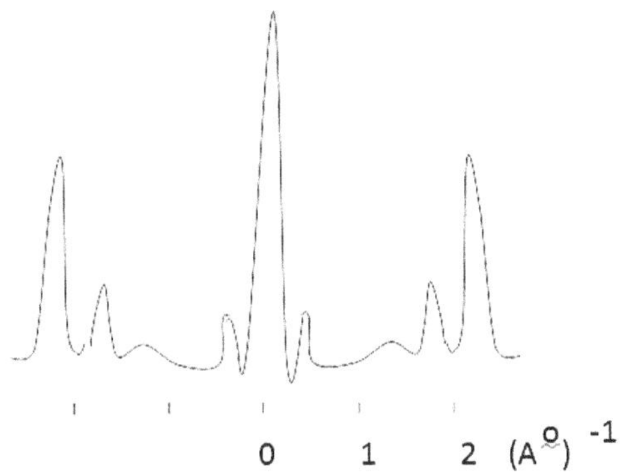

rhombus model of diffraction patterns is not in the center of the regular decagon; this is different from the real diffraction patterns. At a distance from the center of a pentagon, resizing by a geometric progression with a base 1+ τ, results in a hierarchical filling of the plane so that it is filled hierarchical

Figure 8. Fractal from regular pentagons

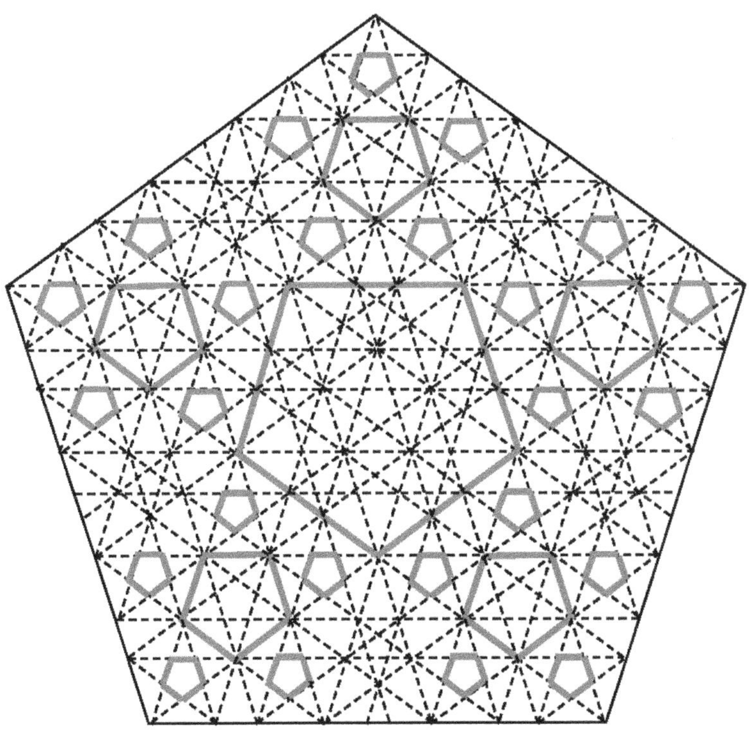

plane (Zhizhin, 2012, 2014a, b, c) in this case a double pentagon. Moreover, such filling can happen, not only for the center of the diffraction patterns, but also for the points uniformly distributed on the diffraction pattern. Regular pentagons form a sectors emanating from the center of the diffraction patterns and similar points, consisting of pentagons, based on the diagonal to each other, so that the size of the parties pentagons increases at geometric progression with the base $1+ \tau / (1+ \tau)$. Since the sequence of line segments division by the golden section leads to the formation of regular pentagons smaller and smaller, it results in a geometric structure of a double fractal. To prove this, one construct a regular pentagon large enough (Figure 8).

Its diagonal form coaxial regular pentagon smaller. Mentally cut out the pentagon (or paint over it with black paint). As can be seen initial pentagon 5 pentagons More of the same size as the central, each having a common vertex with the original pentagon. Diagonals of pentagons formed in each pentagon coaxial with them smaller pentagons. Cut out these pentagons. In each of the pentagons around emitted pentagon has 5 more pentagons smaller, which is also in the future we have to cut. The process can continue indefinitely, so that each of the pentagons emitted around more five smaller pentagons. This process is similar to getting a Sierpinski carpet - fractal geometric figure, which is a square or a triangle (Mandelbrott, 1982; Crownover, 1995). One can calculate the fractal dimension of the regular pentagons. Note that in reducing the size of each side of the pentagon it decreases $(2-\tau)^{-1}$ times. In addition, each number of pentagons downscaling increases by 5 times. Therefore, the fractal dimension D by Hausdorff (1918) can be calculated with the formula:

$$D = \log[f(\lambda x) \,/\, f(x)] \,/\, \log \lambda,$$

where $\lambda = 1 / [(2-\tau)N\,]$ is the scale factor, N is any integer, $f(\lambda x)$ and $f(x)$ are the number of new and the number of initial elements, respectively. Since in this case $f(\lambda x) \,/\, f(x) = 5$, then (Zhizhin, 2014 a, b, c):

$$D = \frac{\log\left[f\left(\dfrac{x}{2-\tau}\right) / f(x)\right]^N}{\log(2-\tau)^{-N}} = \frac{\log 5}{-\log(2-\tau)} = 1.6722756.$$

As already mentioned, there is a regular pentagon double diffraction pattern in the center. Therefore, the geometry of these pentagons (Figure 8) of the diffraction pattern is superimposed on each other rotated by 36°. As opposed to the Sierpinski`s gasket (Crownover, 1995), our considered quasicrystal system can be regarded as a dual irrational ("golden") fractal, of dimension D = 1.6722759 each.

One define the source area of a regular pentagon, which remains after consecutive cut smaller regular pentagons in the "golden" fractal. Let consider the length of the sides of regular pentagons cut off from the initial regular pentagon equal to *d*. Since the length of the sides of decreasing pentagons is given by the golden section (see Figure 5 a), can to find that after each step the scaling of the pentagon side is equal to $d(2-\tau)^N$. Denote by S_0 the original area of the pentagon and by S the area remained from the original pentagon after removing smaller pentagons. Then

$$S = S_0 - \sum_{N=1}^{\infty} 5^N S_N ,$$

$$S_N = \left[d(2-\tau)^N\right]^{2/4} \sqrt{5\left(5+2\sqrt{5}\right)} = S_0(2-\tau)^{2N}$$

is area of the pentagon at the Nth step of the scale. Pulling S_0 in the right side of the equation, can to get

$$S = S_0\left[1-5(2-\tau)^2\left(1+5(2-\tau)^2+5^2(2-\tau)^4+\ldots\right)\right] = \frac{S_0}{1-5(2-\tau)^2}.$$

It is obvious that the formation of fractal regular pentagons is a consequence of repeated division, of step *d* between the lines, on the golden section. In this case, in step *d* a fractal is formed also. Performing the division step *d* in each interval results in two smaller intervals along the edges of the interval *d*, and in the remaining part of the central cut, then the dimension of this fractal, by Hausdorff, is

$$D = \frac{\log\left[f\left(\dfrac{x}{2-\tau}\right)/f(x)\right]^N}{\log(2-\tau)^{-N}} = \frac{\log 2}{-\log(2-\tau)} = 1.6722756.$$

If we assume that $d = 1$, then the cut part of the interval is equal to

$$\sum_{N-1}^{\infty} 2^{N-1}(2\tau - 3)^{N} = \frac{2\tau - 3}{7 - 4\tau} = 0.44721359.$$

Thus, the remaining part of the interval is a little more than half of its original length. This fractal cannot be called the dust of Cantor (Crownover, 1995), whose power is zero; rather it can be called golden dust.

THE SPATIAL STRUCTURE OF QUASICRYSTALS

Since the diffraction pattern is the result of the interaction of the electron beam with the material, the diffraction pattern geometry should be viewed as projections of 3D spatial geometric elements in the 2D plane. However, from 3D space one cannot to receive such projections using convex polyhedrons. Similarly happens with the objects of higher dimensions, particularly the four – dimensional ones. A simple convex correct figure in 4D space is a regular simplex; in projection on the 2D plane, it has the form of a regular pentagon with intersecting diagonals. One can see these pentagons in the diffraction patterns on Figures 1 to 4 and in Figure 6. Thus, a quasicrystal should recognize the four - dimensional body, in which the space is filled with hierarchically regular simplex. Moreover, from each of the right simplex in 4D, as well as from its projection onto the plane 2D, five smaller right simplexes will be formed. Therefore, the regular simplex in the space 4D, as well as their projection on the plane 2D, form doubling fractal with dimension $D = 1.6733759$ each.

Recall that the geometric system of diffraction patterns built by a family of parallel lines is periodically arranged to five directions in the plane 2D. A question arises: what causes this hidden periodicity of the families of parallel lines in 4D space? To answer this question, one turn again to the analysis of the diffraction patterns.

Let choose a rhombus with angles $\gamma = 72^{\circ}$ and $\delta = 108^{\circ}$. Obviously, the plane can be covered by the rhombus without cracks and gaps by translating in two directions x and y parallel to the rhombus edges. In this case, the rhombus vertices form a lattice of nodes with integer values of x and y, if the length of the edge is taken as unity. Through each node in the lattice perpendicular to the plane, the axis of second order is crossed with its rotated by 180°

line to combine the grid to itself. Naturally, each point inside a rhombus or on the edge (side) also is formed the lattice for translation of a rhombus in directions x and y.

If the coordinates of a selected point into the rhombus are

$$x = \alpha, y = \beta, (0 \le \alpha \le 1, 0 \le \beta \le 1),$$

then the coordinates of the new lattice are respectively $\alpha \pm n, \beta \pm n$. Next, on each side of the rhombus defined by the vertices A_1, A_2, A_3, A_4 choose two points, each of which dividing a golden section (Figure 9), i.e.

$$a_1(x = 2 - \tau, y = 0); a_2(x = \tau - 1, y = 0); a_3(x = 2 - \tau, y = 1); a_4(x = \tau - 1, y = 1);$$
$$b_1(x = 0, y = 2 - \tau); b_2(x = 0, y = \tau - 1); b_3(x = 1, y = 2 - \tau); b_4(x = 1, y = \tau - 1);$$
$$\tau = 1.6180...$$

One connect the point A_1 to points a_4,b_4, and point A_3 with points a_1,b_1. Obviously (by symmetry) the points of intersection of lines A_1,a_4,b_1,A_3 and lines A_3,a_1,A_1,b_4 lie on the diagonal of the rhombus A_2,A_4 Line b_1,A_3 is defined by equation $y = x(\tau - 1) + 2 - \tau$, while A_1,a_4 line by $y = x / (\tau - 1)$. The intersection of these lines gives the coordinates of the point

$$a_5 : x_{a_5} = 2 - \tau, y_{a_5} = \tau - 1.$$

Thus, the point a_5 lies on the line a_3a_1 and direct b_2b_4. Similarly, can to find the coordinates of the point a_6, lying on the intersection of the lines a_1A_1 and A_1b_4,

$$x_{a_6} = \tau - 1, y_{a_6} = 2 - \tau.$$

Consequently, the point a_6 lies at the intersection directs a_4a_2,b_1,b_3. Draw further the lines a_3a_1, a_4a_2, b_2,b_4, b_1,b_3 (Figure 9); crossing the line b_1b_3 with line A_1a_4 gives the point a_7 with coordinates

$$x_{a_7} = (\tau - 1)(2 - \tau), y_{a_7} = 2 - \tau;$$

Figure 9. Division sides of the rhombus on the golden section

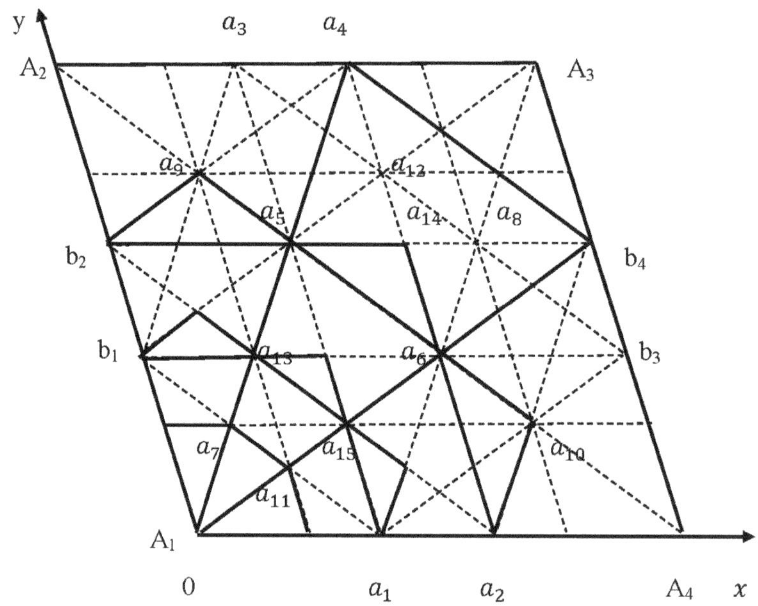

crossing the line b_2,b_4 with the line A_3,a_1, given by the equation

$$\frac{x-1}{y-1}=\frac{x-2+\tau}{y},$$

results in the point a_5 with coordinates

$$x_{a_8}=2(2-\tau), y_{a_8}=\tau-1.$$

Draw now the parallel lines a_3b_1,b_4a_2. The intersection of these lines with a diagonal A_2A_4 gives the points a_9,a_{10}. The equation of the line a_3b_1 is

$$\frac{x-2+\tau}{y-1}=\frac{x}{y-2+\tau};$$

equation of the line b_4a_2 is

$$\frac{x-1}{t-\tau+1}=\frac{x-\tau+1}{y};$$

equation of the line A_2A_4 is $\frac{x}{y-1}=\frac{x-1}{y}$. The point of intersection of the diagonal A_2A_4 with the line a_3b_1 can be found from the system:

$$\frac{x}{y-1}=\frac{x-1}{y},\frac{x-2+\tau}{y-\tau}=\frac{x}{y-2+\tau},$$

i.e. one obtain the coordinates of the point a_9:

$$x_{a_9}=(\tau-1)(2-\tau)=x_{a_7},y_{a_9}=2(2-\tau).$$

The point of intersection with the diagonal line b_4a_2 is found out from the system:

$$\frac{x}{y-1}=\frac{x-1}{y},\frac{x-1}{y-\tau+1}=\frac{x-\tau+1}{y},$$

i.e. we obtain the coordinates of the point a_{10}:

$$x_{a_{10}}=2(2-\tau)=x_{a_8},y_{a_{10}}=2\tau-3.$$

Lines a_3a_1, b_1b_3 intersect in the point a_{13} with coordinates $x_{a_{13}}=y_{a_{13}}=2-\tau$.

Lines a_2a_4, b_2b_4 intersect in the point a_{14} with coordinates $x_{a_{14}}=y_{a_{14}}=\tau-1$.

It is easy to see that the points b_1,a_5,a_6,a_1,A_1 and b_4,a_5,a_6,a_4,A_3 form two (of the same size) regular pentagon with the side 2 - τ. In addition, in Figure 9, one can see a number of regular pentagons with the sides 2τ - 3 and a number of regular pentagons with sides 5 - 3τ.

Note that in the vicinity of vertices A_1 and A_3 the original rhombus is formed by three expanded sectors, consisting of regular pentagons that grow in size so that each subsequent pentagon is built on the diagonal of the previous pentagon (i.e. has edge length equal to the diagonal of the previous pentagon). If connecting points the a_4 with b_4 and a_1 with b_1 one finds that sectors in the vicinity of the vertices A_1A_3 coincide with each other when rotating by 36°. The same two sectors are formed in the vicinity of vertices of A_2,A_4. They coincide with one another by rotation with 36° around the corresponding vertex. It is essential that the described sectors, emanating from each vertex, exist simultaneously in the rhombus. Let now translate the rhombus $A_1A_2A_3A_4$ at coordinates x and y equaling unit; then the vertices of the rhombus on the plane form a lattice. Draw rhombus translation from the basic position one the step to the left, than one the step down and at a time one the step to the left and down. When you translation to the left select sector emanating from the vertex of the rhombus A_4 mainly position; when translated down select sector emanating from the vertex of the rhombus A_2 mainly position; when translated down and left select sector emanating from the vertex of the rhombus A_3 mainly position. The result of these translations is presented in Figure 10.

It can be seen clearly that the point O in Figure 10 has rotation axis of the fifth and tenth order. If one needs to increase the number of translations on the coordinate axes to save the axes of rotation of 5th and 10th order, a system of five families of parallel lines in the initial state of the rhombus can be drawn:

The First Family:

$$A_2A_1, a_3a_1, a_4a_2, A_3A_4, a_9a_{13}, a_8a_{10}.$$

The Second Family:

$$A_4A_1, b_3b_1, b_4b_2, A_3A_2, a_9a_{12}, a_7a_{10}.$$

The Third Family:

$$A_3b_1, b_3a_1, a_4b_2, A_1b_4, a_{14}a_{13}, a_8a_{10}.$$

The Fourth Family:

$$a_4A_1, a_3b_1, b_4a_2, A_3a_1, a_{11}a_{12}, a_{15}a_{14}.$$

The Fifth Family:

$$A_4A_2, a_1b_1, b_2a_2, b_3a_3, a_4b_4.$$

Figure 10. Increasing the number of translations on the coordinate axes to save the axes the axes of rotation of 5th and 10th order in the translated rhombus $A_1A_2A_3A_4$

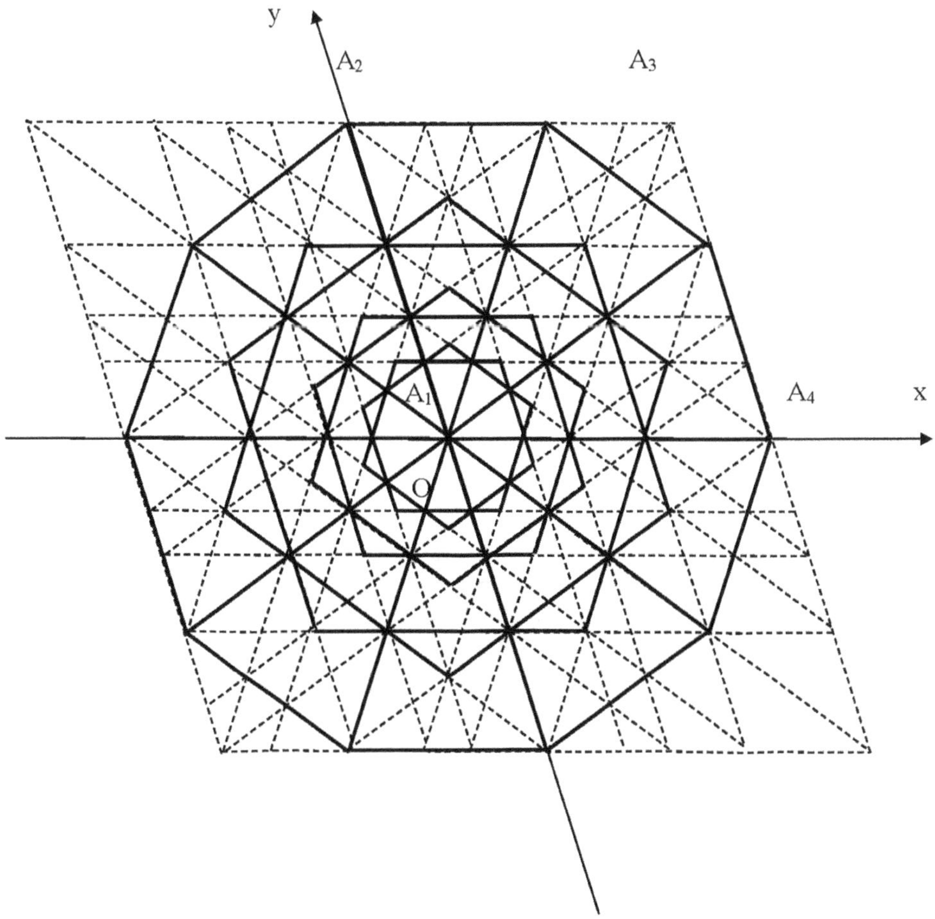

Comparing Figure 10 with the diffraction patterns in Figures 1 to 4, one can see that Figure 10 fully describes the basic elements of the diffraction patterns. A more detailed arrangement of the points marked on the diffraction patterns in Figure 10 regular pentagons one can be obtained using earlier subfamily of parallel lines.

GOLDEN GIPER - ROMBOHEDRON: TRANSLATIONAL BASIS OF QUASICRYSTALS (ZHIZIN, 2014, A, B, C)

What is essential, the grid of families of parallel lines in the four positions of the rhombus in Figure 10, for condition draw lines to subfamilies, has the property of translation (by construction).

The property of translation can be found also in experimental diffraction patterns of quasicrystals (contrary to the claims of Shechtman et al. (1984), for which they were awarded the Nobel Prize). Indeed, Figure 11 shows that through the bright spots of the diffraction pattern in Figure 2, in the vicinity of the central bright spot, a periodic flat grid can be distinguished with nodes coinciding with other bright spots.

Figure 11. Construction of a flat translational lattice on an experimental diffraction pattern of the alloy $Al_{72}Ni_{20}Co_8$

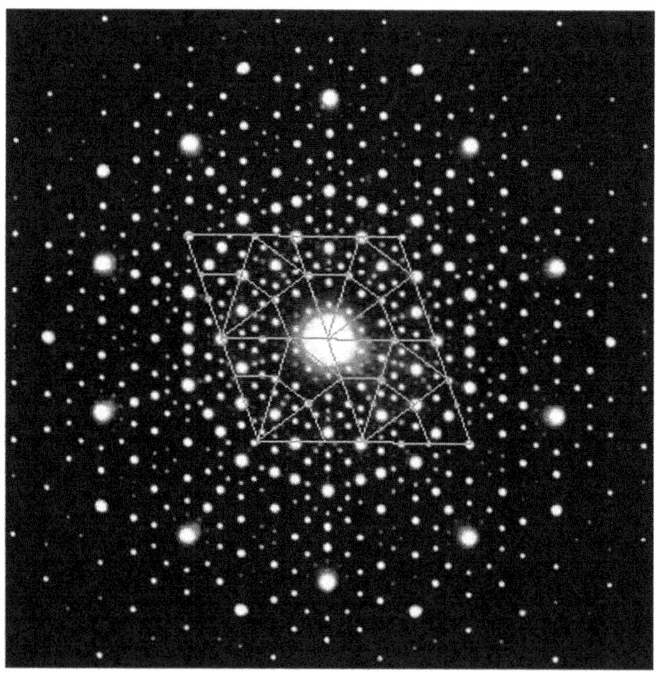

At the same time, the same system of rhombuses allows building an axisymmetric lattice with an axis of the tenth order passing through the center of the diffraction pattern on the experimental diffraction pattern of the alloy $Al_{72}Ni_{20}Co_8$, i.e. through the brightest and largest spot of the diffraction pattern (Figure 12).

Figure 12. Construction of an axisymmetric lattice on an experimental diffraction pattern of the alloy $Al_{72}Ni_{20}Co_8$

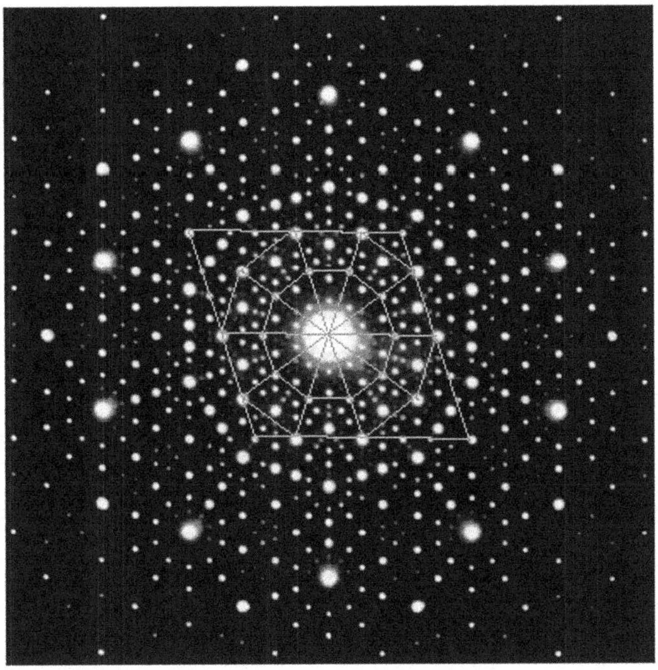

From Figure 12 it follows that the axisymmetric flat lattice is most prominent due to the use of bright spots of the diffraction pattern as nodes. One can say that it is a hierarchical filling of the plane with the original figure in the form of a regular decagon (or a double regular pentagon). With distance from the center of the diffraction pattern, the distance between the nodes is constantly increasing. It should be noted that the axisymmetric lattice requires the use of a smaller number of diffraction pattern spots as compared to the periodic lattice in Figure 11. It is important, however, that these spots are on the diffraction pattern, that is, the diffraction pattern makes it possible to distinguish its translational component. It can be assumed that the axisymmetric component of the diffraction pattern is associated with the axisymmetric conditions of the experiment - the axisymmetric incidence of the beam on the alloy sample.

From a comparison of Figures 11 and 12, it follows that among the luminous spots of diffraction patterns there are spots that are involved in the creation of an axisymmetric lattice and do not participate in the creation of a translational lattice. In order to isolate the translational part from the diffraction pattern, it is necessary to remove from it those spots that ensure only the existence of an axisymmetric lattice. Here it must be remembered that Figure 11 shows the simplest two -dimensional part of the translational lattice. However, it is clear that the lattice must be spatial. To find it, one turn to Figure 10, which represents a diffraction pattern model with all its nodes. Remove from it the nodes that ensure the existence of only an axisymmetric lattice. The remaining nodes belong to the translational lattice (Zhizhin, 2014 c). They are presented in Figure 13.

Figure 13. Translational lattice formed by families of parallel lines diffraction pattern

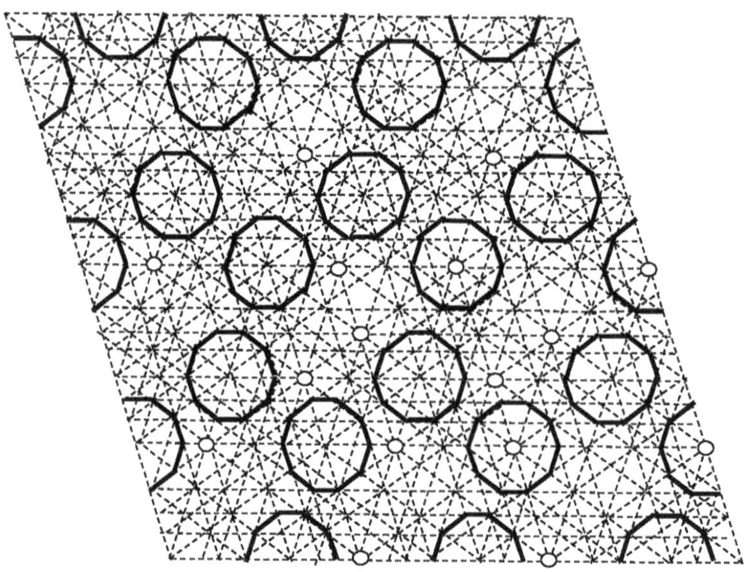

From Figure 13 it can be seen that in the absence of central axial symmetry (in which the lattice, taking into account the spatial dimension, is a hierarchical filling of space with a double simplex with dimension 4), local axes of order 10 are uniformly distributed across the lattice.

The procedure selects from the diffraction patterns of the periodic part of the incommensurate structure of the quasicrystal led to this. Exploring the structure of Figure 13, one can find that it is a projection of the periodic space filling by 4D - polytopes is depicted in Figure 14 (see the outlined circles in Figure 14).

Figure 14. Golden hyper - rhombohedron

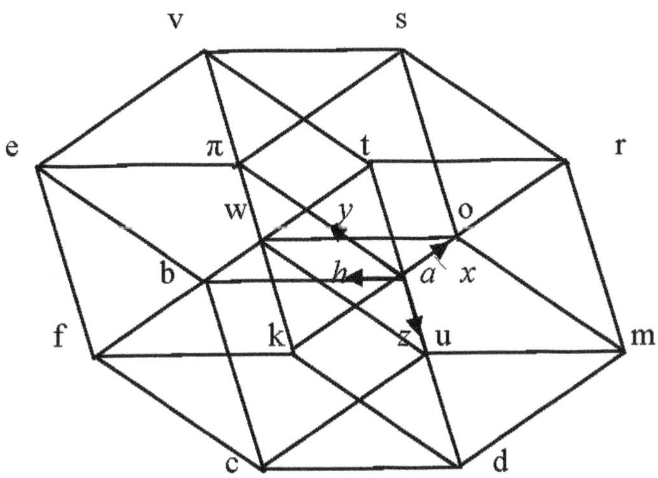

The polytope in Figure 14 consists of 8 – cells (i.e., 3D – polyhedrons) whose faces were built up by the golden section. That is, all two - dimensional faces of the 3D - polytope are the same rhombus, with angles $\varphi = \arccos 1 / 2\tau$, $\gamma = \pi - \varphi$.

Any two 3D - faces (rhombohedrons) share to each other a 2D - face, which is precisely described as a flat rhombus; this is a necessary condition for the existence of a 4D – polytope (Zhizhin, 2018, 2019). In opposition to the 4 – Cube (also having 8 - cells), the set of cells is divided into pairs of rhombohedrons adjacent to each other in two lanes *πauw* and *woab*, whose width is 2 - τ on the length of the edge. Thus, polytope in Figure 14 can be called the "golden hyper - rhombohedron" and assume that the periodic part of the structure of quasicrystals is a translational space filling by 4D golden hyper - rhombohedrons.

Figure 15 shows an example of the location of the golden hyper - rhombohedron on the diffraction pattern of the alloy $Al_{72}Ni_{20}Co_8$.

Interestingly, the coordinates of the vertices of the golden hyper - rhombohedron can be expressed in four - dimensional space in integers. Let`s choose the vertices *a* Figure 14 as beginning of the coordinate system (*x, y, z, h*). Let the axis *x* be directed along the edge *ar*, the axis *y* be directed along

Figure 15. Example of the location of the golden hyper - rhombohedron on the diffraction pattern of the alloy $Al_{72}Ni_{20}Co_8$.

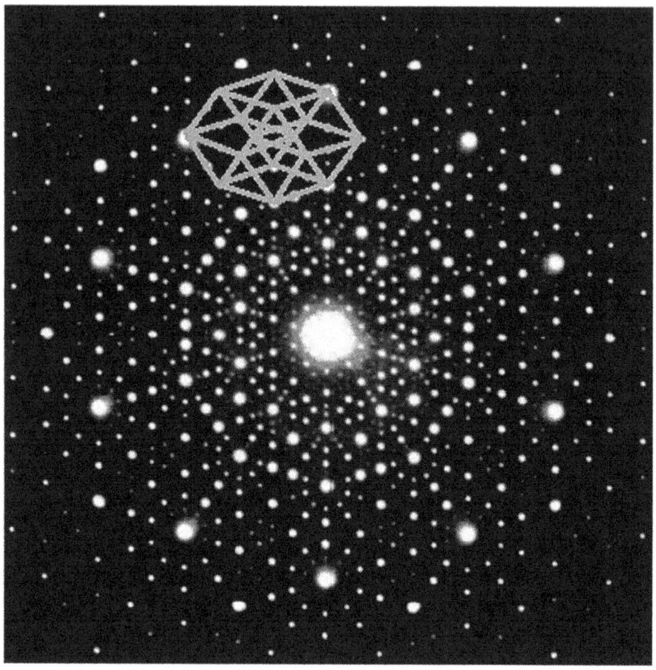

the edges *aπ,* the axis *z* be directed along the edges *ad*, the axis *h* be directed along the edges *ab* (see Figure 14). Then the coordinates of the rhombohedron *aπebdkfc* are expressed in whole numbers: a(0, 0, 0, 0), π(0, 1, 0, 0), e(0, 1, 0, 1), b(0, 0, 0, 1), k(0, 1, 1, 0), f(0, 1, 1, 1), c(0, 0, 1, 1), d(0, 0, 1, 0). To determine the coordinates of the other vertices of the hyper - rhombohedron, one consider the rhombohedron v*twusrom.* From Figure 14 it follows that at the vertex r the value of the variable x is equal to one. In other vertices of this rhombohedron, the value of the variable x cannot change, since the edges of this rhombohedron are not directed along the x coordinate. The coordinates of the vertices of this rhombohedron can to find in Figure 14 taking into account this condition:

v(1, 1, 0, 1), *t*(0, 0, 0, 1), *w*(1, 1, 1, 1), *u*(1, 0, 1, 1),

s(1, 1, 0, 0), *r*(1, 0, 0, 0), *o*(1, 1, 1, 0), *m*(1, 0, 1, 0).

SCALING THE GOLDEN HYPER – RHOMBOHEDRON IN THE HIGHER DIMENSIONAL SPACE

On the diffraction pattern (Figure 15), as an example, the projection of golden hyper - rhombohedron with the edges of a certain size is applied. It is essential that in addition to the already built golden hyper - rhombohedron certain size, on a lattice of vertices can build these polytopes and other dimensions (see. Figure 16), i.e. allows the process of scaling (Zhizhin, 2014a, b, c, e).

Figure 16. Scaling on the lattice of golden hyper - rhombohedron

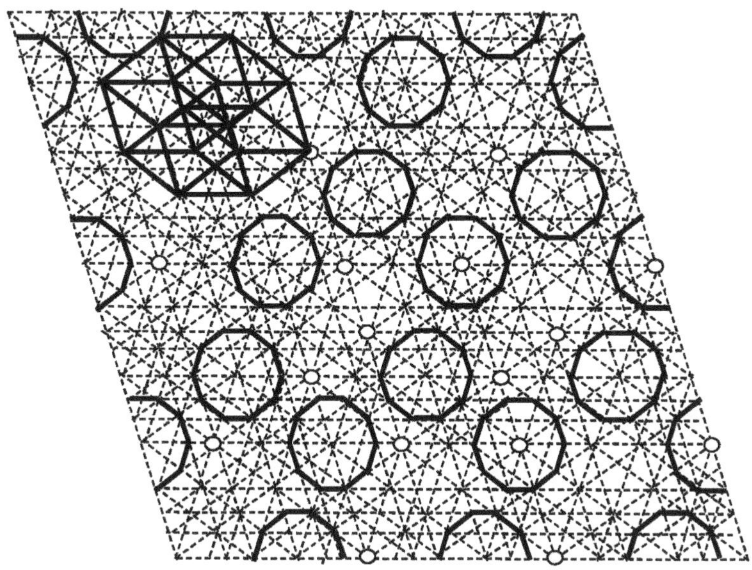

However, in contrast to the ideas of Kadanoff (1966), involving cell merging smaller to larger, here the analysis of lattice vertices golden hyper - rhombohedron it follows that the process of scaling occurs hierarchically, i.e. smaller sell hyper - rhombohedron is not combined with similar, but simply increases in hierarchical mode, absorbing more and more of lattice sites.

In Figure 16, two hyper - rhombohedrons, the smaller one (indicated by thick solid lines) and the larger one (indicated by small circles around the vertices), are in contact with each other along the edges. Moreover, it is clear that the length of the edge of the smaller hyper - rhombohedron is 2 - τ of the length of the edge of the larger hyper - rhombohedron (see Figure 5a). Thus, the dimensions of the hyper - rhombohedrons in the process of scaling are related by the golden section, i.e. the lengths of the edges of hyper - rhombohedrons of different sizes correspond to each other as degrees $(2-\tau)^{\pm n}$, $n = 0, 1, 2, \ldots$ Interestingly, at the vertices of the lattice golden hyper - rhombohedron is possible to construct the projection of the golden hyper - rhombohedron of space even more dimension. Figure 17 is a projection on to the plane 2D golden hyper - rhombohedron of space 5D.

Figure 17. Golden hyper - rhombohedron in space 5D

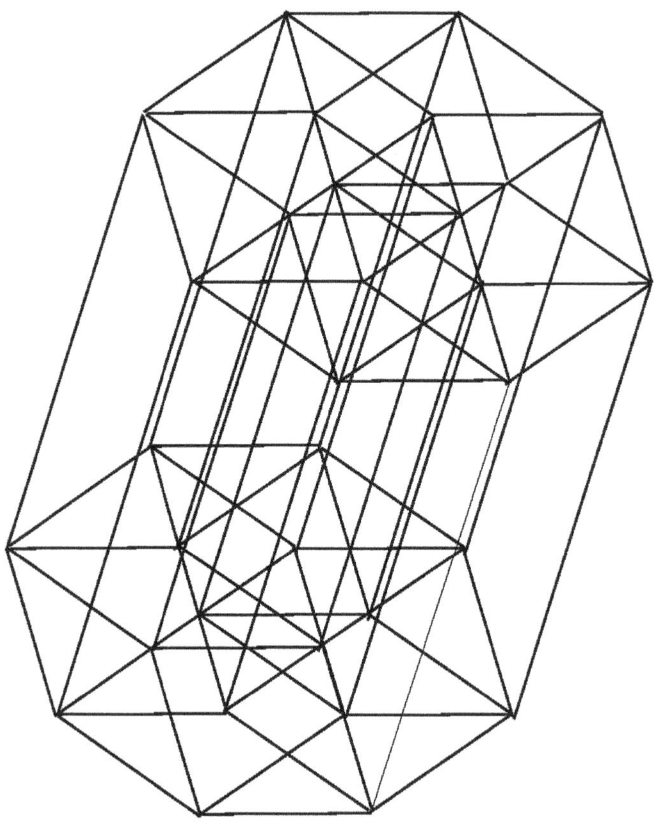

From figure 17 it follows that a hyper - rhombohedron of dimension 5 is a polytopic prism with the bases of a hyper - rhombohedron of dimension 4 (Zhizhin, 2018).

The polytopic prism in Figure 17 is the product of a hyper - rhombohedron 4 D and a segment. One can get the product of a hyper - rhombohedron 4 D on other geometric shapes (triangles, squares, cubes, etc.). This will lead to getting complex lattices with high dimension. The existence of lattices describing the structure of the nanoworld unrelated to the hyper -rhombohedron should not be ruled out. All these lattices can be considered as aggregates of several lattices or several subspaces of different dimensions. Since all the listed products of the shapes are polytopic prismahedrons (Zhizhin, 2015, 2018), they all allow us to construct partitions of *n* - dimensional spaces. The conditions of contact of polytopic prismahedrons in these *n* - dimensional spaces will be studied further.

Figure 18 representing the lattice vertices golden hyper–rhombohedron 4D, as an example, shows how on the vertices of the golden hyper - rhombohedron 4D can obtain projection shapes of Platonic: tetrahedron, cube, octahedron, icosahedron, dodecahedron. Moreover, these figures may have vary in size.

Figure 18. Platonic solids on the lattice golden hyper - rhombohedron

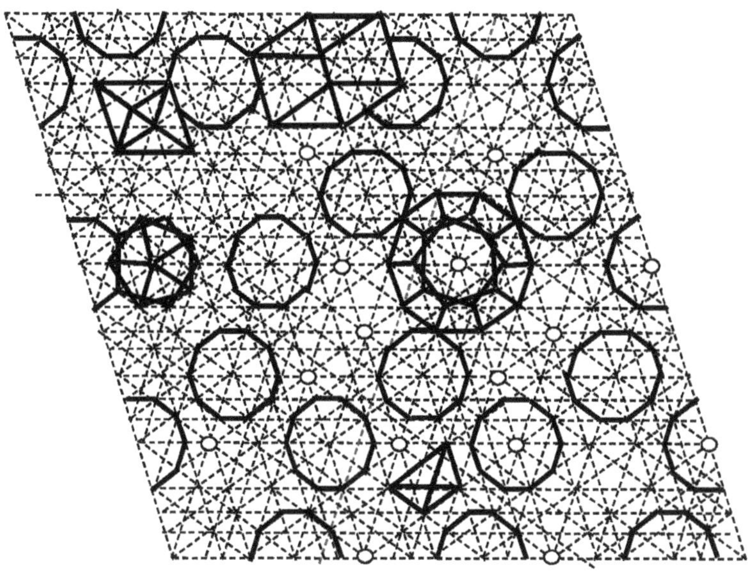

On the grid of vertices golden hyper - rhombohedron can build more complex cells body than Platonic, for example, Archimedean solids, the same as the Bravais cells (Brave, 1974), Delaunay cells and Voronoi cells (Delone, 1937), and even more complex cells. It follows that the golden hyper - rhombohedron have universal value in the description of the process of filling the space of different dimensions.

It is also important to note that on the diffraction patterns of inter metallic one can see not only the latent periodicity of the structure in spaces of higher dimension, but also the hierarchical filling of spaces in the vicinity al almost any point of the structure . It is expressed in an increase in the size of the figures around an arbitrary point of the structure while maintaining the shape of the figure (similar uniform increasing Universe).

Thus, one can conclude that the lattice vertices of golden hyper - rhombohedron in 4D space is flexible enough to describe all kinds of basic cells. The fact that the basis of this lattice is a polytope with irrational ratios characteristic sizes shows the irrationality of the world 4D. This demolish the unfounded allegations of Shevchenko (Shevchenko, 2011) about getting rid of the irrationality of the transition in the space of higher dimension.

CONCLUSION

More Plato suggested that objects of Nature, their quality and operating between the forces, may be the result of exposure to hidden from us the geometric structures (Yau & Nadis, 2010). In recent decades, a shining example of the opening of the geometric structure is string theory. Based on the solutions of Einstein's equations, this theory postulated that the manifold of higher dimension hidden in every point in space (Colabi -Yau manifold of dimension 6). It has vanishingly small size. In the nanoworld, as shown in (Zhizhin, 2014a, b, c, d, e; Zhizhin, 2015; Zhizhin & Diudea, 2016; Zhizhin, Khalaj & Diudea, 2016) is also a hidden dimension but objects (especially molecules), having these dimensions, are tangible sizes. A common basis for understanding the hidden dimensions of the objects of nature, and in string theory, explains the structure of the universe and the structure of elementary particles (Green, 2000; Gepner, 1988), and in the theory of the nanoworld is Riemann geometry (Riemann, 1854), allowing the existence of spaces of higher dimension in the space of lower dimension.

The unit cell of the smallest dimension 4 of the nanoworld structure, determined from the analysis of diffraction patterns of intermetallics, is the golden hyper - rhombohedron. It is a polytopic prismahedron consisting of eight rhombohedrons. On diffraction patterns, cells of a higher dimension can be distinguished, representing works of the golden hyper - rhombohedron and polytopes of arbitrary dimension. Simultaneously with the translational symmetry in the space of the highest dimension in the diffraction patterns of inter-metallics, one can see the central symmetry at each point of the diffraction pattern, like a homogeneous expanding universe.

REFERENCES

Belousov, L. V. (2013). *Symmetry transformations in the development of organisms. In Morphogenesis in individual and historical development: symmetry and asymmetry. A series of "Geo - biological systems in the past"* (pp. 6–21). Moscow: Paleontological Institute RAS.

Bohr, H. (1952). *Collected Mathematical Work. II Almost periodic functions*. Copenhagen: Dansk Mathematisk Forening.

Crownover, M. R. (1995). *Introduction to Fractals and Chaos*. Boston, London: Jones and Bartlett Publishers.

Curie, P. (1966). *Selected works*. Moscow: Science.

Delone, B. (1937). Geometry of positive quadratic forms. *Advances in Mathematical Sciences*, *3*, 16–62.

Eiji, A., Yanfa, Y., & Pennycook, S. J. (2004). Quasicrystals as cluster aggregates. *Nature Materials*, *3*(11), 759–767. doi:10.1038/nmat1244 PMID:15516956

Galicin, U. G., Mansurov, V. G., Marahovka, I. I., & Petrenco, I. P. (1998). Commensurable and incommensurable phases In on surface (III) A In As. *Physics and Technology of Semiconductors, 1*, 89-94.

Gepner, D. (1988). Yukawa Couplings for Calabi -Yau String Compactification. *Nuclear Physics B, 311*(1), 191–204. doi:10.1016/0550-3213(88)90147-2

Gratias, D., & Cahn, J. W. (1986). Periodic and quasi - periodic crystals. *Scripta Metallurgica, 20*(9), 1193–1197. doi:10.1016/0036-9748(86)90030-X

Green, B. (2000). *The Elegant Universe*. New York: Vintage Books.

Ichikawa, T. (1981). Rhee study of In – induced superstructures on Ge (III) surfaces. *Surface Science, 3*(2), 227–259. doi:10.1016/0039-6028(80)90707-4

Izyumov, Yu. A. (1984). Modulated or long-period magnetic metal structures. *Advances in Physical Sciences*, *3*, 439 - 474.

Janssen, T., Chapuis, G., & De Boissieu, M. (2007). *Aperiodic Crystals. From Modulated Phases to Quasicrystals*. Oxford: Oxford University Press. doi:10.1093/acprof:oso/9780198567776.001.0001

Kadanoff, L. P. (1966). Scaling Laws for Ising Models Near T_c^*. *Physics*, *2*(6), 263–272. doi:10.1103/PhysicsPhysiqueFizika.2.263

Landau, L. D. (1937). To Theory of Phase Transitions. *Journal of Experimental and Theoretical Physics*, *7*, 19–38.

Lord, E. A., Mackay, A. L., & Ranganathan, S. (2006). *New Geometry for New Materials*. Cambridge: Cambridge University Press.

Mandelbrott, B. B. (1982). *The Fractal Geometry of Nature*. New York: Freeman.

Mukhopadhyay, N. K. (1993). Diffraction studies of icosahedral phases in $Al_{70}Fe_{20}W_{10}$. *Journal of Non - Crystalline Solids*, *153 – 154*, 1193 – 1197.

Penrous, R. (1979). Pentaplexity: A class of Nonperiodic Tillings of the Plane. *The Mathematical Intelligencer*, *2*(1), 32–37. doi:10.1007/BF03024384

Pouling, L. (1987). So - called Icosahedral Decagonal Quasicrystals Are Twins of an 820-atom cubic crystal. *Physical Review Letters*, *58*(4), 365–368. doi:10.1103/PhysRevLett.58.365 PMID:10034915

Riemann, B. (1854). *On the Hypotheses Which Lie at the Foundations of Geometry*. Gottingen: Gottingen Observatory.

Shechtman, D., & Blech, I. (1985). Article. *Trans. Ser. A.*, *16*, 1005–1009.

Shechtman, D., Blech, I., Gratias, D., & Cahn, J. (1984). Metallic phase with longer angle orientational order and no translational symmetry. *Physical Review Letters*, *53*(20), 1951–1953. doi:10.1103/PhysRevLett.53.1951

Shevchenko, V. (2011). Structural chemistry of nanoworld - a new page of Inorganic Chemistry. *Glass Physics and Chemistry*, *5*, 635–650.

Shevchenko, V., Zhizhin, G., & Mackay, A. (2013 a). On the structure of the quasicrystals in the high dimension space. In M. V. Diudea (Ed.), *Diamond and Related Nanostructures*. Dordrecht: Springer. doi:10.1007/978-94-007-6371-5_17

Shevchenko, V., Zhizhin, G., & Mackay, A. (2013 b). On the structure of the quasicrystals in the high dimension space. *News RAS. Chemical series*, *2*, 269 – 274.

Vernadskiy, V. I. (1965). *Chemical structure of Earth and its vicinity*. Moscow: Science.

Weiering, H., Hesinga, D., & Himba, T. (1992). Structure and growth of epitaxial Pb on Si(III). *Physical Review B: Condensed Matter and Materials Physics*, *45*(11), 5991–6002. doi:10.1103/PhysRevB.45.5991

Zang, X., & Kelton, K. (1993). High-order crystal approximant alloys $Ti_{54}Zr_{26}Ni_{20}$. *Journal of Non-Crystalline Solids*, *153-154*, 114–118. doi:10.1016/0022-3093(93)90325-R

Zhizhin, G. V. (2012, October). *Hierarchical filling of spaces with polytopes*. Paper presented at "St. Petersburg Scientific Forum: Science and Human Progress". 7th St.- Petersburg meeting of Nobel Prize laureates, St. Petersburg, Russia.

Zhizhin, G. V. (2014a). The fractal nature of disproportionate phases. In *10th all-Russian Scientific School "Mathematical research in the natural sciences"*. Apatity: Geological institute KSC RAS.

Zhizhin, G. V. (2014b). Incommensurable and fluctuating structures in the terrestrial space. *Biosphere*, *3*, 211–221.

Zhizhin, G. V. (2014c). *World – 4D*. St. Petersburg: Polytechnic Service.

Zhizhin, G. V. (2014d). On the higher dimension in nature. *Biosphere*, *6*(4), 313–318.

Zhizhin, G. V. (2014e). Phase transitions of the second kind with a fluctuation of the geometric structure. In *10th All-Russian Scientific School "Mathematical research in the natural sciences."* Geological institute KSC RAS.

Zhizhin, G. V. (2015, November). *Polytopic prismahedrons – fundamental regions of the n-dimension nanostructures*. Paper presented at The International conference "Nanoscience in Chemistry, Physics, Biology and Mathematics", Cluj-Napoca, Romania.

Zhizhin, G. V. (2018). *Chemical Compound Structures and the Higher Dimension of Molecules: Emerging Research and Opportunities*. Hershey, PA: IGI Global. doi:10.4018/978-1-5225-4108-0

Zhizhin, G. V. (2019). *The Geometry of Higher – Dimensional Polytopes*. Hershey, PA: IGI Global. doi:10.4018/978-1-5225-6968-8

Zhizhin, G. V., & Diudea, M. V. (2016). Space of Nanoworld. In M. V. Putz & M. C. Mirica (Eds.), *Sustainable Nanosystems, Development, Properties, and Applications* (pp. 214–236). New York: IGI Global.

KEY TERMS AND DEFINITIONS

Diffraction: A wide range of phenomena occurring in the propagation of waves in heterogeneous environments in the space.

Dimension of the Space: The member of independent parameters needed to describe the change in position of an object in space.

Fractal: The set is self-similar, that is, uniformity at different scales.

Golden Hyper-Rhombohedron: Polytope in four-dimensional space with facets as rhombohedron and metric characteristics associated the golden section.

Polytope: Polyhedron in the space of higher dimension.

Quasicrystal: A solid body, characterized by symmetry without translation in three-dimensional Euclidean space.

Chapter 2
Higher Dimensions of Clusters of Intermetallic Compounds

ABSTRACT

The author has previously proved that diffraction patterns of intermetallic compounds (quasicrystals) have translational symmetry in the space of higher dimension. In this chapter, it is proved that the metallic nanoclusters also have a higher dimension. The internal geometry of clusters was investigated. General expressions for calculating the dimension of clusters are obtained from which it follows that the dimension of metallic nanoclusters increases linearly with increasing number of cluster shells. The dimensions of many experimentally known metallic nanoclusters are determined. It is shown that these clusters, which are usually considered to be three-dimensional, have a higher dimension. The Euler-Poincaré equation was used, and the internal geometry of clusters was investigated.

INTRODUCTION

A systematic study of the geometry of the structures of chemical compounds (Zhizhin, 2014a, b, c, d, 2016, 2018, 2019a) showed that almost all elements of the periodic system form molecules of higher dimension. It is natural to assume that clusters, as larger than education molecules, including a large number of atoms, can have a higher dimension. However, until recently, clusters considered as three - dimensional objects (Lord, Mackay, & Ranganathan, 2006; Pauling, 1960). This Chapter discusses clusters of real chemical

DOI: 10.4018/978-1-7998-3784-8.ch002

compounds (as opposed to the book of Diudea, 2018, in which, as the author emphasizes, clusters are not related to real chemical compounds). Moreover, in this work, consideration of clusters is limited to a special type of chemical compounds - intermetallic compounds, since the study of intermetallic compounds has had a significant impact on the development of scientific views in recent decades. In particularly, the discovery of so - called quasicrystals is associated with intermetallic compounds, i.e. crystals supposedly devoid of translational symmetry (Shechtman et al., 1984). Although it was later shown that quasicrystals have translational symmetry, but in the space of higher dimension (Shevchenko, Zhizhin, & Mackay, 2013a, b; Zhizhin, 2014c; Zhizhin, & Diudea, 2016). The ideas about the higher dimension of clusters and the calculation of this dimension should be taken account to in their practical use as objects with valuable physic - chemical properties.

CLASTERS OF MACKAY

Mackay's cluster consists of two icosahedrons of different sizes with a common center (Mackay, 1962). A larger icosahedron is obtained by attaching a number of tetrahedrons and octahedrons to the surface of the smaller icosahedron.

Figure 1. The cluster of Mackay

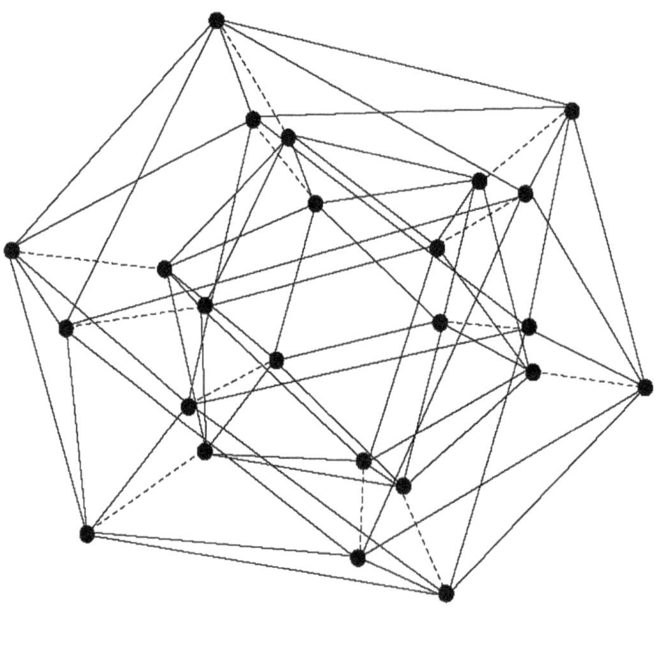

However, to determine the dimension of Mackay's cluster, only the result of this connection is important - the formation of a larger icosahedron. Moreover, each vertex of the larger icosahedron is located on a certain line passing through the common center of the icosahedrons. On the same line is located one of the vertices of the smaller icosahedron (Lord, Mackay, & Ranganathan, 2006). In all vertices of each icosahedron, one atom is located. In addition, there is one more atom in the middle of each edge. Since the icosahedron has 12 vertices and 30 edges, it turns out that Mackay's cluster has 54 atoms (Figure 1).

In this figure, the atoms in the midpoints of the edges of the larger icosahedron are not shown, since in determining the dimension the atoms located on the linear portions of the edges do not matter. Connect the edges of the vertices of both icosahedrons lying on the same line passing through the common center of the icosahedrons (dotted lines in Figure 1). This shows that the space between the icosahedrons is completely filled with triangular prisms, the bases of which are the triangular faces of the smaller and larger icosahedrons (Figure 2).

Figure 2. The triangular prism

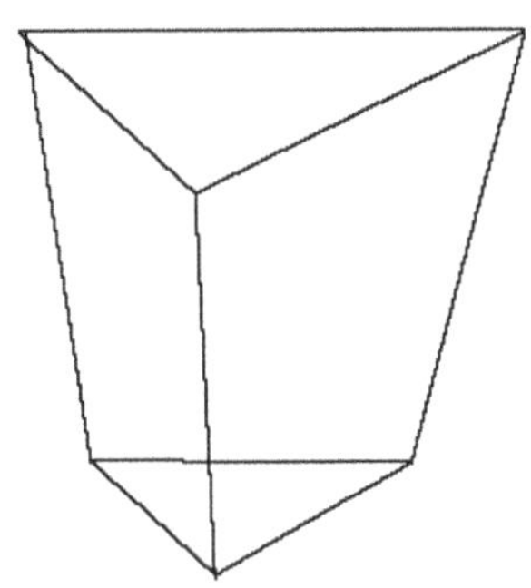

These prisms are adjacent to each other along flat quadrilateral side faces. The number of these prisms is equal to the number of triangular faces of the icosahedron - 20. To determine the dimension of the construction of two icosahedra with a common center, between which there are 20 triangular prisms can be determined by the Euler - Poincaré formula (Poincaré, 1895)

$$\sum_{i=0}^{d-1}(-1)^i f_i(P) = 1 + (-1)^{d-1} . \tag{1}$$

In (1) $f_i(P)$ is the number of faces with dimension i in polytope P with dimension d.

For a given design, we have

$$f_0 = 2 \cdot 12 = 24, f_1 = 2 \cdot 30 + 12 = 72, f_2 = 2 \cdot 20 + 30 = 70, f_3 = 20 + 2 = 22.$$

Substituting these values into equation (1), we have 24 – 72 + 70 – 22 = 0. This proves that the dimension of Mackay's cluster with two icosahedral shells is equal to 4.

To the second icosahedron, as well as to the first one, tetrahedrons and octahedrons can be attached and a third icosahedral shell can be obtained. This process can be continued.

Theorem 1. The dimension d of a cluster of n - icosahedrons with a common center is $2 + n$.

Proof.

If a cluster consists of three shells of an icosahedron with a common center, then

$$f_0 = 3 \cdot 12 = 36, f_1 = 3 \cdot 30 + 2 \cdot 12 = 114, f_2 = 3 \cdot 20 + 2 \cdot 30 = 120, f_3 = 2 \cdot 20 + 3 = 43.$$

The number of four - dimensional figures in this case is equal to $f_4 = C_3^2 = 3$. Substituting these values into equation (1), one have 36 – 114 + 120 – 43 + 3 = 2. This proves that the dimension of Mackay's cluster with three icosahedral shells is $d = 5$.

If a cluster consists of four shells of an icosahedron with a common center, then

$$f_0 = 4 \cdot 12 = 48, f_1 = 4 \cdot 30 + 3 \cdot 12 = 156, f_2 = 4 \cdot 20 + 3 \cdot 30 = 170, f_3 = 3 \cdot 20 + 4 = 64.$$

The number of four - dimensional figures in this case is equal to $f_4 = C_4^2 = 6$. The number of five - dimensional figures in this case is equal to $f_5 = C_4^3 = 4$. Substituting these values into equation (1), one have 48 – 156 + 170 – 64 + 6 - 4 = 0. This proves that the dimension of Mackay's cluster with four icosahedral shells is $d = 6$.

If a cluster consists of five shells of an icosahedron with a common center, then

$$f_0 = 5 \cdot 12 = 60, f_1 = 5 \cdot 30 + 4 \cdot 12 = 198, f_2 = 5 \cdot 20 + 4 \cdot 30 = 220, f_3 = 4 \cdot 20 + 5 = 85.$$

The number of four - dimensional figures in this case is equal to $f_4 = C_5^2 = 10$. The number of five - dimensional figures in this case is equal to $f_5 = C_5^3 = 10$. The number of six - dimensional figures is equal to $f_6 = C_5^4 = 5$. Substituting these values into equation (1), one have 60 – 198 + 220 – 85 +10 – 10 + 5 = 2. This proves that the dimension of Mackay's cluster with five icosahedral shells is $d = 7$.

These constructions can be continued. However, it is now possible to write general expressions for the numbers of elements of different dimensions in a cluster with an arbitrary number n of shells of icosahedra and to give a formula for calculating its dimension. So, in the n - shell cluster of icosahedrons, we have

$$\begin{aligned} &f_0 = n \cdot 12, f_1 = n \cdot 30 + (n-1) \cdot 12, f_2 = n \cdot 20 + (n-1) \cdot 30, \\ &f_3 = (n-1) \cdot 20 + n, f_4 = C_n^2, f_5 = C_n^3, \ldots, f_{n+1} = C_n^{n-1}. \end{aligned} \tag{2}$$

Substituting these values into equation (1) and opening the brackets, one can to see that in this case the left side of the Euler - Poincaré equation (1) takes the form

$$\sum_{i=0}^{d-1} (-1)^i f_i(P) = 2 - n + \sum_{k=2}^{n-1} C_n^k (-1)^k. \tag{3}$$

To calculate the sum on the right side of equation (3), one used the well - known expression for the alternating series of combinations (Vilenkin, 1969)

$$C_n^0 - C_n^1 + C_n^2 - \ldots + (-1)^n C_n^n = 0. \tag{4}$$

From the series (4), taking into account the equalities $C_n^0 = C_n^n = 1, C_n^1 = n$, it follows that the sum $\sum_{k=2}^{n-1} C_n^k (-1)^k$ in the right side of equation (3) is n - 2 for n even and n, if n is odd. Therefore, the right side of equation (3) coincides with the right side of equation (1). This proves that the figures in question are closed convex polytopes and they satisfy the Euler - Poincaré equation.

At the same time, from equation (2), since $f_{n+2} = C_n^n = 1$, it follows that the dimension of a cluster of n - icosahedrons with a common center is equal $d = n + 2$. Q.E.D.

Thus, it is proved that the dimension of a cluster of n icosahedral shells with a common center increases linearly with the number of icosahedral shells (Zhizhin, 2019 b; 2020).

The total number of atoms in a cluster, including the nth layer, is $n(10n^2+ 15n+ 11)/3$ (Lord, Mackay, & Ranganathan, 2006). Clusters of icosahedral shells are a structural feature of many complex alloys.

CLASTERS OF γ - BRASS

The structure of γ - brass was described as early as 1926 in the concept of a lattice of cubic cells (Bradley & Thewlis, 1926). Pauling, considering the structure of γ - brass, noticed that this structure is icosahedral (Pauling, 1960). Nyman & Anderson (1979) described the alloy Mn_5Si_3,Th_6Mn_{23} and γ - brass as a 26 - atomic cluster of identical balls, although it should be noted that atoms are not balls, especially since the atoms in the alloy of γ - brass are different.

Theorem 2. A γ - brass cluster has the dimension $d = 4 + 3n$, where n is the number of cluster shells ($n = 0, 1, 2,...$). The cluster is a d - cross - polytope

Figure 3. The tetrahedron

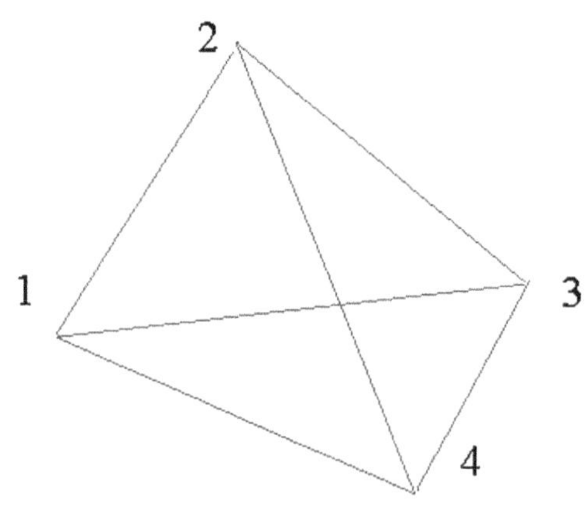

and the number of elements of the dimension i included in the cluster is determined by the ratio $f_i(d) = 2^{1+i} C_d^{d-1-i}$.

Proof. Alloys of γ - brass, as well as other intermetallic alloys, are conveniently considered using tetrahedrons (Lord, Mackay & Ranganathan, 2006). Place four atoms at the vertices of the 1234 tetrahedron (Figure 3).

Then on each flat face of the 1234 tetrahedron one place another a tetrahedrons (Figure 4).

Figure 4. The tetrahedron with tetrahedrons on its faces

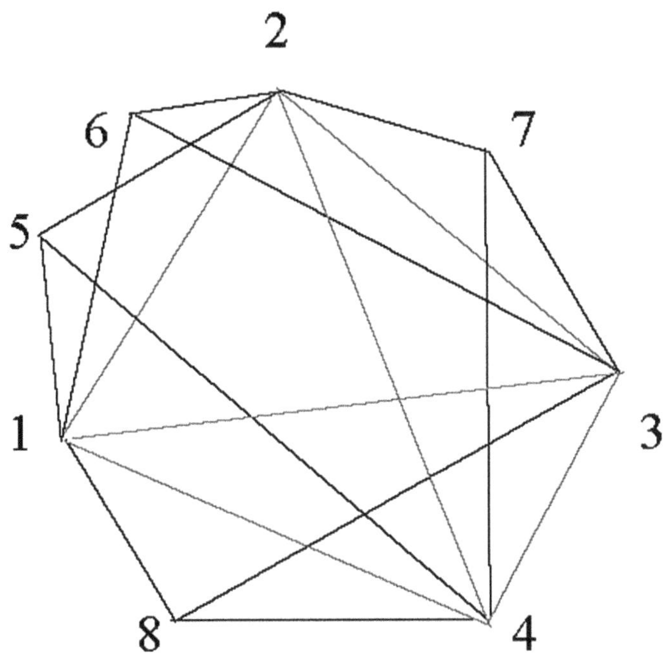

Then a figure appears, including 8 atoms (vertices 1 - 8). Connect the vertices 5, 6, 7, 8 of edges. They also form a tetrahedron. As a result, the resulting figure (Figure 5) is a 4 - cross - polytope (Zhizhin, 2019 a).

In this polytope, in addition to 8 vertices, there are 24 edges, 32 flat triangular faces, and 16 tetrahedrons. Each vertex has no edge connection to some other (opposite) vertex. These unconnected vertices form pairs $\begin{matrix} 1 & 2 & 3 & 4 \\ 7 & 8 & 5 & 6 \end{matrix}$. Each vertex in the top row does not have a connection with the vertex in the bottom row, just below that vertex. This polytope has dimension

Figure 5. The 4 - cross - polytope

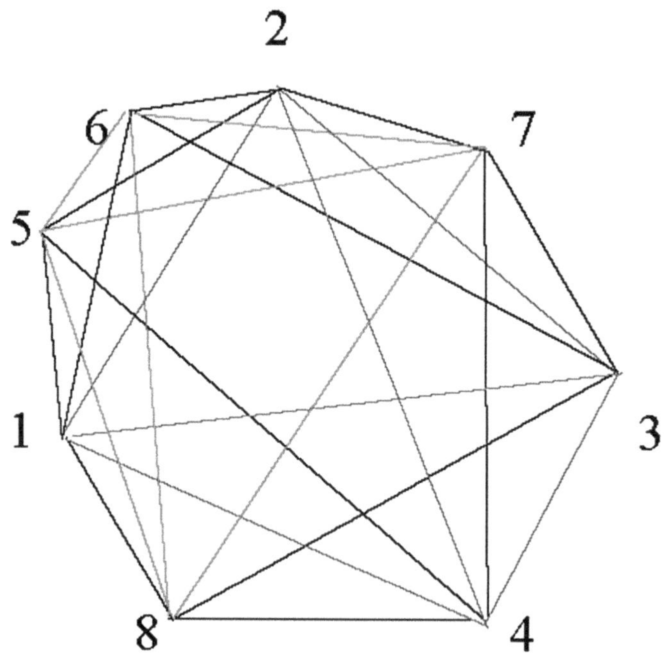

4. Upon further joining of the tetrahedrons to the edges of the original tetrahedron 1234 (two tetrahedrons to the edge), a figure is formed containing 14 vertices. Each newly formed vertex is located opposite one of the edges of the tetrahedron 1234. If we designate the newly formed vertices by a pair of vertices of the corresponding edges of the original tetrahedron 1234, then these are the next vertices (1,2), (1, 3), (1, 4), (2, 3), (2, 4), (3, 4). Among these newly formed vertices, opposite vertices can be distinguished, given the opposite edges of the original 1234 tetrahedron. These opposite vertices also form pairs $\begin{matrix}(2,3)(1,2)(1,3)\\(1,4)(4,3)(2,4)\end{matrix}$. Connecting the remaining vertices with edges, leaving pairs $\begin{matrix}1 & 2 & 3 & 4\\7 & 8 & 5 & 6\end{matrix}$ unconnected, can to get a 7 - cross - polytope. In a topologically equivalent form, this polytope of dimension 7 is shown in Figure 6.

From the general expression for the number of elements of the dimension and in the d - cross - polytope (Zhizhin, 2013, 2014 c, 2018, 2019) $f_i(d) = 2^{1+i} C_d^{d-1-i}$ it follows that

Figure 6. The 7 - cross - polytope

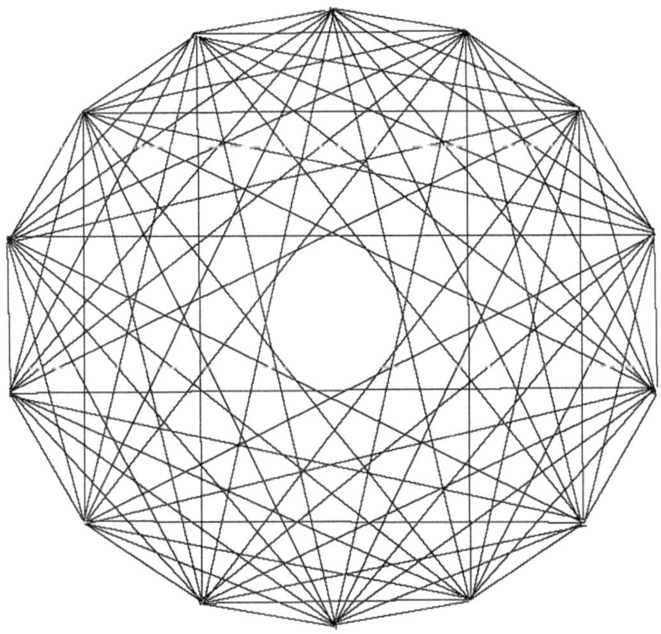

the number of vertices in this polytope $f_0 = 2 \cdot C_7^6 = 14$,
the number of edges $f_1 = 2^2 \cdot C_7^5 = 84$,
the number of triangle faces $f_2 = 8 \cdot C_7^4 = 280$,
the number of tetrahedrons $f_3 = 16 \cdot C_7^3 = 560$,
the number of four - dimension simplexes $f_4 = 32 \cdot C_7^2 = 672$,
the number of five - dimension simplexes $f_5 = 64 \cdot C_7^1 = 448$,
the number of six - dimension simplexes f_6=128.

Connecting a pair of tetrahedrons to the edges of the tetrahedron 5678, we obtain six more vertices (5, 6), (5.7), (5, 8), (6, 7), (6, 8), (7, 8). Opposite ones can be distinguished among these vertices, considering the opposite of the edges of the tetrahedron 5678 $\begin{matrix}(5,6)(5,8)(6,8)\\(8,7)(6,7)(5,7)\end{matrix}$. Connecting the remaining vertices with edges, leaving pairs

$$\begin{matrix} 1 & 2 & 3 & 4 & (2,3)(1,2)(1,3)(5,6)(5,8)(6,8) \\ 7 & 8 & 5 & 6 & (1,4)(4,3)(2,4)(8,7)(6,7)(5,7) \end{matrix}$$

unconnected, one get a 10 - cross - polytope. In this polytope:

the number of vertices is $f_0(10) = 2 \cdot C_{10}^{9} = 20$,
the number of edges is $f_1(10) = 2^2 \cdot C_{10}^{8} = 180$,
the number of triangle faces $f_2(10) = 8 \cdot C_{10}^{7} = 960$,
the number of tetrahedrons $f_3(10) = 16 \cdot C_{10}^{6} = 3360$,
the number of four - dimension simplexes $f_4(10) = 32 \cdot C_{10}^{5} = 8064$,
the number of five - dimension simplexes $f_5(10) = 64 \cdot C_{10}^{4} = 13440$,
the number of six - dimension simplexes $f_6(10) = 128 C_{10}^{3} = 15360$,
the number of seven - dimension simplexes $f_7(10) = 256 C_{10}^{2} = 11520$,
the number of either - dimension simplexes $f_8(10) = 512 C_{10}^{1} = 5120$,
the number of nine - dimension simplexes $f_9(10)=1024$.

Continuing to attach tetrahedrons to a cluster of 20 vertices, keeping the order of attachment, as in the previous steps, a cluster of 26 atoms can be obtained. This will be the 13 - cross - polytope. In this polytope the number of vertices is $f_0(13) = 2 \cdot C_{13}^{12} = 26$, the number of tetrahedrons is $f_3(13) = 2^4 \cdot C_{13}^{9} = 45760$. Thus, instead of a cluster in the form of four interpenetrating icosahedrons in three - dimensional space, the image of this cluster in the space of dimension 13 in the form of a convex standard cross - polytope can serve. Two such clusters make up an elementary cell of γ - brass.

Thus, it was proved that the addition of tetrahedrons to a γ - brass cluster of 8 atoms, having the form of a 4 - cross - polytope, leads to the creation of a number of shells, and the dimension of the cluster when each shell is attached increases by three. The number of elements of different dimensions in a cluster for any shell number n ($n = 0, 1, 2, ...$) is determined by the formula established earlier for d - cross - polytopes.

CLUSTERS OF BERGMEN, SAMSON, R – PHASES

In many metal alloys there are clusters in which the initial element is an icosahedron with a central atom (in contrast to Mackey's clusters). The extension of the tetrahedrons to the outer surface of the icosahedron leads to the formation of the next convex shell. This process can be continued and a series of larger convex hulls can be obtained. The location of the vertices on each convex hull is often determined by the author, based on the author's commitment to some of the ideas prevailing at a given time. There were clusters, designated by the name of the authors. However, to determine the dimension of these clusters with a central atom, it is important that an icosahedral surface can be distinguished on all these outer shells. The sizes of triangles on these surfaces naturally increase as one move to more and more distant shells. Such clusters will be called icosahedral complexes with a central atom.

Theorem 3. The dimension d of icosahedral complexes with a central atom equal $n + 3$, where n is the number of the shells in the complexes.

Proof. The icosahedral with a central atom has 13 vertices (f_0=13), 42 edges (f_1=42), 50 flat triangle faces (f_2=50), 21 three - dimensional faces (20 triangle pyramids and 1 icosahedron) (f_3=21). Substituting the values in the equation (1) we have $13 - 42 + 50 - 21 = 0$. This proves that the icosahedron with its center has a dimension of 4.

If one continue the edges going from the center of the icosahedron (Figure 7), and on these edges at the appropriate distance arrange more atoms forming

Figure 7. Two icosahedrons with a common center

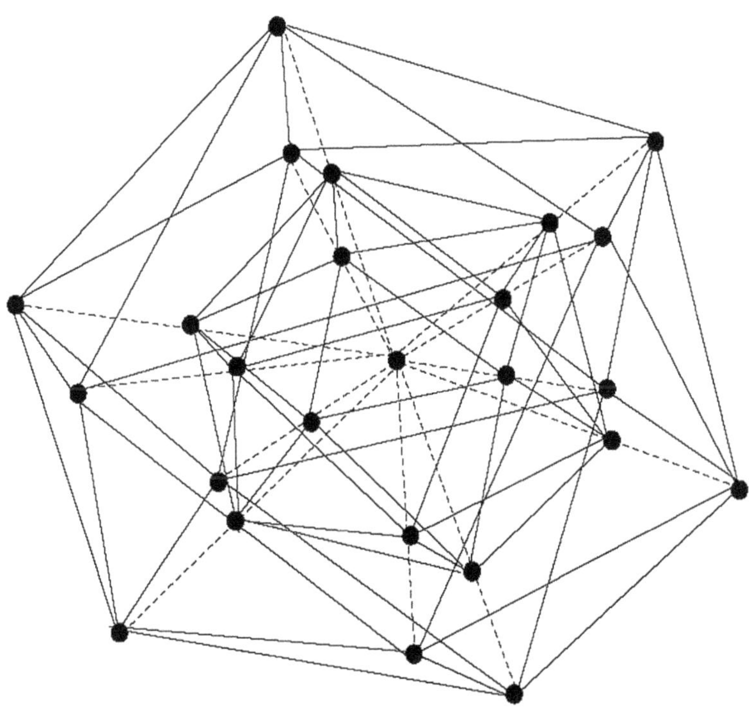

the second icosahedron of a larger size, such a construction of two icosahedra with a common center will have 25 vertices (f_0=25), 84 edges (f_1= 2•30+ 2•12= 84), 100 flat faces (f_2= 2•20+ 2•30= 100), 42 three - dimensional faces (f_3=2•20+2=42), 3 four - dimensional faces ($f_4 = C_3^2 = 3$). Substituting these values in the equation (1) you can get 25 - 84 + 100 – 42 + 3 = 2. This proves that the two icosahedrons with common center has a dimension 5.

Note that the number of four - dimensional faces in this case includes a smaller icosahedron with a center, a larger icosahedron with a center, and the space between two icosahedrons. Indeed, for this space one have 24 vertices (f_0=24), 72 edges (f_1= 2•30+ 12= 72), 70 flat faces (f_2= 2•20+ 30= 70), 22 three - dimensional faces (f_3= 20+ 2= 22). Substituting these values in the equation (1), can get 24 - 72 + 70 – 22 = 0. This proves that the space between two icosahedrons has a dimension 4. The total number of four - dimensional elements in this case is determined by the number of combinations between the three characteristic elements of the construction (the center and two icosahedral surfaces) of two elements. Any of these combinations gives a four - dimensional element.

If one continue again the edges coming from the center of the icosahedrons, and on these edges one construct the third icosahedron of a still larger size, then such a construction will have 37 vertices (f_0= 25+ 12= 37), 126 edges (f_1= 3•30+ 3•12= 126), 150 flat faces (f_2= 3•20+ 3•30= 150), 63 three - dimensional faces (f_3=3•20+3=63), 6 four - dimensional faces ($f_4 = C_4^2 = 6$), 4 five - dimensional faces ($f_5 = C_4^3 = 4$). Substituting these values into the equation (1), can get 37 - 126 + 150 – 63 + 6 - 4 = 0. This proves that the tree icosahedrons with common center has a dimension 6.

The total number of four - dimensional elements in this case is determined by the number of combinations between the four characteristic elements of the construction (the center and three icosahedral surfaces) of two elements. Any of these combinations gives a four - dimensional element.

The total number of five - dimensional elements in this case is determined by the number of combinations between the three characteristic elements of the construction (the center and three icosahedral surfaces) of three elements. Any of these combinations gives a five - dimensional element.

These constructions can be continued. However, it is now possible to write general expressions for the numbers of elements of different dimensions in a cluster with a center for arbitrary number n of shells of icosahedra and to give a formula for calculating its dimension. So, in the n - shell cluster of icosahedrons with a center, we have

$$f_0 = n \cdot 12 + 1, f_1 = n \cdot 30 + n \cdot 12, f_2 = n \cdot 20 + n \cdot 30,$$
$$f_3 = n \cdot 20 + n, f_4 = C_{n+1}^2, f_5 = C_{n+1}^3, \ldots, f_{n+2} = C_{n+1}^n. \tag{5}$$

Substituting these values into equation (1), one can to see that in this case the left side of the Euler - Poincaré equation (1) takes the form

$$\sum_{i=0}^{d-1} (-1)^i f_i(P) = 1 - n + \sum_{k=2}^{n} C_{n+1}^k (-1)^k. \tag{6}$$

To calculate the sum on the right side of equation (6), one use the expression (4) for the alternating series of combinations (Vilenkin, 1969).

From the series (4), taking into account the equalities $C_n^0 = C_n^n = 1, C_n^1 = n$, it follows that the sum $\sum_{k=2}^{n} C_{n+1}^k (-1)^k$ in the right side of equation (6) is *n* + 1 for *n* even and *n* - 1 if *n* is odd. Therefore, the right side of equation (6) coincides with the right side of equation (1). This proves that the figures in question are closed convex polytopes and they satisfy the Euler – Poincaré equation.

At the same time, from equation (5), since $f_{n+3} = C_{n+1}^{n+1} = 1$, it follows that the dimension of a cluster from *n* - icosahedrons with an atom in a common center is equal $d = n + 3$.

Thus, as in the Mackay clusters, the dimension of clusters of several icosahedrons with one common central atom increases linearly with the number of shells. The dimension of these clusters is greater than the dimension of the corresponding Mackay clusters by one due to the presence of an atom in a common center.

From Theorem 3, for example, it follows that the dimension of a giant palladium cluster (Vargaftik et al., 1985), containing 561 atoms in five shells, is 8. The claims of some authors that the palladium cluster in this case is an E_8 lattice (Shevchenko, 2011) are groundless. The lattice E_8, as you know (Conway & Sloane, 1988), is a collection of points in an eight-dimensional space with coordinates (±1, ±1, 0, 0, 0, 0, 0, 0), where units can stand anywhere on the line with arbitrary signs, as well as points with coordinates

$$\left(\pm\frac{1}{2}, \pm\frac{1}{2}, \pm\frac{1}{2}, \pm\frac{1}{2}, \pm\frac{1}{2}, \pm\frac{1}{2}, \pm\frac{1}{2}, \pm\frac{1}{2}\right).$$

Obviously, this lattice has nothing to do with the structure of a cluster consisting of five icosahedral shells, although it is for such a cluster that the number of atoms gives 561 (Lord et al., 2006). It is assumed (Coxeter, 1963) that the lattice E_8 corresponds to the polytope of Gosset (1900), which draws from simplexes and cross-polytopes. But from the previous it follows that the polytope corresponding to the giant palladium cluster does not include either simplexes or cross-polytopes.

The outer surface of the clusters of Bergman, Samson, and R - phases (Lord, Mackay, & Ranganathan, 2006) outwardly seems to be different from the icosahedron. However, this difference is not significant. For example, in the Bergman cluster of 45 atoms, the outer surface is supposed to consist of rhombuses. Can note by construction it is not, it is the icosahedron. For example, the clusters from $Mg_{32}(Al,Zn)_{49}$ was specifically decided to deform in order to bring it closer to Pauling's three - contohedron (Bergman et al., 1952, 1957). In the Samson cluster of 105 atoms of alloy Mg_6Pd (Samson, 1972), the outer surface is a truncated icosahedron. It is easy to get an icosahedron from this surface, connecting the centers of the pentagons, especially since the initial construction is based on tetrahedrons. The surfaces of the R - phase clusters from $Mg_{32}(Al,Zn)_{49}$, Mo-Cu-Cr,Al_5CuLi_3 consist of elements of icosahedral surfaces (Bergman et al., 1952, 1957; Komura et al., 1960; Audier et al., 1988). Therefore, the dimensions of these clusters can be calculated from the formulas obtained in this work, taking into account the number of shells in these clusters. Since the Bergman cluster has two icosahedral shells, according to Theorem 3 its dimension is 5. Since the Samson cluster has three shells, its dimension is 6. Since the R - phase cluster has four shells, its dimension is 7.

THE DIMENSION OF METALLIC CLUSTERS OF SEVERAL SHELLS IN THE FORM OF PLATO`S BODIES

The icosahedron is only one of Plato's bodies found in the skeleton of metal clusters (Gubin, 1987). Suppose there is a cluster, each shell of which is a convex regular three - dimensional polyhedron (Plato's body) with some possible number of vertices n and the same for all shells of this cluster. The flat sides of the shells are regular m - corner. All shells of this cluster differ only in size and have a common center. Assume that the number of flat edges in the shell is j. One denoted S_j a shell satisfying the indicated conditions. Then the following statement is true.

Theorem 4. The dimension d of a cluster of N shells with a common center is $N + 2$, if there is no atom in the common center, and is equal to $N + 3$, if there is an atom in the common center.

Proof. Denote by the symbol t the number of edges emanating from each vertex of the shell S_j. Then the total number of edges of the shell is $nt/2$, and the number of faces of the shell is $j=nt/m$. The number j in Plato's body also gives the form of the shell, i.e. sets the numbers n, m, t (Table 1).

Table 1. Defining Plato bodies by the number of two - dimensional faces of the outer shell

J	n	m	T	**Polyhedron**
4	4	3	3	**Tetrahedron**
6	8	4	3	**Cube**
8	6	3	4	**Octahedron**
12	20	5	3	**Dodecahedron**
20	12	3	5	**Icosahedron**

Let a shell with the number of faces j and be given. Therefore, the number of vertices n_j, the number of sides m_j at the flat face, the number of edges t_j emanating from each vertex are given. Consider two shells of different sizes with a common center, arranged so that every two corresponding vertices of both shells are on the same straight line connecting them with a common center (there is no atom in the center). Then the space between the shells is filled with prisms, the bases of which are flat faces of the larger and smaller shells. The number of these prisms is equal to the number of faces of the shell, i.e. j. The number of vertices in a polytope of two shells is $2n_j=f_0$. The number of edges of this polytope, taking into account the edges connecting the corresponding vertices of the shells, is $n_jt_j+n_j=f_1$. The number of flat edges in a polytope is equal to twice the number of flat edges in each shell and the number of edges in one of the shells, i.e. $2j+\frac{n_jt_j}{2}=f_2$. The number of three - dimensional figures in the polytope is equal to $j+2=f_3$. Substitute these numbers in the equation (1) Euler - Poincare

$$2n_j - n_j(t_j + 1) + 2j + \frac{n_j t_j}{2} - j - 2 = \left(n_j - \frac{n_j t_j}{2} + j\right) - 2 = 0.$$

There $\left(n_j - \frac{n_j t_j}{2} + j\right) = 2$ on the equation (1) for the polytope of dimension 3.

This proves that a polytope composed of two shells S_j with a common center (in the absence of an atom in the center) has dimension 4 for any possible j.

If the cluster consists of three shells S_j with a common center, then $f_0=3n_j$. The number of edges in this cluster, taking into account the edges connecting of the corresponding vertices in shells S_j, is equal to $f_1 = 3\frac{n_j t_j}{2} + 2n_j$. The number of two - dimensional faces is $3j+n_j t_j=f_2$. The number of three - dimensional faces is $2j+3=f_3$. The number of four - dimensional faces is $f_4 = C_3^2 = 3$. Substitute these numbers in the equation (1) Euler - Poincare

$$3n_j - n_j\left(\frac{3}{2}t_j + 2\right) + 3j + n_j t_j - (2\text{j} + 3) + 3 = n_j - \frac{n_j t_j}{2} + j = 2.$$

This proves that dimension of the cluster of tree shells S_j is equal 5.

One can write a general expression for the numbers of elements of different dimensions in a cluster with an arbitrary number N of shells S_j and get a formula for calculating its dimension. So, in the N - shell cluster of shells S_j, we have

$$\begin{aligned} &f_0 = Nn_j, f_1 = N\frac{n_j t_j}{2} + (N-1)n_j, f_2 = Nj + (N-1)\frac{n_j t_j}{2}, \\ &f_3 = (N-1)\,\text{j} + N, f_4 = C_N^2, f_5 = C_N^3, \ldots, f_{N+1} = C_N^{N-1}. \end{aligned} \tag{5}$$

Substituting the values (5) in the equation (1) and opening the brackets, one can see that in this case the left side of the Euler - Poincaré equation (1) takes the form

$$\sum_{i=0}^{d-1}(-1)^i f_i(P) = 2 - N + \sum_{k=2}^{N-1} C_N^k (-1)^k. \tag{6}$$

To calculate the sum on the right side of equation (6), we use the expression for the alternating series of combinations (Vilenkin, 1969)

$$C_N^0 - C_N^1 + C_N^2 - ... + (-1)^N C_N^N = 0. \tag{7}$$

From the series (7), taking into account the equalities $C_N^0 = C_N^N = 1, C_N^1 = N$, it follows that the sum $\sum_{k=2}^{N-1} C_N^k (-1)^k$ in the right side of equation (6) is N - 2 for N even and N for N odd. Therefore, the right side of equation (6) coincides with the right side of equation (1). This proves that the figures in question are closed convex polytopes and they satisfy the Euler - Poincaré equation.

At the same time, from equation (5), since $f_{N+2} = C_N^N = 1$, it follows that the dimension of a cluster of N shells S_j with a common center if at center not atoms is equal $d = N + 2$.

Now consider a cluster of several shells S_j with a common center in which the atom is located. The shell S_j with a central atom has n_j+1=f_0 vertices, $\frac{n_j t_j}{2} + n_j = f_1$ edges, $\frac{n_j t_j}{2} + j = f_2$ flat faces, j+1=f_3 three – dimensional faces. Substituting the values in the equation (1) one can get

$$n_j + 1 - \left(n_j + \frac{n_j t_j}{2} \right) + j + \frac{n_j t_j}{2} - (j - 1) = 0.$$

This proves that the shell S_j with its center has a dimension of 4.

If one continue the edges going from the center of the shell S_j to its vertices and at the appropriate distance arrange more atoms forming the second a shell S_j of a larger size. Such a construction of two the shells S_j with a common atom in center will have $2n_j$+1=f_0 vertices, $n_j t_j$+$2n_j$=f_1 edges, $2j$+$n_j t_j$=f_2 flat faces, $2j$+2=f_3 three - dimensional faces, $f_4 = C_3^2 = 3$ four - dimensional faces. Substituting these values in the equation (1), one can get

$$2n_j + 1 - 2n_j(1 + t_j / 2) + 2j + n_j t_j - (2j + 2) + 3 = 2.$$

This proves that the two shells S_j with common atoms in the center has a dimension 5.

If one continue again the edges coming from the center of the shells S_j, and on these edges one construct the third shell S_j of a still larger size, then such a construction will have $f_0=3n_j+1$ vertices, $f_1 = 3\frac{n_j t_j}{2} + 3n_j$ edges, $f_2 = 3j + 3\frac{n_j t_j}{2}$ flat faces, $f_3=3j+3$ three - dimensional faces, $f_4 = C_4^2 = 6$ four - dimensional faces, $f_5 = C_4^3 = 4$ five - dimensional faces. Substituting these values into the equation (1), one can get

$$3n_j + 1 - 3\left(n_j + \frac{n_j t_j}{2}\right) + 3\left(j + \frac{n_j t_j}{2}\right) + 3(j+1) + 6 - 4 = 0.$$

This proves that cluster of the tree shells S_j with common atom at center has a dimension 6.

Can write a general expression for the numbers of elements of different dimensions in a cluster with a center for arbitrary number N of shells S_j and give a formula for calculating its dimension. Then in the cluster of N shell S_j with atom in a common center, we have

$$\begin{aligned} &f_0 = Nn_j + 1, f_1 = N\frac{n_j t_j}{2} + Nn_j, f_2 = Nj + N\frac{n_j t_j}{2}, \\ &f_3 = Nj + N, f_4 = C_{N+1}^2, f_5 = C_{N+1}^3, \ldots, f_{N+2} = C_{N+1}^N. \end{aligned} \tag{8}$$

Substituting these values into equation (1), one can see that in this case the left side of the Euler - Poincaré equation (1) takes the form

$$\sum_{i=0}^{d-1} (-1)^i f_i(P) = 1 - N + \sum_{k=2}^{N} C_{N+1}^k (-1)^k. \tag{9}$$

To calculate the sum on the right side of equation (6), we use the expression (7). From the series (7), taking into account the equalities $C_N^0 = C_N^N = 1, C_N^1 = N$, it follows that the sum $\sum_{k=2}^{N} C_{N+1}^k (-1)^k$ in the right side of equation (9) is $N+1$ for N even and N -1for N odd. Therefore, the right side of equation (9) coincides with the right side of equation (1). This proves that the figures in question are closed convex polytopes and they satisfy the Euler - Poincaré equation.

At the same time, from equation (8), since $f_{N+3} = C_{N+1}^{N+1} = 1$, it follows that the dimension of a cluster of N shells S_j with an atom in a common center is equal $d = N + 3$.

CLUSTERS OF THE HYPER-ROMBOHEDRONS

When analyzing the diffraction pattern of intermetallic compounds of Al_6Mn (Shechtman et al., 1984), $Ti_{54}Zr_{26}Ni_{20}$ (Zang & Kelton, 1993), $Al_{70}Fe_{20}W_{10}$ (Mukhopadhyay et al., 1993), $Al_{72}Ni_{20}Co_8$ (Eiji, Yanfa, & Pennycook, 2004) it was found (Zhizhin, 2014c, 2018, 2019a), that the unit cell of the structure created by the luminous points of the diffraction patterns is a figure of dimension 4, called the golden hyper - rhombohedron. It consists of eight rhombohedrons, with angles in the flat edges defined by the golden section, contains 16 vertices (Figure 8).

Figure 8. The golden hyper - rhombohedron

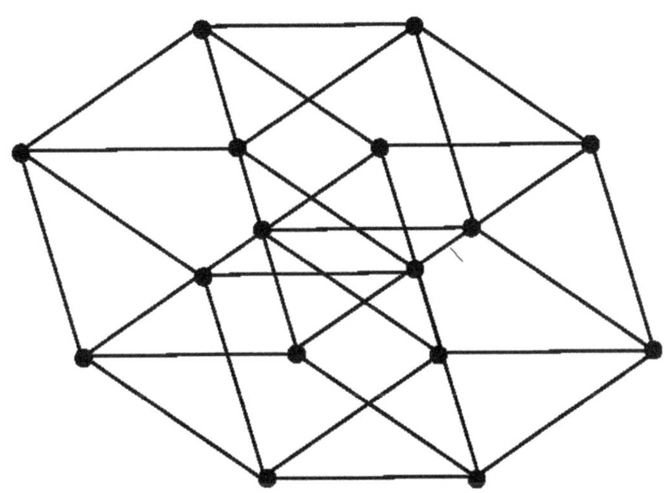

This figure provides the translational symmetry of quasicrystals in the space of higher dimension. The golden hyper - rhombohedron can be considered a cluster of intermetallic compounds, since the luminous points in the diffraction patterns are lattice nodes, which reflect the rays passing through the metal alloy.

The products of the golden hyper - rhombohedron on other geometric figures lead to the formation of clusters of the higher dimension. Definitions and research the works on various geometric shapes are given in the monographs of the author (Zhizhin, 2018, 2019a). In Figure 9 and Figure 10, as an example, the products of the golden hyper - rhombohedron on a triangle and a tetrahedron are shown, correspondingly.

Figure 9. The product of golden hyper - rhombohedron on a triangle

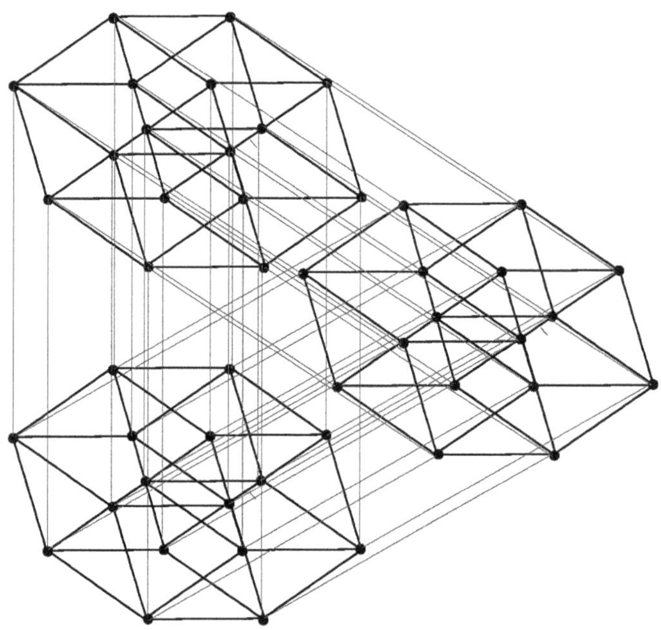

Since the dimension of the product of the figures is equal to the sum of the dimensions of the factors, the dimension of the product of the golden hyper - rhombohedron by a triangle is 6, and the dimension of the product of the golden hyper - rhombohedron by the tetrahedron is 7. The number of vertices in the product of the figures is equal to the product of the numbers of the vertices of the factors. Therefore, the number of vertices in the product of the golden hyper - rhombohedron on a triangle is equal to 48, and the number of vertices in the product of the golden hyper - rhombohedron on a tetrahedron is 64.

Figure 10. The product of golden hyper - rhombohedron on a tetrahedron

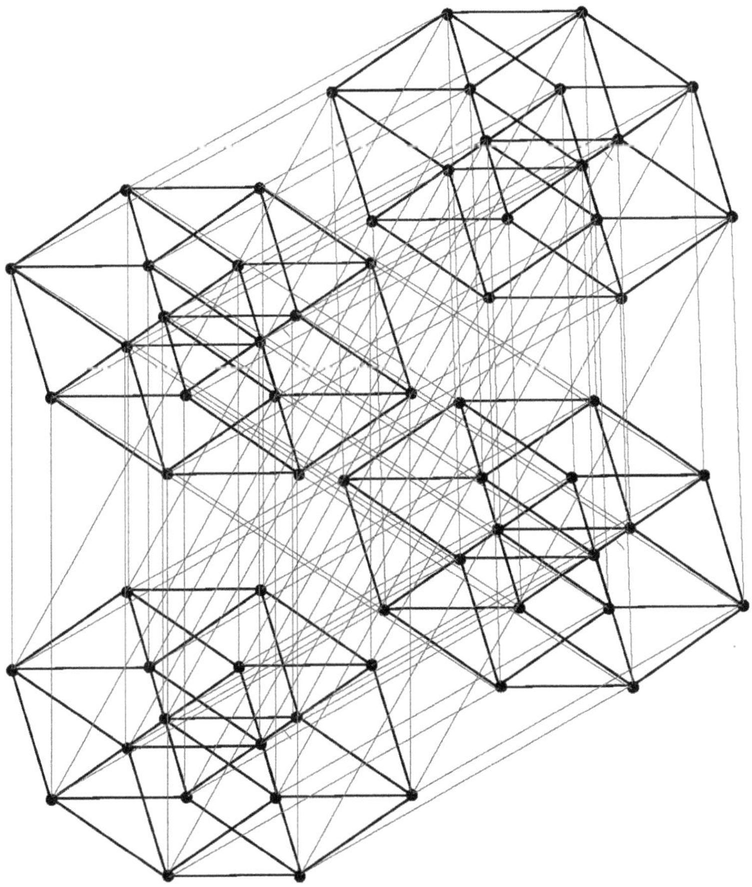

INTERMETALLIC COMPOUNDS AS ASSEMBLIES OF NANOCLUSTERS

The term nanocluster is used currently in the wide sense as a nanometric set of connected atoms, stable either in isolation state or in building unit of condensed matter (Shevchenko, 2011). Metallic nanoclusters are particularly important objects in the nanoworld, and their structural chemistry is a new chapter of modern inorganic chemistry and materials science. The former is characterized by the great diversity in composition and structure, and often is the primary unit in the self - assembly of nanostructured materials. Nanoclusters existing in solids and interacting with each other (so - called quantum dots) are of special interest as well as the process of the self – organization of nanoclusters from isolated atoms in some environment. Quantum processes are still important in large semiconducting nanoclusters. Nanoclusters can be obtained by gas aggregation, vaporization from surfaces, dispersion, laser ablation, etc.

In terms of nanoclusters was studied chemical compounds Rh_7Mg_{44},$ZrZn_{22}$ (Shevchenko & Mackay, 2008; Shevchenko, Blatov, & Ilyushin, 2009), $NaCd_2$ (Samson, 1962, 1964; Shevchenko, Blatov, & Ilyushin, 2009; Bergman, 1996), Mg_2Al_3 (Samson, 1965), Cu_3Cd_4 (Samson, 1967), $MgCu_2$ (Shevchenko, Blatov, & Ilyushin, 2009), $MgZn_2$ and $MgNi_2$ (Shevchenko, Blatov, & Ilyushin, 2009). It was established that practically in all these compounds has two types of nanoclusters A and B are involved simultaneously, each of which contains a large number of atoms. Geometrically, icosahedrons, Friauf polyhedrons, and other polyhedrons are involved in the compounds. Determining the dimension of complicated nanoclusters containing a large number of atoms is a difficult task. However, it follows from the preceding that one can say with certainty, that the dimension of these nanoclusters is more than three. Details of the study of the structure of nanoclusters intermetallic compounds can be found in the cited literature.

CONCLUSION

For the first time, based on previously developed ideas about the multidimensionality of the nanoworld, an analytical expressions are obtained for calculating the dimension of multi - shell clusters depending on the number of shells. These expressions are applicable to clusters consisting of shells in the form of an icosahedron with a common center, if an atom is located or dislocated in this center. From the obtained analytical formulas in it follows that in both cases the dimension of the clusters increases linearly with an increase in the number of shells. Thus, for the clusters of Mackay (1962), Bergman (Bergman et al., 1952, 1957, 1996), Samson (1972) known in the literature, which can be approximately described by such multi - shell models, expressions are obtained for calculating their dimensions and proved that they have a higher dimension. It should be noted an interesting fact that the clusters of γ - brass studied for a long time turned out to have geometry of a high - dimensional cross - polytope. Consequently, not only extended alloys of intermetallic compounds, as was shown in the previous chapter, have a higher dimension, but also isolated compounds of intermetallic compounds in the form of clusters have a higher dimension. Clusters, connecting with each other, lead to the formation of new clusters with an even higher dimension. This also applies to the hyper - rhombohedron, as a cluster of quasicrystals. This is demonstrated in this chapter. In the case of complex clusters of intermetallic compounds containing a large number of atoms and formations in various geometric forms, it is difficult to determine their dimension. However, it is safe to say on the basis the conducted research, that they have a higher dimension.

REFERENCES

Audier, M., Pannetier, J., Leblanc, M., Janot, C., Lang, L.M. & Dubost, B. (1988). An approach to the structure of quasicrystals: A single crystal X - Ray and neutron diffraction study of the R – Al_5CuLi_3 phase. *Physics B,* 136 - 142.

Bergman, G. (1996). A refinement was performed on single - crystal X - ray diffraction data in a ninefold. *Acta Crystallographica. Section B, Structural Science*, *52*, 54–58. doi:10.1107/S0108768195007737

Bergman, G., Waugh, J. L. T., & Pauling, L. (1952). Crystal structure of the intermetallic compound Mg32(Al,Zn)49. *Nature*, *169*(4312), 1057–1058. doi:10.1038/1691057a0

Bergman, G., Waugh, J. L. T., & Pauling, L. (1957). The crystal structure of the metallic phaseMg32(Al,Zn)49. *Acta Crystallographica*, *10*(4), 254–259. doi:10.1107/S0365110X57000808

Bredley, A. J., & Thewlis, J. (1926). The structure of γ - brass. *Proceedings of the Royal Society of London. Series A, Containing Papers of a Mathematical and Physical Character*, *112*(762), 678–681. doi:10.1098/rspa.1926.0134

Conway, J. H., & Sloane, N. J. A. (1988). *Sphere Packings, Lattices, and Groups*. New York: Springer – Verlag. doi:10.1007/978-1-4757-2016-7

Coxeter, H. S. M. (1963). *Regular Polytopes*. New York: Dower.

Diudea, M. V. (2018). *Multi - shell Polyhedral Clusters*. Springer. doi:10.1007/978-3-319-64123-2

Eiji, A., Yanfa, Y., & Pennycook, S. J. (2004). Quasicrystals as cluster aggregates. *Nature Materials*, *3*(11), 759–767. doi:10.1038/nmat1244 PMID:15516956

Gosset, T. (1900). On the regular and semi-regular figures in space of n dimensions. *Messenger of Mathematics., 29*, 43–48.

Gubin, S. P. (1987). *Chemistry clusters. Basics of classification and structure*. Moscow: Science.

Komura, Y., Sly, W. G., & Shoemaker, D. P. (1960). The structure of the R phase, Mo - Cu - Cr. *Acta Crystallographica, 13*(8), 575–585. doi:10.1107/S0365110X60001394

Lord, E. A., Mackay, A. L., & Ranganathan, S. (2006). *New Geometry for New Materials*. Cambridge: Cambridge University Press.

Mackay, A. L. (1962). A dense non - crystalline packing of equal spheres. *Acta Crystallographica, 15*(9), 1916–1918. doi:10.1107/S0365110X6200239X

Mukhopadhyay, N.K. (1993). Diffraction studies of icosahedral phases in $Al_{70}Fe_{20}W_{10}$. *Journal of Non - Crystalline Solids, 153 – 154*, 1193 - 1197.

Nyman, H., & Andersson, S. (1979). On the structure of Mn_5Si_3,Th_6Mn_{23}. *Acta Crystallographica. Section A, Crystal Physics, Diffraction, Theoretical and General Crystallography, 35*(4), 580–583. doi:10.1107/S0567739479001364

Poincaré, A. (1895). Analysis situs. *J. de é Ecole Polytechnique, 1,* 1 - 121.

Pouling, L. (1960). *The Nature of the Chemical Bond and the Structure of Molecules and Crystals: An Introduction to Modern Structural Chemistry* (3rd ed.). Ithaca, NY: Cornell University Press.

Samson, S. (1962). Crystal Strucrure of $NaCd_2$. *Nature, 195*(4838), 259–262. doi:10.1038/195259a0

Samson, S. (1964). The crystal structures of semiconductors and a general valence rule. *Acta Crystallographica, 17*, 49–51.

Samson, S. (1965). The crystal structure of the phase β Mg2Al3. *Acta Crystallographica, 19*(3), 401–413. doi:10.1107/S0365110X65005133

Samson, S. (1967). The Crystal Structure of the Intermetallic Compound Cu_4Cd_3. *Acta Crystallographica, 23*(4), 586–594. doi:10.1107/S0365110X67003251

Samson, S. (1972). Complex cubic A_6B compounds. II. The crystal structure of Mg_6Pd. *Acta Crystallographica. Section B, Structural Crystallography and Crystal Chemistry, 28*(3), 936–945. doi:10.1107/S0567740872003437

Shechtman, D., Blech, I., Gratias, D., & Cahn, J. (1984). Metallic phase with longer angle orientational order and no translational symmetry. *Physical Review Letters*, *53*(20), 1951–1953. doi:10.1103/PhysRevLett.53.1951

Shevchenko, V., Zhizhin, G., & Mackay, A. (2013a). On the structure of the quasicrystals in the high dimension space. In M. V. Diudea (Ed.), *Diamond and Related Nanostructures*. Dordrecht: Springer. doi:10.1007/978-94-007-6371-5_17

Shevchenko, V., Zhizhin, G. & Mackay, A. (2013b). On the structure of the quasicrystals in the high dimension space. *News RAS. Chemical Series*, *2*, 269 - 274.

Shevchenko, V. Y. (2011). *Search in Chemistry, Biology and Physics of the Nanostate*. Saint Petersburg: Lema.

Shevchenko, V. Y., Blatov, V. A., & Ilushin, G. D. (2009). Structural Chemistry of Metal Microclusters: Questions and Answers. *Glass Physics and Chemistry*, *35*(1), 1–12. doi:10.1134/S1087659609010015

Shevchenko, V. Y., & Mackay, A. L. (2008). Geometrical principles of the self - assembly of nanoparticles. *Glass Physics and Chemistry*, *34*(1), 1–8. doi:10.1134/S108765960801001X

Vargaftik, M. N. (1985). Article. *Reports of the Academy of Sciences of the USSR*, *284*(4), 896–899.

Vilenkin, N. V. (1969). *Combinatorics*. Moscow: Science.

Weiering, H., Hesinga, D., & Himba, T. (1992). Structure and growth of epitaxial Pb on Si(III). *Physical Review B: Condensed Matter and Materials Physics*, *45*(11), 5991–6002. doi:10.1103/PhysRevB.45.5991

Zang, X., & Kelton, K. (1993). High - order crystal approximant alloys $Ti_{54}Zr_{26}Ni_{20}$. *Journal of Non-Crystalline Solids*, *153-154*, 114–118. doi:10.1016/0022-3093(93)90325-R

Zhizhin, G. V. (2013). Images of convex regular and semiregular n - dimensional polytopes. In *Proceedings of the 9th All-Russian Scientific School "Mathematical Research in the Natural Sciences"*. Geological Institute KSC RAS.

Zhizhin, G. V. (2014a). The fractal nature of disproportionate phases. In *10th all-Russian Scientific School "Mathematical research in the natural sciences"*. Apatity: Geological institute KSC RAS.

Zhizhin, G. V. (2014b). Incommensurable and fluctuating structures in the terrestrial space. *Biosphere*, *3*, 211–221.

Zhizhin, G. V. (2014c). *World - 4D*. St. Petersburg: Polytechnic Service.

Zhizhin, G. V. (2014d). On the higher dimension in nature. *Biosphere*, *6*(4), 313–318.

Zhizhin, G. V. (2016). The structure, topological and functional dimension of biomolecules. *J. Chemoinformatics and Chemical Engineering*, *5*(2), 44–58.

Zhizhin, G. V. (2018). *Chemical Compound Structures and the Higher Dimension of Molecules: Emerging Research and Opportunities*. Hershey, PA: IGI Global. doi:10.4018/978-1-5225-4108-0

Zhizhin, G. V. (2019a). *The Geometry of Higher - Dimensional Polytopes*. Hershey, PA: IGI Global. doi:10.4018/978-1-5225-6968-8

Zhizhin, G. V. (2019b). Higher Dimensions of Clusters of Intermetallic Compounds. *International Journal of Applied Nanotechnology Research*, *4*(1), 8–25. doi:10.4018/IJANR.2019010102

Zhizhin, G. V. (2020). The Geometry of Higher – Dimensional Multi – Shell Clusters with Common Center and Different Center: The Geometry of Metal Clusters with Ligands. *Int. Journal of Applied Nanotechnology Research.*

Zhizhin, G. V., & Diudea, M. V. (2016). Space of Nanoworld. In M. V. Putz & M. C. Mirica (Eds.), *Sustainable Nanosystems, Development, Properties, and Applications* (pp. 214–236). New York: IGI Global.

Zhizhin, G. V., Khalaj, Z., & Diudea, M. V. M.V. (2016). Geometrical and topological dimensions of the diamond. In Distance, symmetry and topology in carbon nanomaterials. London: Springer.

KEY TERMS AND DEFINITIONS

Diffraction: A wide range of phenomena occurring in the propagation of waves in heterogeneous environments in the space.

Dimension of the Space: The member of independent parameters needed to describe the change in position of an object in space.

Fractal: The set is self-similar, that is, uniformity at different scales.

Golden Hyper-Rhombohedron: Polytope in four-dimensional space with facets as rhombohedron and metric characteristics associated the golden section.

Nanocluster: A nanometric set of connected atoms, stable either in isolation state or in building unit of condensed matter.

Polytope: Polyhedron in the space of higher dimension.

Quasicrystal: A solid body, characterized by symmetry without translation in three-dimensional Euclidean space.

Chapter 3
Dimension of Chemical Compounds of Atoms Metals With Atoms No Metals

ABSTRACT

The structures of compounds of a metal atom with ligands were studied by sequentially changing the groups and subgroups of the periodic system of elements in which the metal atom is located. It is shown that all metals from the first to the eighth groups form chemical compounds of a higher dimension. The formation of molecules of higher dimension occurs due to the chemical bonds of the metal atom with ligands both due to the influence of electron pairs and due to the attraction of ions. Moreover, the apparent valence of the metal atom, as a rule, exceeds the value of the valence determined by the location of the metal in the periodic table of chemical elements.

INTRODUCTION

In previous chapters, compounds of atoms of the same or different metals with each other were considered. The structures of these compounds and the dimensions of the elementary cells of a solid array of material or the dimensions of their nanoparticles (clusters of metal compounds of each other) were determined. However, in nature and in technology, the combination of metals with other chemical elements, their compounds and groups is very important. In coordination chemistry, elements or groups of elements attached

DOI: 10.4018/978-1-7998-3784-8.ch003

to a metal atom are called ligands, and the compounds themselves are called complex compounds. However, there is still no exact (unambiguous) definition of the concept of a complex compound covering all possible types of such compounds (Morozov, 2008). A number of terms and definitions have been introduced that have a narrow scope (Werner, 1936; Greenberg, 1986). In particular, the concepts of the main and secondary valence of chemical elements, etc., are introduced.

In the monograph (Zhizhin, 2018), relying on the works of Gillespie (Gillespie, 1972; Gillespie & Hargittai, 1991), an analysis of almost all elements of the periodic system was carried out and it was shown that most of the molecules of the chemical compounds of these elements have a higher dimension. At the same time, it was not necessary to use special definitions of coordination chemistry, since the ball is sufficient to have ideas about divided and unshared electron pairs. There is for the purpose of determining the dimension of a molecule of a chemical compound, it is important to take into account the arrangement of the atoms of this molecule in space. A priori the equality of the dimension of space by three was not given. In addition, when determining the dimension of molecules, the nature of the chemical bond fades into the background. The result is important - the arrangement of atoms in space. In this case, there is no need to solve accurately the complex problems of determining the degree of covalence and ionization of a chemical bond. It is important that this bond is strong enough to ensure a stable configuration of the molecule. Weak bonds (for example, hydrogen bonds) are not taken into account.

To determine the dimension, the molecule is modeled by a convex figure (polyhedron or polytope), the edges of which are both chemical bonds between atoms and edges that have only a geometrical meaning. They serve to make the molecule look like a closed geometric figure. The dimension of the molecule is determined by the Euler – Poincaré formula (Poincaré, 1895), substituting the numbers of elements of different dimensions that make up the polytope. In the author's works (Zhizhin, 2016a, 2018, 2019, 2020), when considering the geometry of biomolecules, the concept of the functional dimension of biomolecules was introduced, taking complex groups of atoms attached to the central atom to be functional groups, denoting them as vertices of a polytope. In this case, the polytope is a simplified model of the molecule. The concept of functional groups and functional dimensions will also be used here when considering complex molecules.

THE STRUCTURE OF THE COMPOUNDS M_2O

If the anomalous elements has one electron in the outer orbital *s* and subshell *d* completely (or almost completely) filled, then the element at the expense of *s* electron forms a linear molecule, such as a linear molecule oxide M_2O, where M is the atom of metal (anomalous elements Cu, Pd, Ag, Pt, Au, Rg). However, due to the donor - acceptor bond linear molecule can form complex structures in the space. Can choose element Cu from second group of anomalous elements. Figure 1 shows an exemplary structure formed by linear molecules Cu_2O

Figure 1. The structure of the compound Cu_2O. A black small circle is oxygen atom. A brown circle is copper atom.

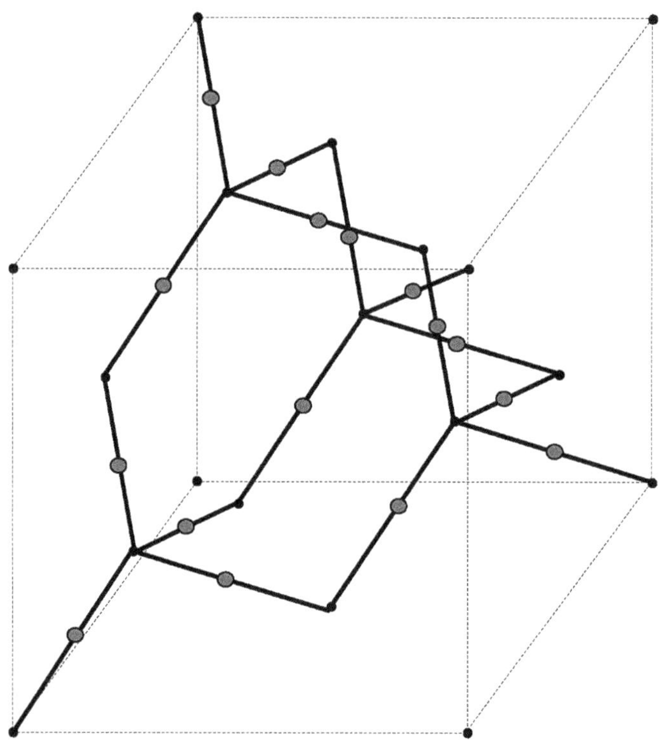

Each oxygen atom in the structure of Figure 1 bonded to four metal atoms (Cu). Two covalent bonds due to the formation of electron pairs divided: one *s* - electron metal atom and a *p* - electron atom of oxygen. In addition, there are two more donor - acceptor chemical bond due to the transfer of two electrons from the *s* - orbital and two electrons from the *p* - orbitals of the oxygen atom to vacant quantum cell of orbital metal. Thus, oxygen atom acquires valence equal four. In addition, each metal atom is linearly between two oxygen atoms.

In this structure, oxygen atoms (except oxygen atoms located at the vertices of a cube) form structure is topologically equivalent to the structure of carbon atoms in the molecule of adamantane. As shown in the article of Zhizhin (2014a) on the basis the monograph of Zhizhin (2014b), the dimension of this molecule is 4. However, the two molecules comprising 10 oxygen atoms have free unallocated space. Therefore, if one set the task of finding the unit cell structure of copper oxide without filling cracks and gaps to help translation the entire space, one need to build polytopic prismahedron Zhizhin (2015b), with bases in the form polytopes corresponding to these molecules. Taking a line segment equal to the length of the edge of the cube, in which is inscribed the structure including 10 oxygen atoms, multiply the polytope corresponding to this structure for this segment. We obtain polytopic prismahedron of dimension 5. With this polytopic prismaehdron can fill space without gaps and clearances.

From the third group of anomalous elements we choose gold (Au). The outer shell of gold atom has one 6*s* - electron and a completely filled 5*d* - orbital. In the compound chlorine triphenylphosphine $(Ph_3P)AuCl$ of gold atom, giving one *s* - electron to chlorine atom, forms ion $Au(Ph_3P)_3^+$ with trigonal coordination (Gillespie, 1972; Perrin, Armarego & Perrin, 1980). Following of Zhizhin (2016 b) can denoted phosphine molecule as a functional group of the compound.

Then ion $Au(Ph_3P)_3^+$ represented in the form of three tetrahedrons with the center, having one common vertex - a gold atom. In the center of each tetrahedron is located phosphorus atom and the remaining vertices of the tetrahedrons are occupied introduced functional groups Ph_3 (Figure 2).

The functional dimension of each tetrahedron with center is still equal to 4. Thus, the ion $Au(Ph_3P)_3^+$ is a collection of three polytopes of dimension 4, having a common vertex. The assertion is proven

Theorem 1. The ion $Au(Ph_3P)_3^+$ has dimension 5.

Proof. To prove the necessity of the three tetrahedrons with a common vertex to form a convex shape. Connect the vertices of a_3, a_4, a_7 by line segments, forming a triangle $a_3\, a_4\, a_7$. Connect also the three centers of the tetrahedrons with each other, forming a triangle $a_7\, a_{11} a_{13}$. Connect the center

Figure 2. Ion $Au(Ph_3P)_3^+$. A white circle is gold atom. A black small circle is phosphorus atom. A black big circle is functional groups Ph_3

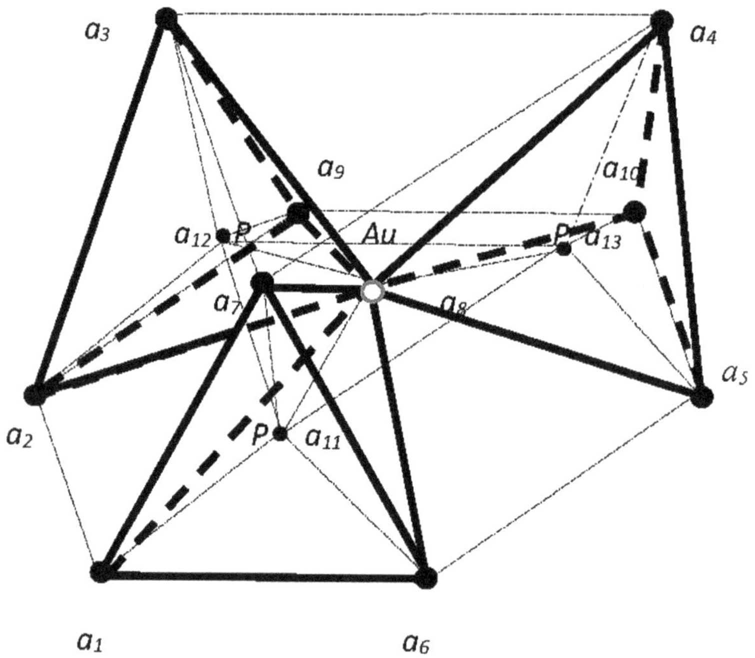

of the tetrahedrons with vertices corresponding of the tetrahedrons and vertices in the grounds of the tetrahedrons, forming a hexagon a_1 a_2 a_9 a_{10} a_5 a_6 (thin lines on Figure 2). Define dimension polytope in Figure 5 on the Euler- Poincaré equation (Poincaré, 1895)

$$\sum_{i=0}^{n-1}(-1)^i f_i(P) = 1-(-1)^i, \tag{1}$$

f_i is the number of the elements with the dimension i at polytope P; n is dimension of the polytope P.

To calculate the number of elements of large dimensions we turn first to a simple polytope, a part of a polytope in Figure 2. Temporarily excluded from Figure 2 the centers of the tetrahedrons – a_{11}, a_{12}, a_{13}, and all edges emanating from these vertices. Then, the polytope has 13 vertices, i. e. f_0=10. The number of edges of the polytope is sum the number edges of three tetrahedrons (6•3=18), the number of edges connecting tetrahedrons at the base figure (3), the number of the edges connecting vertices of the tetrahedrons at top of Figures 2. Thus, the number of edges polytope without centers of the tetrahedrons equal 24, i.e. f_1=24. The number of flat elements is sum of the flat faces of tetrahedrons (4•3=12), 1 hexagon, 3 triangles between tetrahedrons at base of figure, 3 lateral tetragons, 4 triangles of tetrahedron at top of figure. Thus, the number of flat elements is 23, i.e. f_2=23. The number of three - dimension elements is sum of 4 tetrahedrons, 3 figures between tetrahedrons, 1 hexagon at base and figure composed from boundary flat faces. Thus, the number of three - dimension elements is 9, i.e. f_3=9. Substituting the values f_i in equation (1), you can see that it holds for $n = 4$

$$10 - 24 + 23 - 9 = 0.$$

Therefore, three tetrahedrons with common vertices is polytope with dimension 4.

For add centers in tetrahedrons the number of the vertices becomes equal 13, i.e. $(f_0)_c$=13. For this there add the number of the edges: 4•3=12 edges in tetrahedrons with centers, and 3 edges connecting centers. Thus, common number of edges on Figure 5 equal to 39, i.e. $(f_1)_c$=39. The number flat faces there increases on 18 triangles in the tetrahedrons, 4 triangles in tetrahedron $a_{11}a_{12}a_{13}a_8$, 6 tetragons with vertices part which are centers of the tetrahedrons. Thus, common number of flat faces on Figure 2 equal to 51, i.e. $(f_2)_c$=51. For add centers the number of three-dimensions faces increases on 4•3=12 tetrahedrons into tetrahedrons with centers,

tetrahedron $a_{11}a_{12}a_{13}a_8$,

figure $a_1\ a_2\ a_9\ a_{10}\ a_5\ a_6\ a_{11}a_{12}a_{13}a_8$,

prism $a_{11}a_{12}a_{13}a_3a_4a_7$,

3 pyramids with vertex a_8 $(a_8a_1a_2a_{12}a_{11}, a_8a_5a_6a_{11}a_{13}, a_8a_9a_{10}a_{12}a_{13})$,

3 prism $(a_1a_2a_3a_7a_{11}a_{12}, a_5a_6a_4a_7a_{11}a_{13}, a_9a_{10}a_3a_4a_{12}a_{13})$.

Thus, common number of three - dimension faces on Figure 2 equal to 30, i.e. $(f_3)_c$=30.

It is known from the preceding that the Figure 2 has polytopes of dimension 4. Each tetrahedron with center there is polytope of dimension 4 and 3 tetrahedrons without center, but with a common vertex, there is a polytope of dimension 4. In addition, in Figure 2 between any two tetrahedrons with the center is polytope dimension 4. Obviously, such polytopes are 3. That to proof this statement one consider any polytope from them. For example, the polytope $a_4a_5a_6a_7a_{11}a_{13}$ (Figure 3).

Figure 3. The polytope $a_4a_5a_6a_7a_{11}a_{13}$

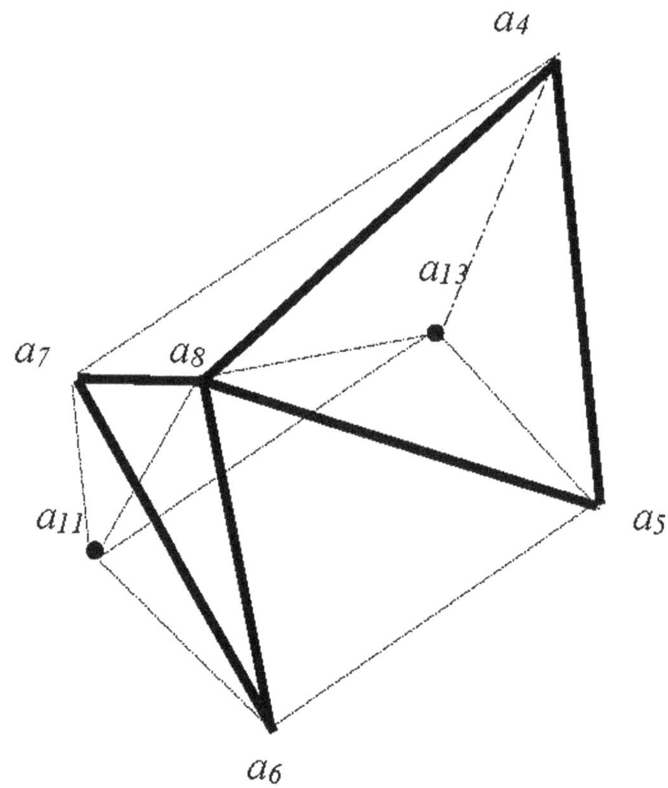

Figure 3 has 7 vertices (f_0=7);

15 edges

$a_{11}a_7$, $a_{11}a_8$, $a_{11}a_6$, $a_{11}a_{13}$, a_7a_6, a_7a_8, a_7a_4, a_8a_4, a_8a_{13}, a_8a_5, a_6a_5, $a_{13}a_5$, a_4a_{13}, a_4a_5, a_8a_6;

14 flat faces

$a_{11}a_8a_6$, $a_{11}a_7a_8$, $a_6a_7a_8$, $a_{11}a_7a_6$, $a_8a_5a_{13}$, $a_8a_{13}a_4$, $a_5a_{13}a_4$, $a_8a_5a_4$, $a_6a_8a_5$, $a_{11}a_8a_{13}$, $a_7a_8a_4$, $a_6a_{11}a_{13}a_5$, $a_6a_5a_7a_4$, $a_{11}a_7a_{13}a_4$;

6 three-dimension faces

$a_6a_{11}a_8a_7$, $a_8a_5a_{13}a_4$, $a_7a_8a_{11}a_{13}a_4$, $a_6a_{11}a_8a_{13}a_5$, $a_{11}a_6a_7a_{13}a_5a_4$, $a_6a_7a_8a_4a_5$.

Therefore, for Figure 3 are

$f_0 = 7, f_1 = 15, f_2 = 14, f_3 = 6$.

Substituting the values f_i in equation (1), you can see that it holds for $n = 4$

$7 - 15 + 14 - 6 = 0$.

This proof that Figure 3 has dimension 4.

As the figures in polytope on Figure 2 is 3, so common number polytopes with dimension 4 in Figure 2 equal to 7, i.e. $(f_4)_c=7$. Substituting the values $(f_i)_c$ in equation (1), you can see that it holds for $n = 5$

$13 - 39 + 51 - 30 + 7 = 2$.

THE DIMENSION OF THE WURTZITE

In the structure of wurtzite every atom of one component has a tetrahedral environment of the atoms of the other component. This results to arrangement of tetrahedrons with center so that vertex one tetrahedron is the center of another tetrahedron (Figure 4).

If to carry construction of atoms in Figure 4 on this principle, can obtain a spatial lattice, the unit cell the lattice is a convex shape it is shown in Figure 5.

This figure is the unit cell structure of the wurtzite. In Figure 5, solid lines represent chemical bonds of the atoms, and the dotted lines are only geometric sense outlining contours of the figure. One define the dimension of this figure by the Euler-Poincaré equation (1). The number of vertices of this figure is equal to 14, i.e. $f_0=14$. The number of edges is equal to 29, i.e.

Figure 4. The tetrahedral coordination atoms in wurtzite. A white circle is atom of one component. A black circle is atom of other component

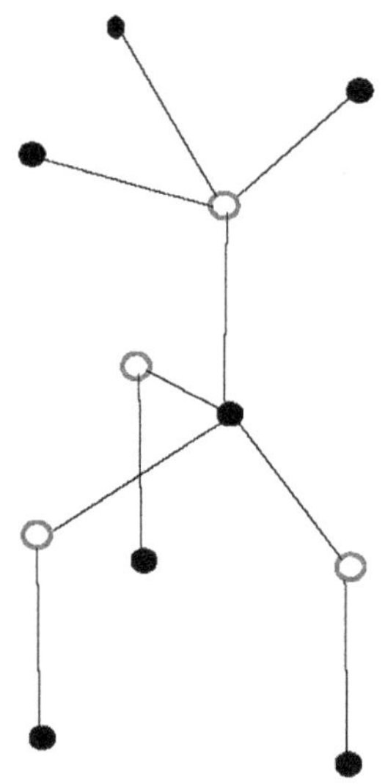

Figure 5. The unit cell of the wurtzite

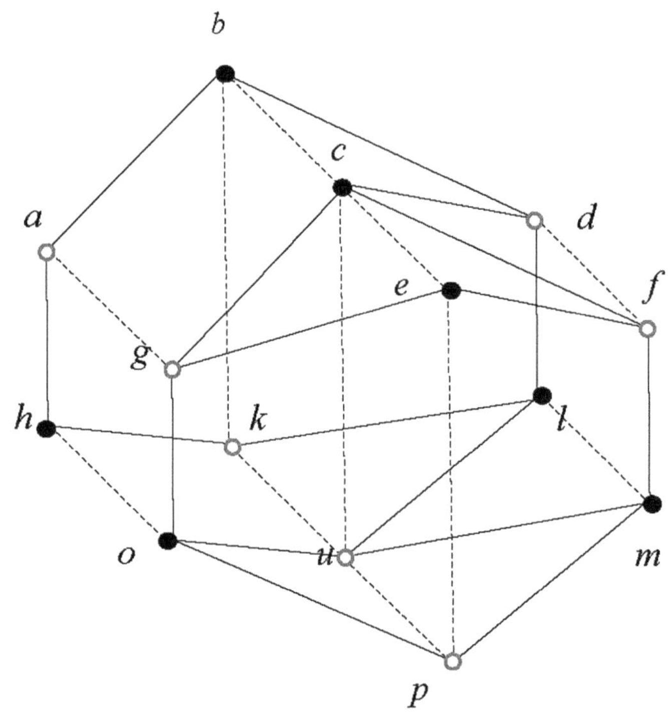

f_1=29. The number of two - dimensional faces is the sum of the number of triangles (8) and number of quadrangles (13), i.e. f_2=21. The number of three - dimensional faces is equal to 6. The figures are *abcghkon, gceruo, cefump, cdfulm, bcdklu,* and all shape on Figure 5 without inner partitions, i.e. f_3=6. Substituting these values f_i, (i=0,1,2,3) in the Euler-Poincare equation (1), one can find that it is satisfied for $n = 4$

$$14 - 29 + 21 - 6 = 0.$$

Thus, the dimension of polytope on Figure 5 is equal to 4, i.e. the unit cell structure of the wurtzite has dimension 4.

THE DIMENSION OF THE FLUORITE

On example of compound $MnCl_2$ one look at the structure of fluorite. Isolate magnesium atoms lying at the centers of the cube faces (•), and chlorine atoms (▲), forming a smaller cube inside the bigger cube, which are located at the vertices of magnesium atoms (Figure 6).

Figure 6. The unit cell of the fluorite

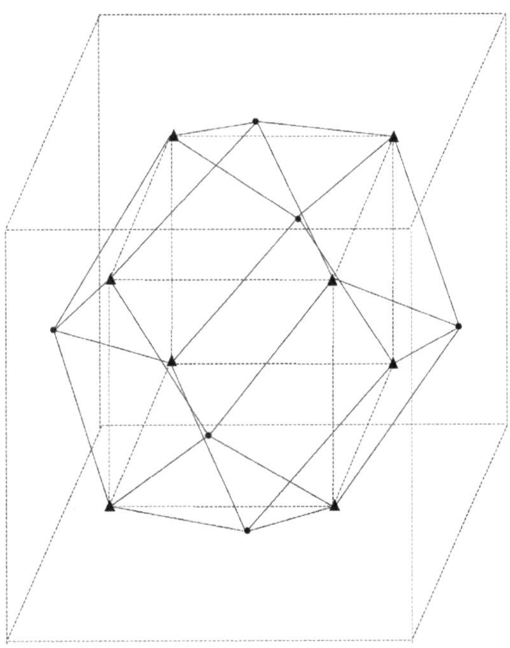

From Figure 6 it follows that the number of vertices is 14, i.e. f_0=14, the number of edges is 36, i.e. f_1=36, the number of flat faces is sum from number of triangles (24) and number of rectangles (6) (smaller cube faces), i.e. f_2=30. The number three - dimension shape to sum up from smaller cube (1), pyramids on its faces (6) and figure (22) without inner parts (1), i.e. f_3=8. Substituting these values f_i, (i=0,1,2,3) in the equation (1), you can find that it is satisfied for $n = 4$

$$14 - 36 + 30 - 8 = 0.$$

Thus, the dimension of polytope on Figure 6 is equal to 4, i.e. the unit cell structure of the fluorite has dimension 4.

THE STRUCTURE AND HIGHER DIMENSION OF ALKALINE METALS

In the vast majority of compounds involving alkali metals (elements of first group of the Mendeleev table) the chemical bond is preferably ionic. Alkali metals have an external electron shell ns^1. They easily give bake one electron exhibiting a degree of oxidation +1. Salts of alkali metals in the condensed state usually have a cubic lattice, forming a structure of type rock salt structure.

However, Oganov and his co - workers have found that under high pressure the structure of many compounds, including alkali metal compounds, acquires new unexpected properties (Zhung et al., 2013; Zhou et al., 2012; Zhu, Oganov & Lyakhov, 2013). In particular, it is shown that the structure of the sodium - chlorine compounds varies significantly (Zhung et al., 2013). The elementary cell of this structure is a cube with sodium atoms at its vertices, and an icosahedron with a center with chlorine atoms at its vertices is located inside the cube. This compound is denoted Pm3-$NaCl_x$.

Theorem 2 (Zhizhin, 2016b). The dimension of the unit cell Pm3-$NaCl_x$ is 5.

In the proof of Theorem 2 one shell use equation Euler - Poincare (1).

Figure 7 shows the structure of this compound, where sodium atoms are located at the vertices of the cube 13 -19, and chlorine atoms are located at the vertices 1 - 12, 21.

Figure 7. The structure of compound sodium and chlorine at high pressure

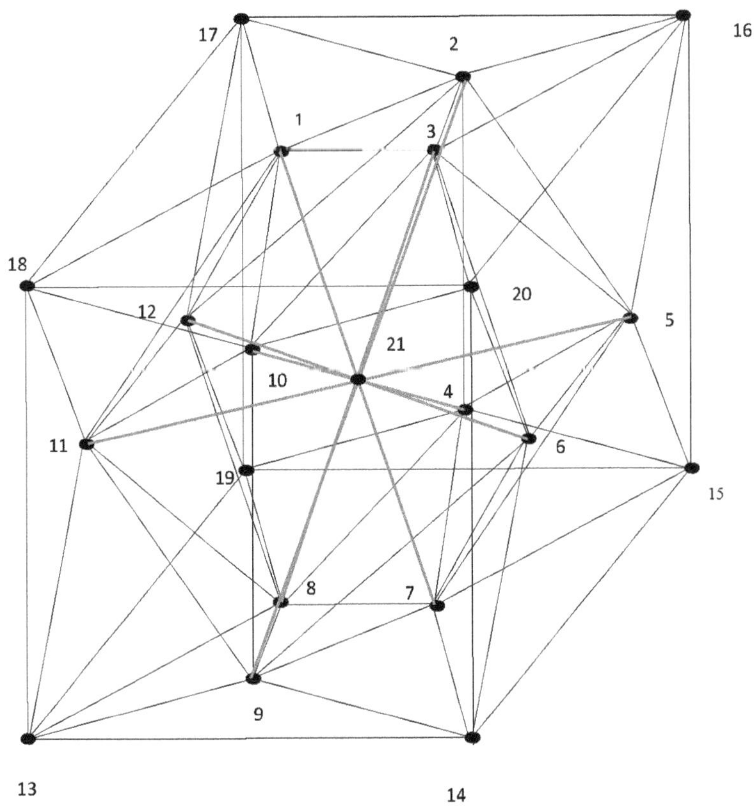

Note that the icosahedron with the center already has dimension 4. Indeed, the icosahedron has 12 vertices, 30 edges, 20 flat faces. If you enter a center in the icosahedron (point 21), then f_0=13, f_1=42, f_2 = 50 for it. In addition, 20 tetrahedrons are added, taking into account that they are located in the icosahedron, we obtain f_3=21. Substituting the obtained values of f_i into equation (1), you can find that it is satisfied for $n = 4$

$$13 - 42 + 50 - 21 = 0.$$

This proves that the icosahedron with the center has dimension 4.

From it follows that the dimension of the polytope with 21 vertices in Figure 7 is greater than 4. To determine this dimension, let us calculate the number of elements of different dimension entering into this polytope. Thus, for this polytope f_0=21. The number of edges is the sum of the number of edges of the icosahedron (30), the number of edges issuing from the center to the vertices of the icosahedron (12), the number of edges of a cube (12), the number of edges issuing from the vertices of the cube to the vertices of the icosahedron (24). Hence, f_1=78.

Two - dimensional elements include trapezoids:

13 - 8 - 7 - 14, 1 - 3 - 17 - 16, 1 - 3 - 18 - 20, 19 - 8 - 7 - 15, 11 - 12 - 18 - 17, 11 - 12 - 19 - 13, 6 - 5 - 20 - 16, 5 - 6 - 14 - 15, 18 - 10 - 9 - 13, 9 - 10 - 20 - 14, 17 - 2 - 4 - 19, 2 - 4 - 15 - 16.

Total number of trapezoids is 12.

In the number of two - dimensional elements includes also triangles:

1. The triangles of the outer surface of the icosahedron (20).
2. The triangles inside the icosahedron ((20 • 3) / 2 = 30).
3. Triangles of tetrahedrons resting on the faces of icosahedrons (with the exception of the icosahedron faces themselves) having common vertices with a cube:

13 - 8 - 9, 13 - 11 - 9, 13 - 8 - 11, 14 - 7 - 9, 14 - 7 - 6, 14 - 9 - 6, 15 - 4 - 5, 15 - 4 - 7, 15 - 5 - 7, 19 - 4 - 8, 19 - 12 - 8, 19 - 12 - 4, 17 - 1 - 12, 17 - 1 - 2, 17 - 12 - 2, 16 - 3 - 2, 16 - 2 - 5, 16 - 3 - 5, 18 - 1 - 11, 18 - 1 - 10, 18 - 10 - 11, 20 - 3 - 10, 20 - 3 - 6, 20 - 6 - 10;

all these triangles is 24.

4. Triangles of pyramids resting on the cube's faces

13 - 1 - 18, 17 - 12 - 19, 18 - 17 - 1, 20 - 3 -16, 16 - 5 - 15, 20 - 6 - 14, 19 - 8 - 13, 14 - 7 - 15, 18 - 10 - 20, 14 - 7 - 15, 17 - 2 - 16, 19 - 4 – 1;

all of these triangles is 12.

The total number of triangles is

20 + 30 + 24 + 12 = 86.

Also, in the number of two - dimensional elements includes 6 squares of cube faces. The total number of two - dimensional elements is

$f_2 = 12 + 86 + 6 = 104.$

In the number of the three - dimensional elements includes:

- 1 cube.
- 1 icosahedron.
- 20 tetrahedrons in the icosahedron.

8 tetrahedrons at the tops of the cube

17 - 1 - 2 - 12, 16 - 2 - 3 - 5, 15 - 4 - 5 - 7, 14 - 6 - 7 - 9, 13 - 11 - 8 - 9, 18 - 1 - 12 - 11, 19 - 12 - 4 - 8, 20 - 3 - 6 – 10.

6 pyramids with a base face of the cube

13 - 11 - 18 - 19 - 12 - 17, 19 - 17 - 4 - 2 - 15 - 16, 16 - 15 - 5 - 6 - 14 - 20, 13 - 18 - 10 - 9 - 20 - 14, 18 - 17 - 1 - 3 - 20 - 16, 13 - 19 - 8 - 7 - 14 – 15.

12 pyramids on the trapezes of these pyramids

18 - 17 - 11 - 12 - 1, 13 - 18 - 11 - 10 - 9, 11 - 12 - 13 - 19 - 8, 20 - 14 - 10 - 9 - 6, 17 - 16 - 1 - 3 - 2, 18 - 20 - 1 - 3 - 10, 13 - 14 - 9 - 8 - 7, 19 – 15 - 8 - 7 - 4, 14 - 15 - 7 - 6 - 5, 20 - 16 - 6 - 5 - 3, 16 - 15 - 5 - 2 - 4, 19 - 17 - 12 - 2 - 4.

The total number of three - dimensional elements is

$f_3 = 2 + 20 + 8 + 6 + 12 = 48.$

The four - dimensional element, as already proved, is an icosahedron with a center. There are other four - dimensional elements. The second such element is the polytope in Figure 7 after removing the center. Indeed, in this case

$f_0 = 21 - 1 = 20, f_1 = 78 - 12 = 66, f_2 = 104 - 30 = 74, f_3 = 48 - 20 = 28.$

Substituting these values of f_i in the equation (1), you can find that it is satisfied for $n = 4$

$20 - 66 + 74 - 28 = 0.$

This proves that the figure in Figure 7 after removing the center is a polytope of dimension 4. One can find three more elements of dimension 4 in Figure 7 if separate the prism connections on the parallel faces of the cube: upper and lower, right and left, front and back. Since these constructions are compatible with a rotation by 90^0, we prove the desired equality for only one of these constructions, for example, for the upper and lower faces of the cube. This construction is shown in Figure 8.

Figure 8. Four - dimensional part of the unit cell of the compound sodium and chlorine at high pressure

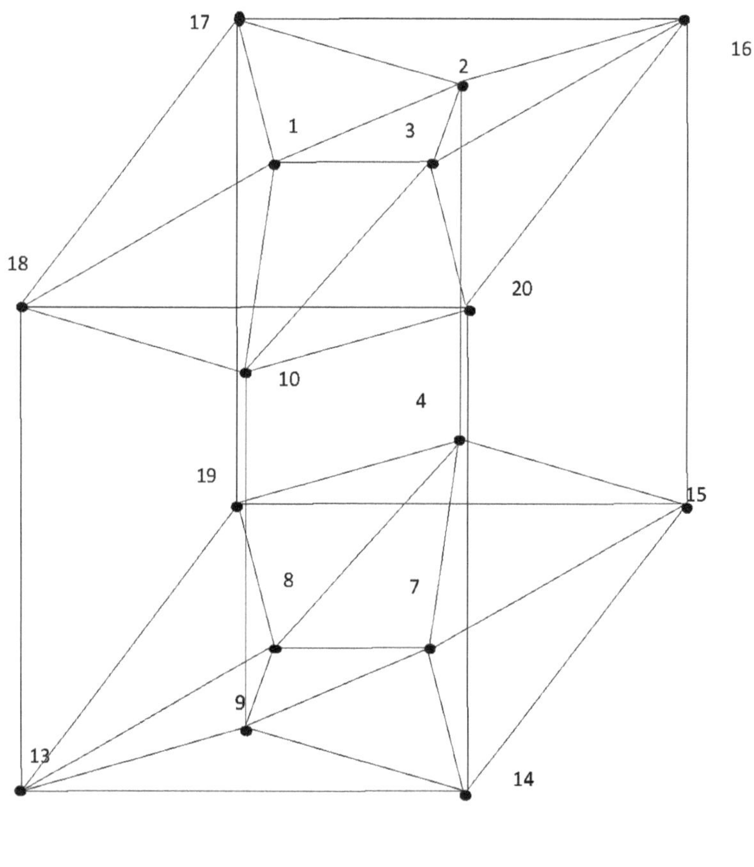

In it the number of vertices is 16, i. e. $f_0 = 16$. The number of edges consists of the number of edges of the cube (12); the number of edges of the cube in the upper part (without the edges of the cube) is

17 - 2, 2 - 16, 17 - 1, 1 - 2, 2 - 3, 1 - 3, 3 - 20, 3 - 10, 1 - 10, 1 - 18, 18 - 10, 10 - 20, 3 - 16,

i.e. 13 edges; the same number of edges (13) in the lower part; 2 connecting (vertical) edges 2 - 4, 10 - 9. Thus, the total number of edges is $f_1 = 40$.

In the number of two - dimensional elements includes:

- 6 faces of the cube.
- 10 facets of the pyramids in the upper part of the structure and 2 triangles from the pyramid with a crouching upper bound of the cube, i.e. 12 two - dimensional elements.
- 12 two - dimensional elements in the lower part of the structure.
- 2 vertical trapezes at the back wall of the cube and 2 vertical trapezes near the front wall of the cube.

Thus, the total number of two - dimensional elements is $f_2 = 34$.

In the number of the three - dimensional elements in Figure 8 includes:

- 2 pyramids with a trapezoidal base and one pyramid with a cube face at the top of the structure.
- 3 the same pyramids in the lower part of the structure.
- 2 pyramids with the bases of the back and front walls of the cube.
- 1 cube.

the figure left from the cube after deducting all the pyramids from it.

Thus, the total number of three - dimensional elements $f_3 = 10$. Substituting the values of f_i, determined for the polytope in Figure 8, in the equation (1) we find that it is satisfied for $n = 4$

$$16 - 40 + 34 - 10 = 0.$$

This proves that the construction in Figure 8 has dimension 4. Taking into account the existence of two more similar constructions and the impossibility of the existence of other similar constructions, we conclude that for the polytope in Figure 7 $f_4 = 5$. Substituting the values of f_i, determined for the polytope in Figure 7, in the equation (1), we find that it is satisfied for $n = 5$

$$21 - 78 + 104 - 48 + 5 = 2.$$

This proves that the polytope in Figure 7 has dimension 5.

It should be expected that other alkali metal salts at high pressure have structures with elementary cells of higher dimension. Besides atoms of alkali metals enters into complex compounds with many of different elements. We shall see later that these compounds also have higher dimensionality at normal pressures.

THE STRUCTURE AND HIGHER DIMENSION OF COMPOUNDS ELEMENTS OF SECOND GROUP

The elements of the second group of the periodic table have on the outer shell two of the valence electrons and can form linear molecules. For example, beryllium forms linear molecules with halogens BeF_2, $BeCl_2$, $BeBr_2$, BeI_2. However, in the vast majority of its compounds, beryllium manifests around itself tetrahedral coordination. An important example of such a compound is beryllium oxide. This compound is a valuable optical material. It is used in nuclear power engineering, microelectronics, and laser technology. The structure of this compound is a wurtzite structure. It shows tetrahedral coordination both around the beryllium atom and around the oxygen atom. It turned out to be equal to 4. Thus, the dimension of the beryllium oxide molecule is 4. For beryllium oxide, atoms of beryllium are located at the vertices *b, c, e, h, l, m, o* of the polytope Figure 2, and oxygen atoms are located at the vertices *a, d, g, k, p, f*. The nature of the chemical bond in beryllium oxide is still open. Estimates of the relative contribution of the ionic and covalent bonds are contradictory (Sholl & Walter, 1969; Hidaka, 1976). In any case, it is impossible to explain the observed structure of beryllium oxide by any distribution of electrons, divided or unshared electron pairs over elementary quantum cells of the adopted system of electronic orbitals. However, this also applies to compounds of the wurtzite type.

When the beryllium oxide is an extended crystalline body, then in order to build a model of such a body, the polytope Figure 2 need to multiply by an edge and to obtain a polytopic prismahedron of dimension 5.

In crystalline beryllium fluoride, the linearity of the combination of beryllium atoms with fluorine atoms and the tetrahedral coordination of beryllium atoms with one another surprisingly are combined. In this case, a structure is formed topologically equivalent to the adamantane molecule. The difference from the adamantane molecule is the linear arrangement between the beryllium atoms of the fluorine atoms (Figure 9). Since the adamantane molecule has a dimensionality of 4 (Zhizhin, 2014a), the unit cell dimension in crystalline beryllium fluoride is also 4.

Figure 9. The structure of crystalline beryllium fluoride

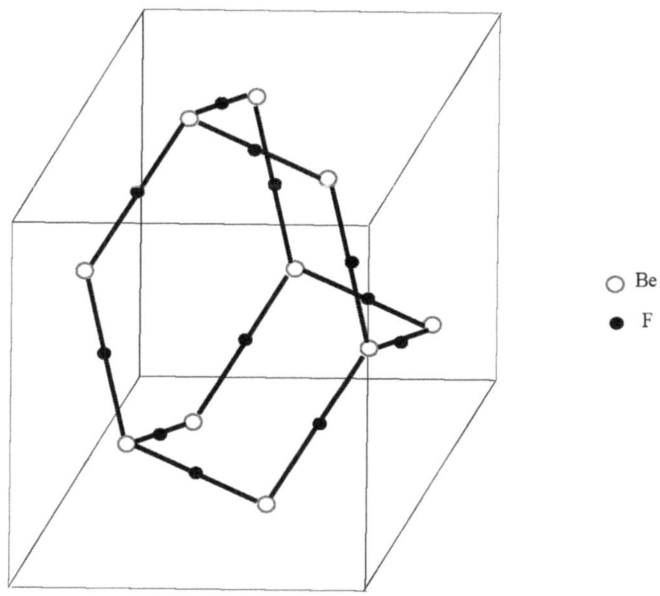

The second element in the second group of the periodic table after beryllium is the element of magnesium. Like beryllium, in accordance with the arrangement in the second group, it has two valence *s* - electrons on the outer layer. However, in contrast to beryllium, it has a completely filled pre - extrinsic layer of electrons. This layer includes two *s* - electrons and six *p* -electrons. These four electron pairs, starting from each other, create tetrahedral coordination around the magnesium atom. Taking vacant quantum cells of ligands, they increase the possible value of valence of magnesium to six. This gives magnesium more chemical activity especially important for living organisms. It participates in all metabolic processes in living organisms. Magnesium is one of the basic elements of the cell. It stimulates the work of enzymes that break down proteins and other nutrients. Magnesium participates in the harmonious work of all body systems, especially the central and peripheral nervous system, affects the growth of estrogen hormones and blood coagulability.

Even for the chemical bonds of magnesium with valence 2, compounds of higher dimension are formed. Consider a molecule of bis (neopentyl) magnesium $Mg(C_5H_{11})_2$ (Gillespie & Hargittai,, 1991). Magnesium in this molecule exhibits a valence of 2. In each group C_5H_{11}, the carbon atoms form the geometric form of a tetrahedron centered. This already gives the dimension of this form equal to 4. In addition, around each carbon atom there is also a tetrahedral coordination of other atoms (hydrogen and carbon). Each group C_5H_{11} can be represented in the form of a tetrahedron with a center in which its vertices contain functional groups CH_3, and in the fourth (attached to the magnesium atom) is a functional group CH_2. At the center of the tetrahedron is a carbon atom. Then the bis (neopentyl) magnesium molecule has the form of two tetrahedrons with a center connected to each other by a magnesium atom (Figure 10). Functional groups CH_3 are located in the vertices $a_1, c_1, d_1, a_2, c_2, d_2$; functional groups CH_2 are located in the vertices b_1, b_2; at the points o_1, o_2 are carbon atoms; at the point o there is a magnesium atom.

Figure 10. The structure of bis(neopentyl) magnesium molecule

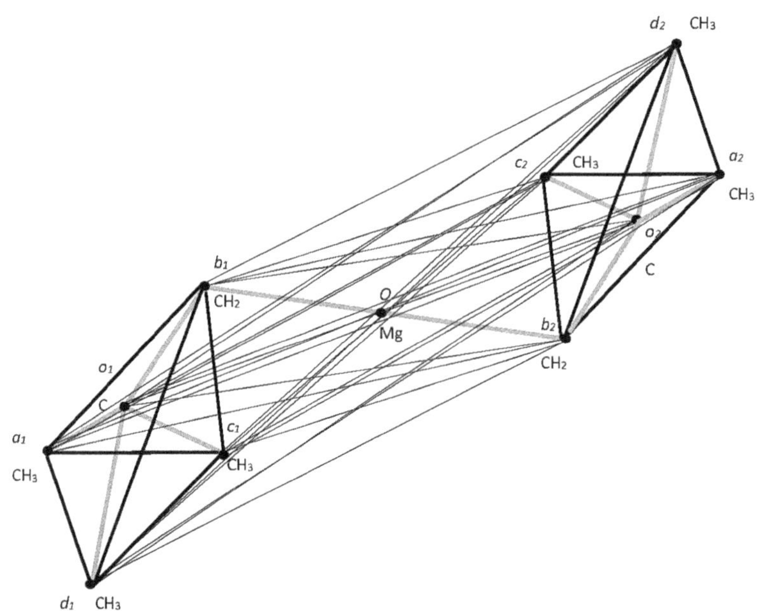

Valentine bonds are indicated in Figure 10 with a brown color. The remaining edges (black) serve to form a convex figure (polytope), the dimension of which must be established.

Theorem 3. The dimension of bis(neopentyl) magnesium molecule equal to 6.

Proof. For proof of theorem 3 we noted that polytope on Figure 10 is 5 – cross – polytope with centrum (Figure 11).

Figure 11. The 5 – cross – polytope with centrum

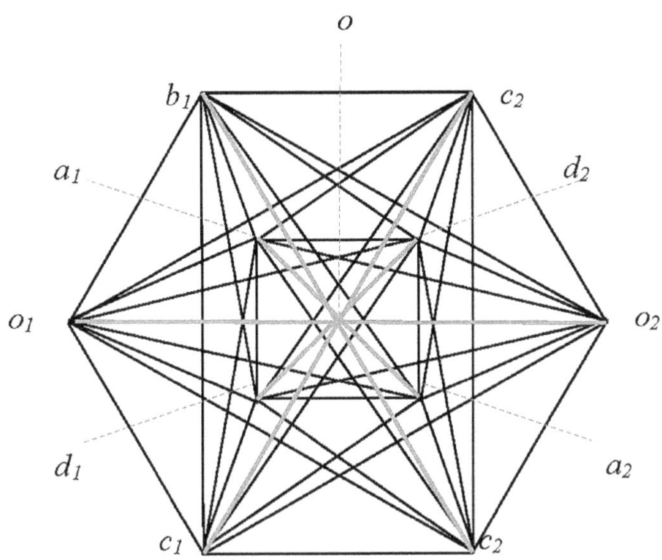

Comparing Figures 10 and 11, we see that these figures are topologically equivalent, that is, in Figure 11, the same vertices are shown as in Figure 10. Moreover, each of the corresponding vertices in Figure 11 is incidental to the number of edges as in Figure 10 and the connection of vertices by edges in Figure 11 is topologically the same as in Figure 10. If we denote in Figure 11 the edges issuing from vertex O to other vertices in green, the remaining figure, as can be seen, is the 5 - cross - polytope, given in the monograph by Zhizhin, (2014b). In addition, the vertex O is the center of 5 - cross - polytope. As follows from Zhizhin (2014b) 5 - cross -polytope has 10 vertices (f_0=10), 40 edges (f_1= 40), 80 triangular faces (f_2= 80), 80 tetrahedrons (f_3= 80), 32 4 - cross - polytopes (f_4= 32). The introduction of the center into the 5 - cross - polytope adds, according to Figure 11, 10 edges

$$\left(oa_1, ob_1, oc_1, od_1, oo_1, ob_2, oa_2, oc_2, od_2, oo_2\right),$$

24 triangular faces

$$\left\{\begin{matrix} o_1b_1o, b_1a_1o, b_1oa_2, b_1od_2, b_1oc_2, b_1oc_1, c_2d_2o, c_2od_1, c_2oa_2, c_2oo_2, o_2d_2o, o_2oa_2, o_2ob_2, \\ b_2a_2o, b_2od_2, b_2oc_2, b_2oc_1, b_2d_1o, c_1od_1, c_1od_2, c_1oa_2, c_1oo_1, a_1o_1o, o_1d_1o \end{matrix}\right\},$$

28 tetrahedrons

$$\left(\begin{matrix} b_1od_2a_2, b_1c_1oo_1, b_1d_1oa_1, b_1a_1d_2o, b_1od_1a_2, b_1oa_2c_2, b_1od_1c_1, c_2oa_1d_1, c_2b_2oo_2, \\ c_2a_2od_2, c_2a_1d_2o, c_2oa_2d_1, c_2od_1b_1, c_2oa_2b_2, c_1oa_2d_2, c_1a_1od_1, c_1d_1a_2o, c_1oa_1d_2, c_1od_2b_2, c_1od_1b_1, \\ o_1a_1d_1o, b_2od_1a_1, b_2d_2oa_2, b_2a_2d_1o, b_2od_2a_1, b_2oa_1c_1, b_2od_2c_2, o_2d_2a_2o \end{matrix}\right),$$

18 4 - simplexes

$$\left(\begin{matrix} b_1a_1od_2c_2, b_1c_1d_1a_1o, b_1o_1a_1oc_2, b_1od_2o_2c_2, b_1od_2a_2o_2, c_1d_1oa_2b_2, c_1o_1d_1b_2o, \\ c_1oa_2o_2b_2, c_1oo_2d_2o_2, c_2b_2a_2d_2o, c_2o_2d_2ob_1, c_2oa_1o_1b_1, c_2oa_1d_1o_1, b_2a_2od_1c_1, b_2od_1o_1c_1, \\ b_2od_1a_1o_1, o_2c_2od_2a_2b_2 \end{matrix}\right),$$

6 5 - simplexes

$$\left(o_1b_1a_1od_1c_1, c_2od_2a_2o_2b_2, c_1od_2a_2o_2b_2, b_2oa_1o_1d_1c_1, c_2od_1a_1o_1b_1, b_1oa_2o_2d_2c_2\right).$$

Adding the obtained quantities of geometric figures of different dimensions connected with the center of the 5 - cross - polytope to the corresponding numbers of figures not connected with the center of the 5 - cross - polytope, we obtain the total number of geometric figures of different dimensions in the 5 - cross - polytope with center:

$f_0=11, f_1=50, f_2=104, f_3=108, f_4=50, f_5=7$

Substituting these values in the equation (1), we find that the Euler-Poincare equation is satisfied for $n = 6$

$$11 - 50 + 104 - 108 + 50 - 7 = 0.$$

This proves that a 5 - cross - polytope with center has dimension 6.

It should be noted that the above evidence accurately lists (in view of the work of Zhizhin, 2014b) all the 108 three - dimensional figures included in the 6 - dimensional 5 - cross - polytope with the center. This is significantly different from the proof of the existence of 4 - dimensional 100 - cell and 600 - cell cells in Coxeter's work (Coxeter, 1963), for which the direct enumeration of three - dimensional figures included in these polytopes has not yet been received. Therefore, the question of proving the existence of these 4 - dimensional polytopes remains open.

If the electron pairs of magnesium at the second energy level enter into a chemical bond, then its valence is more than two. For example, in Grignard reagent the magnesium valence is 4 and in the vicinity of magnesium atom there is tetrahedral coordination. While the nearest neighborhood of the magnesium atom has a dimension of 4, and with the account of the attached groups of atoms this dimension is even higher. An interesting example is the complex magnesium ion $Mg(OAsMe_3)_5^{2+}$, Me=CH_3. In this compound, magnesium exhibits a valence of 5. In this case, the nearest environment of magnesium is of dimension 5. Indeed, the nearest environment of magnesium by oxygen atoms has the form of a 4 - simplex with a center in the magnesium atom (Figure 12). At the vertices *a, b, c, d, e* of the polytope, in Figure 12, there are oxygen atoms, in the vertex *o* there is a magnesium atom. The valence bonds are indicated in Figure 12 with a brown color, the other edges (black) are needed to create a convex figure in space. The vertices together with the connecting ribs form a 4 - simplex. The addition of a magnesium atom and valence bonds converts this polytope into a 4 - simplex with a center.

In Figure 12 can to indicate 6 vertices ($f_0 = 6$); 15 edges ($f_1 = 15$);

20 trigonal faces

(*abc, aeb, abo, abd, bcd, bco, bce, aeo, aed, aec, edo, edc, edb, dco, dca*), $f_2 = 20$;

Figure 12. The 4 - simplex with centrum

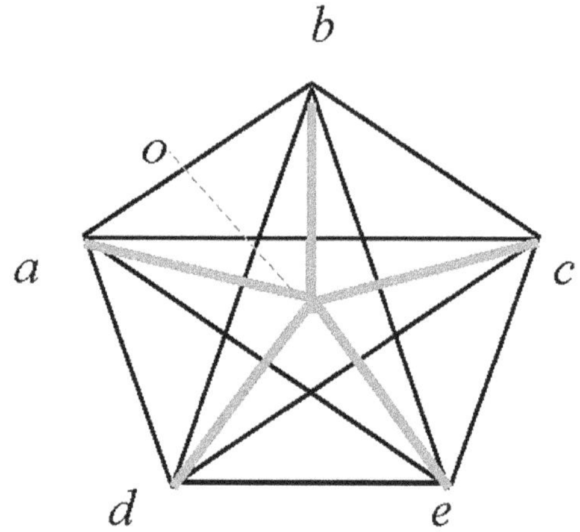

15 tetrahedrons

(*abed, abec, abcd, dbce, aecd, obcd, oecd, aoed, aoeb, aobc, boed, coae, doeb, eobc, aocd*), $f_3 = 15$;

6 4 - simplexes

(*abcde, abedo, abeco, abcdo, dbceo, aecdo*), $f_4 = 6$.

Substituting values f_i into equation (1), we find that the Euler - Poincaré equation is satisfied for $n = 5$

$6 - 15 + 20 - 15 + 6 = 2$.

This proves that a 4 - simplex with center has dimension 5. If we take into account the presence of other atoms in the ion $Mg(OAsMe_3)_5^{2+}$, then its dimension will be even higher.

Such compounds can form other alkaline - earth elements, i. e. calcium and barium.

THE STRUCTURE AND HIGHER DIMENSION OF COMPOUND ELEMENTS OF THE GROUP THREE (a)

The first element of group 3a of the element table is boron. Like all elements of this group, it has two *s* - electrons and one *p* - electron on the outer layer. Boron does not have vacant *d* - and *f* - orbitals, and there are not several electron pairs on the pre - existing layer, as, for example, for the atoms of alkaline - earth elements. However, the property of the collective interaction of electron pairs is also manifested here, but in a slightly different way compared to magnesium. Here pairs of electrons of the second energy level of several boron atoms interact, creating (repelling from each other) tetrahedral coordination of boron atoms. Therefore, in the compound B_4Cl_4, the boron atom has an effective valence of 4, and not three, which would correspond to the group number (Figure 13).

Figure 13. The structure of the B_4Cl_4 molecule

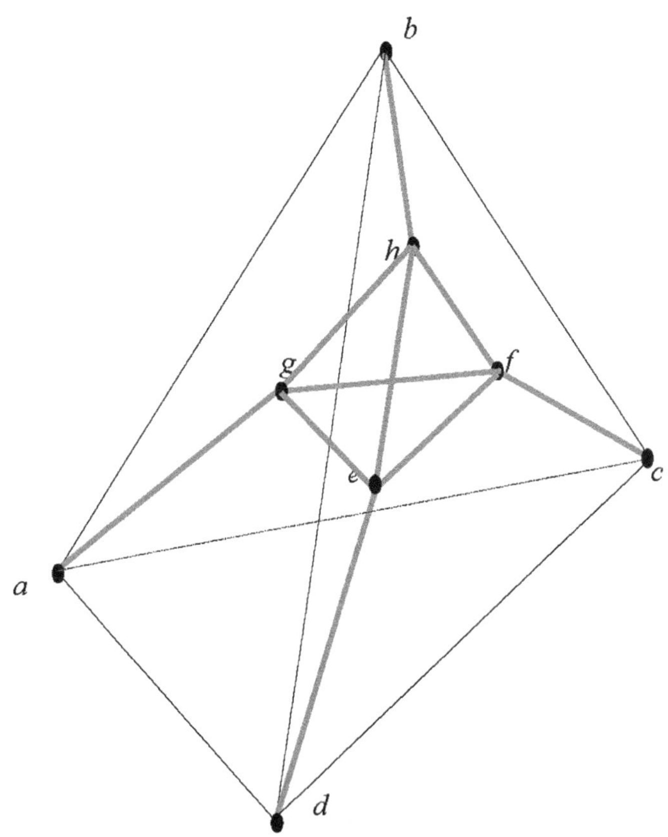

At the vertices *g, h, f, e* in Figure 13 boron atoms are located, and at the vertices *a, b, c, d* chlorine atoms arranged.

Theorem 4. The B_4Cl_4 molecule has dimension 4.

Proof. In Figure 13 there is eight vertices, $f_0 = 8$. The number of edges is 16

(*ab, bc, cd, ad, bd, ac, gh, hf, ef, he, gf, eg, bh, fc, ed, ag*), $f_1 = 16$.

The number of elements of dimension 2 is 14

(triangles *abd, bcd, abc, acd, ghe, hef, ghf, gfe*

and

quadrangles *aghb, aged, hbed, hbfc, efcd, hfbc*), $f_2 = 14$.

The number of elements of dimension 3 is 6

(tetrahedrons *abcd, ghef* and prismatoides *abhged, hbedbfc, aghbfc, gefadc*), $f_3 = 6$.

On Figure 13 the edges correspondent of chemical bounds is indicated brown, remain edges (black) it is need for creating convex body. Substituting the values of the number of elements of different dimension in the equation Euler - Poincaré (1), we obtain

$$8 - 16 + 14 - 6 = 0 .$$

We find that it holds for $n = 4$. This proves that the figure who projection is shown in Figure 13 there is polytope of dimension 4.

Due to the interaction of electron pairs of several atoms, formation of other compounds is also possible. For example, in Figure 14. The image of the B_6Cl_6 molecule is shown. Here, also, the edges corresponding to the chemical bonds is indicated in brown, the remaining edges are necessary for obtaining a convex figure.

In the compound, both the boron atoms and the chlorine atoms have octahedral coordination. The effective valence of boron in this compound is 5. In the polytope in Figure 14, boron atoms are located at the vertices $a_1, b_1, c_1, d_1, e_1, f_1$ and hydrogen atoms are located at the vertices $a_2, b_2, c_2, d_2, e_2, f_2$.

Theorem 5. The B_6Cl_6 molecule has dimension 4.

Figure 14. The structure of the B_6Cl_6 *molecule*

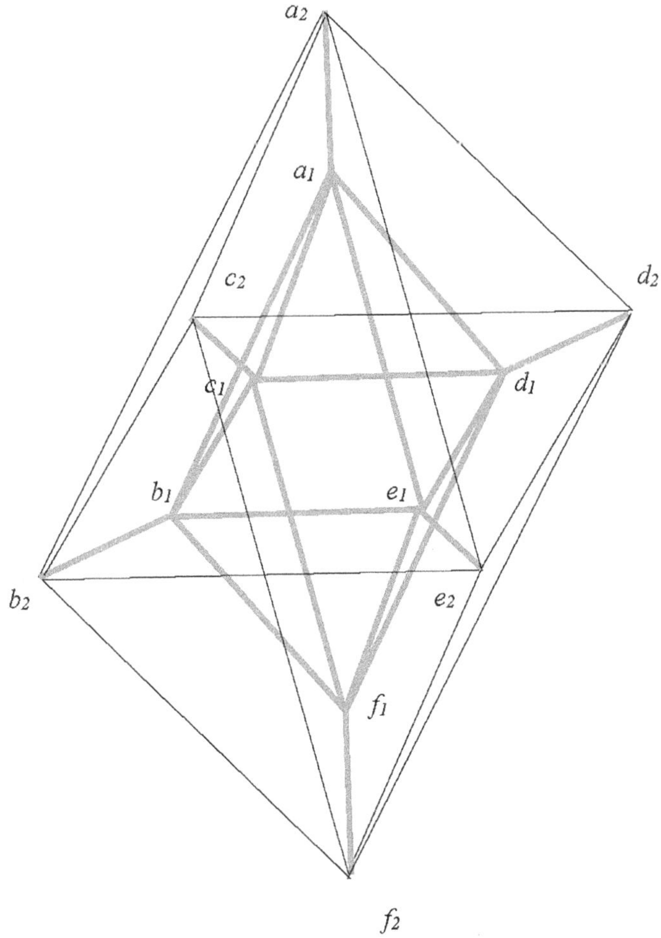

Proof. In this case the number of elements of zero dimension is f_0 = 12. The number of elements of dimension one is

$$f_1 = 12 + 12 + 6 = 30.$$

The number of elements of dimension 2 is sum of the number small triangles 8 and big triangles 8, add 12 quadrangles, i. e. $f_2 = 28$. The number of elements of dimension 3 is sum two octahedrons and eight prisms, i.e. $f_3 = 10$. Substituting the values of numbers of elements of different dimensions in the equation (1), we obtain

$12 - 30 + 28 - 10 = 0.$

We find that it holds for $n = 4$. This proves that the figure 8 is polytope of dimension 4.

Elements Al, Ga, In and Tl have vacant d - and f - orbitals and tend to supplement their valence shell to 6 electron pairs, and in several compounds In and Tl have more than 6 electron pairs. These elements in many compounds exhibit tetrahedral coordination in the vicinity of the atom. Taking into account the possible addition of other elements to tetrahedral coordination, complex compounds with high dimensionality can arise. For example, aluminum (a biogenic element) forms a cyclic compound $[(CH_3)_2AlF]_4$ (Figure 15).

Figure 15. A cyclic compound $[(CH_3)_2AlF]_4$

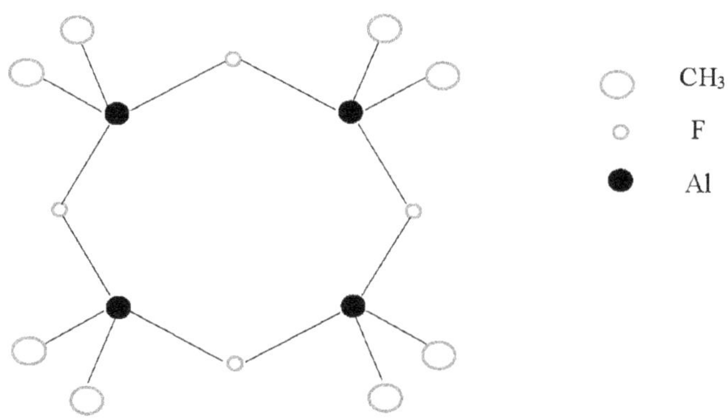

If we form a convex figure from Figure 15, we get the polytope shown in Figure 16. At the vertices of a_1, a_4, a_7, a_{10} fluorine atoms are located. At the vertices $a_{13}, a_{14}, a_{15}, a_{16}$ aluminum atoms are located. Functional groups CH_3 are located in the $a_2, a_3, a_5, a_6, a_8, a_9, a_{11}, a_{12}$ vertices.

Theorem 6. The polytope of cyclic compound $[(CH_3)_2AlF]_4$ has dimension 5.

Proof. The polytope in Figure 16 has 16 vertices, $f_0 = 16$; 52 edges, $f_1 = 52$. In addition, it has 4 polytopes of dimension 4 each (tetrahedrons with a center)

$$a_1a_2a_3a_4a_{14},\ a_4a_5a_6a_7a_{15},\ a_7a_8a_9a_{10}a_{16},\ a_{10}a_{11}a_{12}a_{13}.$$

Figure 16. The convex polytope of cyclic compound $[(CH_3)_2AlF]_4$

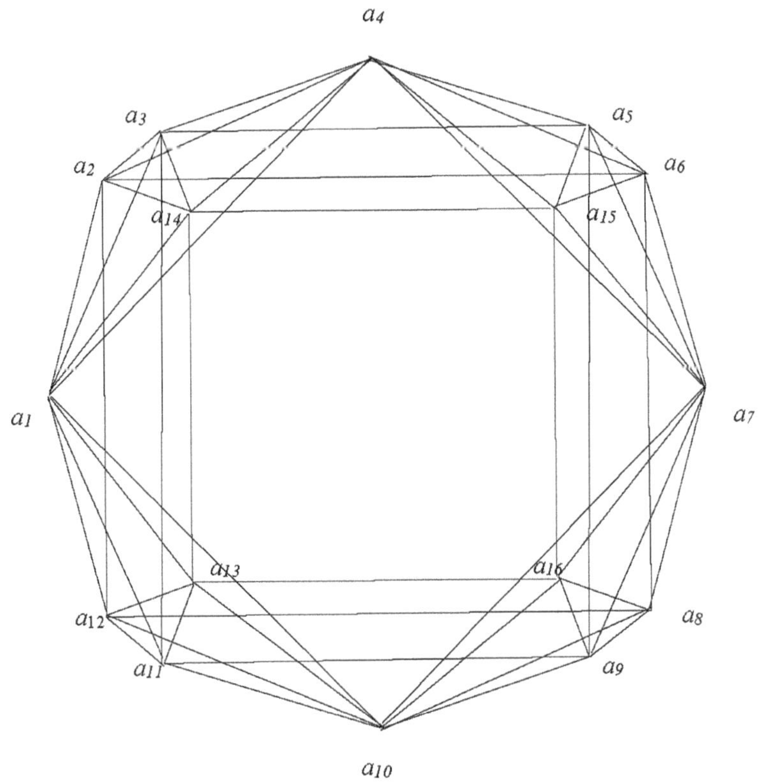

Each tetrahedron with a center has 10 triangular faces. This gives 40 triangular faces in the polytope 10. In addition, three triangular faces are formed at the vertices a_1,a_4,a_7,a_{10} with horizontal and vertical sides. This gives another 4 • 3 = 12 triangles. There are 4 more rectangular faces

$(a_{13}a_{14}a_{15}a_{16}, a_6a_2a_8a_{12}, a_{11}a_5a_3a_9, a_1a_7a_{10}a_4)$

and 12 trapezoids

$(a_3a_5a_{15}a_{14}, a_3a_2a_5a_6, a_{15}a_2a_6a_{14}, a_6a_8a_{15}a_{16}, a_6a_8a_5a_9, a_{15}a_5a_9a_{16},$

$a_{11}a_9a_8a_{12}, a_9a_{11}a_{13}a_{16}, a_{12}a_8a_{13}a_{16}, a_2a_{12}a_{14}a_{13}, a_{12}a_2a_3a_{11}, a_3a_{11}a_{13}a_{14})$.

Thus, the total number of two - dimensional faces 68, $f_2 = 68$. Each tetrahedron with a center has 5 tetrahedrons. Therefore, the total number of tetrahedrons in Figure 10 is 5 • 4 = 20. Each of the vertices a_3, a_7, a_4, a_{10} is the vertex of the three pyramids. The total number of these pyramids is 12:

$$a_5a_{15}a_3a_4a_{14}, a_6a_2a_3a_4a_5, a_4a_2a_6a_{15}a_{14}, a_1a_2a_3a_{12}a_{11}, a_1a_{14}a_3a_{11}a_{13},$$

$$a_{10}a_{12}a_{11}a_8a_9, a_{10}a_{12}a_{13}a_8a_{16}, a_{10}a_{11}a_{13}a_{16}a_9, a_7a_8a_9a_5a_6, a_7a_9a_{16}a_5a_{15}, a_7a_8a_{16}a_6a_{15}.$$

There are four triangular prisms:

$$a_2a_3a_{14}a_5a_6a_{15}, a_{15}a_5a_6a_8a_9a_{16}, a_{11}a_{12}a_{13}a_8a_9a_{16}, a_2a_3a_{14}a_{11}a_{12}a_{13},$$

and six quadrangular prisms:

$$a_{13}a_{14}a_{15}a_{16}a_3a_5a_9a_{11}, a_{13}a_{14}a_{15}a_{16}a_2a_6a_8a_{12}, a_{13}a_{14}a_{15}a_{16}a_1a_4a_7a_{10},$$
$$a_2a_6a_8a_{12}a_3a_5a_9a_{11}, a_1a_2a_4a_6a_7a_8a_{10}a_{12}, a_1a_3a_4a_5a_7a_9a_{10}a_{11}.$$

Then, the total number of three -dimensional figures is 42, $f_3 = 42$.

In addition to the 4 tetrahedrons mentioned with the center, as four - dimensional figures, there are another four - dimensional figures. In particular, this is a figure (F), shown in Figure 17. Indeed, this figure has 12 vertices, $f_0(F) = 12$; 24 edges, $f_1(F) = 24$; 19 two - dimensional faces, $f_2(F) = 19$; and 7 three - dimensional figures, $f_3(F) = 7$. Substituting these values in the Euler - Poincaré equation (1), we obtain that it is satisfied for $n = 4$

$$12 - 24 + 19 - 7 = 0.$$

This proofs that polytope F has dimension 4.

Four identical polytopes of dimension 4 exist in a neighborhood of each of the vertices a_1, a_4, a_7, a_{10}. One of these polytopes (L) is depicted in Figure 18. It has 7 vertices, $f_0(L) = 7$; 15 edges, $f_1(L) = 15$; 14 two - dimensional faces, $f_2(L) = 14$; and 6 $3D$ facets, $f_3(L) = 6$. Substituting these values into the Euler - Poincaré equation (1), we obtain

$$7 - 15 + 4 - 6 = 0,$$

Figure 17. The 4 - dimension polytope F included in Figure 16

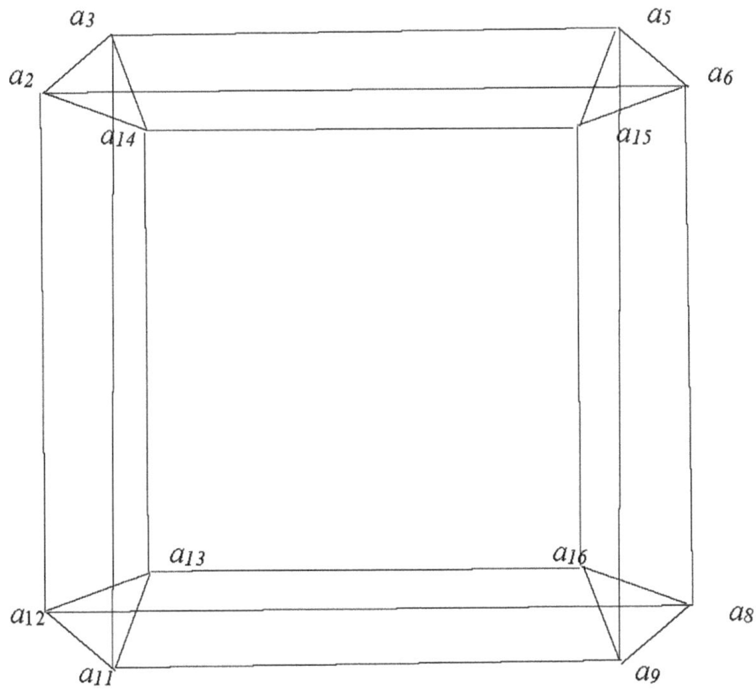

Figure 18. The 4 - dimension polytope L included in Figure 16

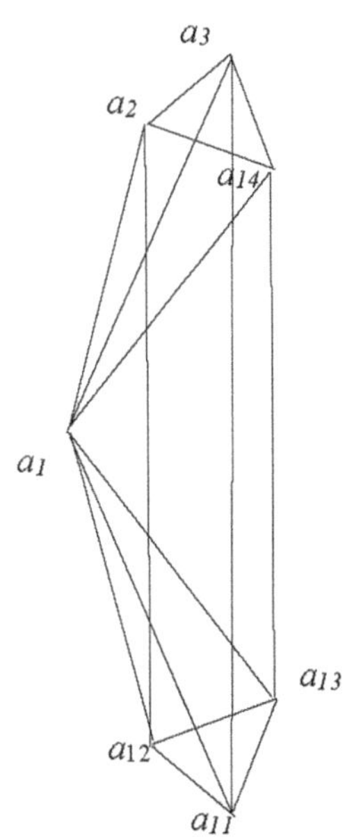

i.e. the equation (1) hold for $n = 4$ and all the polytopes L has dimension 4.

Three more topologically equivalent polytopes of dimension 4 can be distinguished from Figure 16. Each of these polytopes consists of a rectangular prism and four tetrahedrons connected to each other in a cycle along the vertices of a_1, a_4, a_7, a_{10}. These are polytopes

$$a_1a_4a_3a_{14}a_5a_{15}a_7a_{16}a_8a_{10}a_{13}a_{12},\ a_1a_2a_{14}a_4a_{15}a_6a_7a_{16}a_9a_{10}a_{11}a_{13},\ a_1a_2a_3a_4a_5a_6a_7a_8a_9a_{10}a_{11}a_{12}.$$

One of them (polytope K) is shown in Figure 19.

Figure 19. The 4 - dimension polytope K included in Figure 16

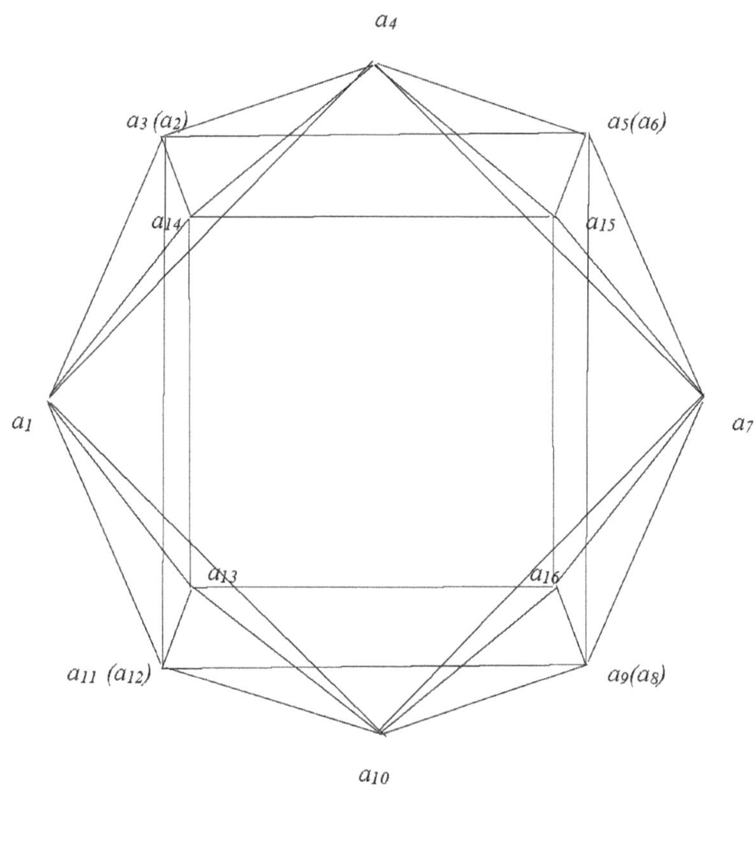

The K polytope has 12 vertices, $f_0(K) = 12$; 32 edges, $f_1(K) = 32$; 31 two - dimensional faces, $f_2(K) = 31$; and 11 $3D$ facets, $f_3(K) = 11$. Substituting these values into the Euler - Poincaré equation (1), we obtain

$$12 - 32 + 31 - 11 = 0,$$

i. e. the equation (1) hold for $n = 4$ and all the polytopes K has dimension 4.

Thus, the polytope on Figure 16 has 11 polytopes of dimension 4. Therefore, for polytope on Figure 16 the Euler - Poincaré equation (1) has face

$$16 - 52 + 68 - 42 + 11 = 2,$$

i.e. it holds for $n = 5$.

THE DIMENSION OF THE FE – PORPHYRIN

In the center of myoglobin globule (Lehninger, 1982) is hemo - group containing Fe - porphyrin (iron atom surrounded by five nitrogen atoms).

Theorem 7. The dimension of the Fe - porphyrin before joining the oxygen atom is equal to 5.

Proof. Consider the first coordination sphere of the iron atom in the center of the porphyrin (Zhizhin, 2015a), since only in the first coordination sphere of atoms are linked by a covalent bond, and in the following focal areas of intermolecular bonds between atoms. Before joining of the oxygen atom the first coordination sphere of Fe - porphyrin may be represented as a plane projection (Figure 20), at the vertices *a, c, d, f* of which the nitrogen atoms of the porphyrin are located, an iron atom is located at the vertex *g*, and the nitrogen atom of the nearest histidine residue is located at the vertex *b*. The deflection of vertex *g* from the center of the rectangle *acdf* corresponds to a certain "dome" character of porphyrin (Steed & Atwwod, 2007; Lehn, 1998). The projection in Figure 12 represents some polytope (let`s denote *A* - polytope).

The *A* – polytope has six elements with dimension 0, $f_0(A)=6$. There are vertices *a, c, d, f, g, b*. The number of elements with dimension 1 is $f_1(A) = C_6^2 = 15$. It are edges

ab, bc, bd, bf, bg, ac, cd, fd, af, fc, ad, ag, gc, fg.

The number of elements with dimension 2 is $f_2(A) = C_6^3 = 20$. It are triangles

Figure 20. The first coordination sphere of Fe - porphyrin before binding oxygen

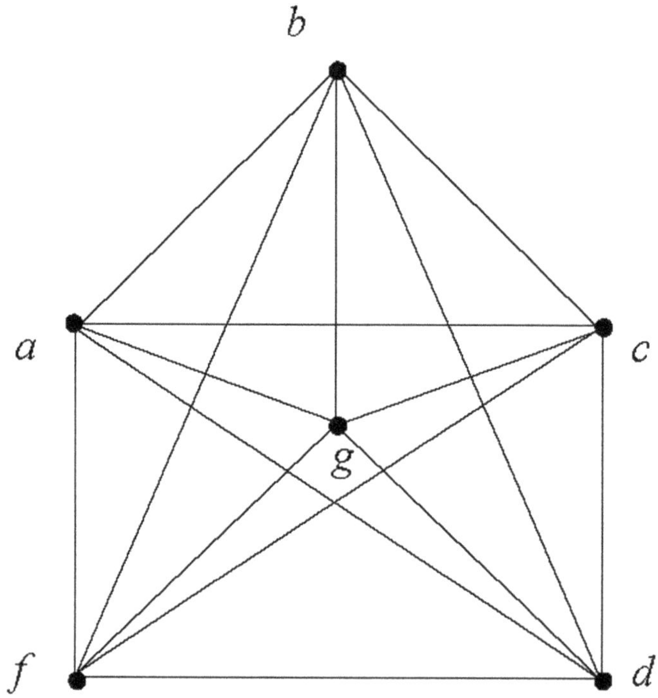

abf, bfg, bgd, dbc, bga, bgc, agc, dfg, adc, acf, fcd, bgd, fbg, agd, fgc, fbc, abd, afg, gcd, afd).

The number of elements with dimension 3 is $f_3(A) = C_6^4 = 15$. It are tetrahedons

abgf, bsgd, abfc, abcd, bfcg, abdg, acfg, abdf, acdg, bfdg, abgc, fbcd, fgcd, afgd, afcd).

The number of elements with dimension 4 is $f_4(A) = C_6^4 = 6$. It are simplexes

abcdf, adcdg, abdfg, abcfg, bcdfg, acdfg.

Substituting the received numbers of elements of different dimensions in the equation (1) at a value of $n = 5$, we obtain

6 - 15 + 20 - 15 + 2 = 2,

i.e. the Euler - Poincare equation is satisfied for A - polytope with $n = 5$. This is a simplex of dimension 5.

Theorem 8. The dimension of Fe - porphyrin after joining of oxygen atom is 6.

Proof. The first coordination sphere after joining oxygen atoms is complemented by one vertex e (Figure 21).

Figure 21. The first coordination sphere of Fe - porphyrin after joining of oxygen atom

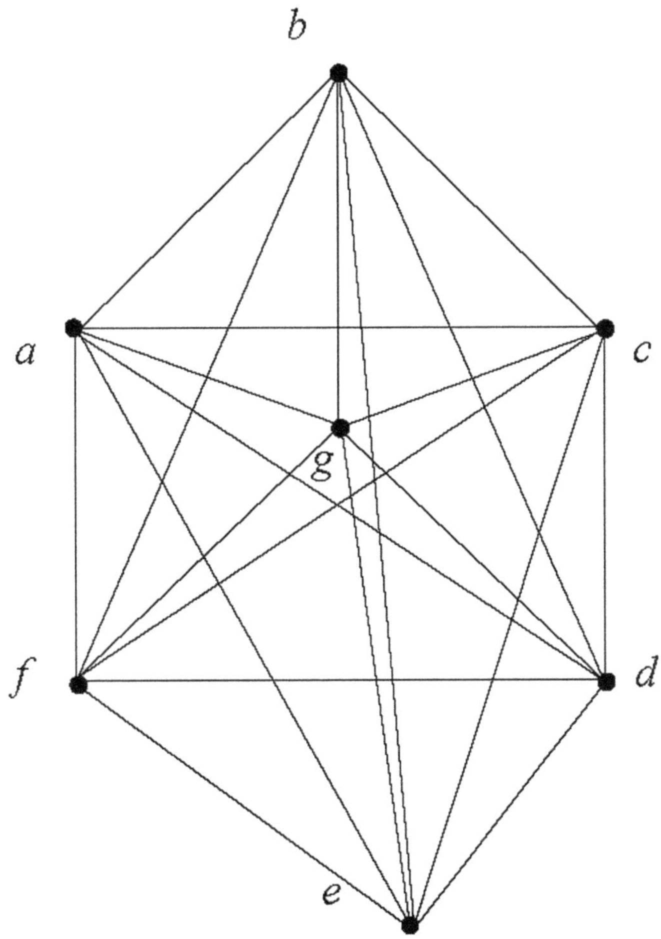

"Dome" character of Fe - porphyrin after joining of oxygen atom decreases, but it is not possible to affirm that it disappears completely (Steed & Atwood, 2007). Therefore, the deflection of vertex *g* from the center of rectangle in Figure 21 quality is maintained qualitatively. Taking into account the significant difference between the geometry and mass of the groups attached to the iron atom at the top and bottom, it is shown in Figure 21 that the vertices *e* and *b* do not lie on the same line. In the polytope in Figure 21 (*B* - polytope) the number of elements of zero dimension is increased compared with to the *A*-polytope by one vertex *e*, $f_0(B)=7$. This leads to the increase in the dimension of the polytope by 1, as the number of edges issuing from each top also increases by 1. In the polytope *B* the number of elements of dimension 1 is $f_1(B) = C_7^2 = 21$ (edges). The number of elements with dimension 2 is $f_2(B) = C_7^3 = 35$ (triangles). The number of three-dimensional figures is $f_3(B) = C_7^4 = 35$ (tetrahedrons). The number of elements with dimension 4 is $f_4(B) = C_7^5 = 21$, (simplexes of dimension 4). The number of elements with dimension 5 is $f_5(B) = C_7^6 = 7$, (simplexes of dimension 5). Substituting the numbers of the elements of different dimensions in equation (1) with $n = 6$, can to get

$$7 - 21 + 35 - 35 + 21 - 7 = 0,$$

i.e. the Euler - Poincaré equation for *B* - polytope is satisfied when $n = 6$. Therefore, *B* - polytope is a simplex of dimension 6. This proves theorem 8.

The dimensions of molecules increase with an increase of its energy again. It is shown that myoglobin is associated coil circuit elements of higher dimension (4) and in the center of the coil is a group of atoms even greater dimension.

CONCLUSION

The structures of compounds of a metal atom with ligands were studied by sequentially changing the groups and subgroups of the periodic system of elements in which the metal atom is located. It is shown that all metals from the first to the eighth groups form chemical compounds of a higher dimension. The images of these compounds in the spaces of higher dimension were constructed and presented, the values of the dimensions of various specific compounds were determined. The formation of molecules of higher dimension occurs due to the chemical bonds of the metal atom with ligands both due to the influence of electron pairs, and due to the attraction of ions. Moreover, the apparent valence of the metal atom, as a rule, exceeds the value of the valence, determined by the location of the metal in the periodic table of chemical elements. The concept of secondary valency, used in coordination chemistry, loses its meaning under these conditions. Consideration of the geometry of the compound of a single metal atom with ligands allows us to proceed to the analysis of the geometry of clusters of several atoms with ligands in the subsequent chapters.

REFERENCES

Coxeter, H.S.M. (1963). *Regular Polytopes.* New York: John Wile & Sons, Inc.

Gillespie, R. J. (1972). *Molecular Geometry*. New York: Van Nostrand Reinhold Company.

Gillespie, R. J., & Hargittai, I. (1991). *The VSEPR Model of Molecular Geometry*. London: Allyn & Bacon.

Grinberg, A. A. (1986). *Introduction to the chemistry of complex compounds.* Moscow: Chemistry.

Hidaka, T. (1976). Bond - ioniciti anisotropy of the distorted wartzite - tipe crystals. *Physics*, *BC84*(30), 345–352.

Lehn, J. M. (1998). *Supramolecular chemistry. Concept and perspectives.* Novosibirsk: Science.

Lehninger, A. L. (1982). *Principles of Biochemistry* (Vol. 1-3). New York: Worth Publisher, Inc.

Morozov, A. N. (2008). *The theory of the structure of coordination compounds.* Rostov: Southern Federal University.

Perrin, D. D., Armarego, W. L. F., & Perrin, D. R. (1980). *Purification of Laboratory Chemical.* New York: Pergamon.

Poincaré, A. (1895). Analysis situs. *J. de é Ecole Polytechnique, 1,* 1 – 121.

Sholl, C. A., & Walter, J. A. (1969). Covalent and ionic models for the electric field gradient in BeO. *Journal of Physical Chemistry*, *2*, 2301–2309.

Steed, J. V., & Atwood, J. L. (2000). *Supramolecular chemistry*. Chichester: John Wiley and Sons.

Werner, A. (1936). *New appearances in the field of inorganic chemistry*. Moscow: ONTI.

Zhang, W., Oganov, A. R., Goncharov, A. F., Zhu, Q., Boulfelfel, S. E., Lyakhov, A. O., ... Konopkova, Z. (2013). Unexpected Stable Stereochemical of Sodium Chlorides. *Science*, *342*(6165), 1502–1505. doi:10.1126cience.1244989 PMID:24357316

Zhizhin, G. V. (2014a). On the higher dimension in nature. *Biosphere*, *6*(4), 313–318.

Zhizhin, G. V. (2014b). *World – 4D*. St. Petersburg: Polytechnic Service.

Zhizhin, G. V. (2015a). The dimensions of supramolecular compounds. *Biosphere*, *7*(2), 149–154.

Zhizhin, G. V. (2015b, November). *Polytopic prismahedrons – fundamental regions of the n-dimension nanostructures*. Paper presented at The International conference "Nanoscience in Chemistry, Physics, Biology and Mathematics", Cluj-Napoca, Romania.

Zhizhin, G. V. (2016a). From the multidimensional phase space to a multidimensional space in nature. *Biosphere*, *8*(3), 258–267.

Zhizhin, G. V. (2016b). The structure, topological and functional dimension of biomolecules. *J. Chemoinformatics and Chemical Engineering*, *5*(2), 44–58.

Zhizhin, G. V. (2018). *Chemical Compound Structures and the Higher Dimension of Molecules: Emerging Research and Opportunities*. Hershey, PA: IGI Global. doi:10.4018/978-1-5225-4108-0

Zhizhin, G. V. (2019). *Attractors and Higher Dimensions in Population and Molecular Biology: Emerging Research and Opportunities*. Hershey, PA: IGI Global. doi:10.4018/978-1-5225-9651-6

Zhizhin, G. V. (2020). The Structure od DNA taking into account the Higher Dimension of its Components. In Encyclopedia of Organizational Knowledge, Administration, and Technologies. Hershey, PA: IGI Global.

Zhou, X. F. (2012). First Principles Determination of the Structure of Magnesium Borohydride. American Physical Society, 109, 245503-1 – 245503-5.

Zhu, Q., Oganov, A. R., & Lyakhov, A. O. (2013). Novel stable compounds in the Mg – O system under high pressure. *Phys. Chem., 15*, 7696–7700. PMID:23595296

KEY TERMS AND DEFINITIONS

Anomalous Elements: Transitional chemical elements in which the filling of the orbitals by electrons occurs with a violation of the experimental series of an increase in the energy of orbitals.

Divided Electron Pair: The binding electron pair, which simultaneously belongs to two atoms in the molecule.

Geometrical Image of a Chemical Compound: The geometrical image of a chemical compound (molecule) is a convex polytope, at the vertices of which atoms (or functional groups) are located. The edges of the polytope connecting the vertices correspond to the chemical bonds of the compound. The part of edges only carries a geometric function. They are necessary to give the molecule the image of a convex geometric figure. The dimension of the polytope is determined by the Euler-Poincare equation.

N-Cross-Polytope: The convex polytope of dimension n in which opposite related of centrum edges do not have connection of edge.

N-Simplex: The convex polytope of dimension n in which each vertex is joined by edges with all remain vertices of polytope.

S- and P-Elements: The chemical elements in which is filling with electrons s- and p-orbitals of atoms.

Tetrahedral Coordination of Electron Pairs: The location of the electronic pairs of the outer and the pre-outer electron layer at the vertices of the tetrahedron.

Transitional Elements: Chemical elements in which electrons fill *d*- and *f*-orbitals of an atom (*d*- and *f*-elements).

Undivided Electron Pair: A non-bonding electron pair belonging to one atom in a molecule.

Chapter 4
Chains of Metallic Clusters With Ligands

ABSTRACT

This chapter geometrically investigated the structure of clusters, the core of which represent the metal chains (linear or curved) of both identical and different elements. It was shown that the dimension of the structures of these clusters is more than three. To create a model of these chains in a higher dimension space, a new geometric approach has been developed that allows us to construct convex, closed polytopes of these chains. It consists of removing part of the octahedron edges necessary for constructing the octahedron and adding the same number of new edges necessary to build a closed polytope chain while maintaining the number of metal atoms and ligands and their valence bonds. As a result, it was found that metal chain polytopes consist of polytopes of higher dimension, adjacent to each other along flat sections.

INTRODUCTION

Models containing polymetallic chains formed by metal atoms with ligands attached to them represent a special class of multicore clusters. Experimentally (Gubin, 2019) obtained a significant number of such compounds in the form of metal chains (usually three - link chains). Metal chains can be both homo - element and hetero - element, and the structure of the metal chain can be linear and non - linear (curved). Binuclear compounds with a metal - metal bond is partially consecrated in the monograph Cotton, Walton (1982). From short

DOI: 10.4018/978-1-7998-3784-8.ch004

chains, as from bricks, longer metal chains can be constructed. In addition, the formation of polymeric compounds is possible, the main chain in which is built of metal atoms linked together.

In this monograph, it is of interest to determine the dimensionality of the chains of metal clusters with ligands. The methods for analyzing the geometry of high - dimensional polytopes, developed by the author (Zhizhin, 2018, 2019a, b), are used.

LINEAR HOMO - ELEMENT METAL CHAINS

An example of a linear homo - element metal chain is shown in Figure 1 (Smart, Cook, & Woodward, 1977; Evans, Okrasinski, Pribula, & Norton, 1976; Gochin & Moss, 1980).

Figure 1. Schematic structure of cluster connection $Os_3(CO)_{12}R_2$

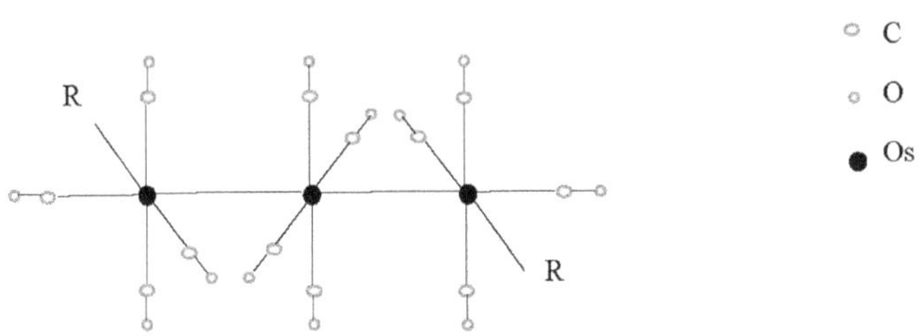

In this example, we illustrate the sequence of actions to calculate the dimension of the metal chain. When determining the dimension of metal chains, atoms that are linearly located and do not have branching of the chain at their location do not matter (Zhizhin, 2018). At the same time, not only metal atoms, but also ligand atoms are of importance for determining the dimension of a compound. Therefore, Figure 1 can be represented as Figure 2.

In Figure 2, the red edges correspond to chemical bonds, and the black edges serve only to create a convex closed figure. It can be seen that each link of the metal chain creates an octahedron with the center as a convex figure. Moreover, the center of each octahedron is simultaneously the vertex of another neighboring octahedron or two neighboring octahedrons. In Figure

Figure 2. The geometric structure of cluster connection $Os_3(CO)_{12}R_2$

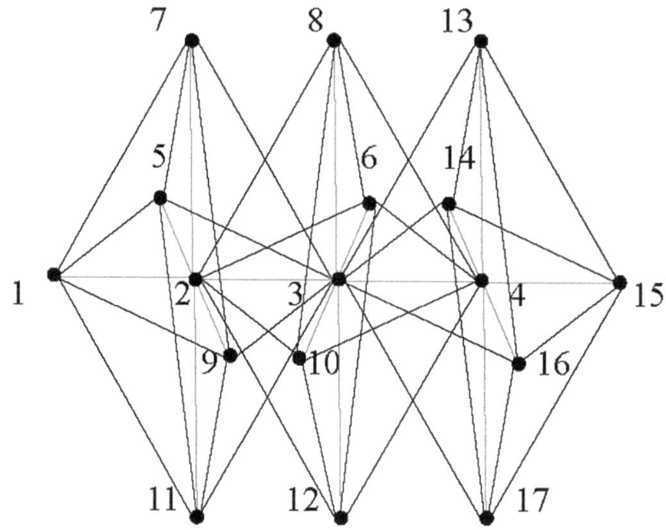

2 at the centers of octahedrons are disposed Os atoms (vertices 2, 3, 4). At the vertices

1, 6, 7, 8, 9, 10, 11, 12, 13, 14, 15, 17

Figure 3. Octahedron with a center

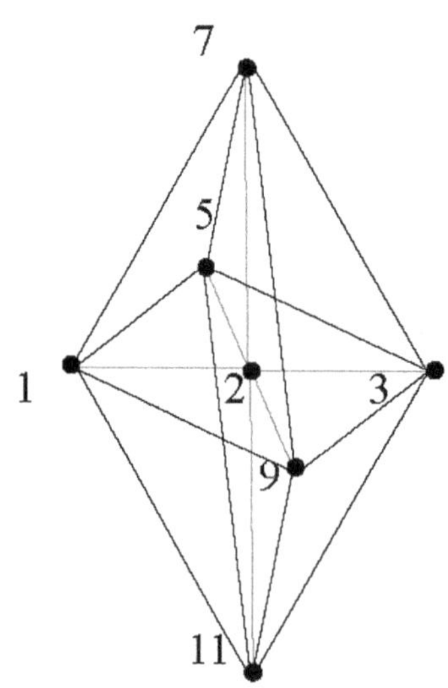

are disposed the functional group CO, at the vertices 5, 16 are disposed the functional group R (Zhizhin, 2016). It is easy to prove that an octahedron with a center has a dimension of 4. To do this, refer to Figure 3, which shows separately the first of the chain octahedron.

Polytope on Figure 3 has 7 vertices (f_0=7), 18 edges (f_1=18)

1 - 2, 2 - 3, 2 - 5, 2 - 9, 1 - 5, 5 - 3, 3 - 9, 1 - 9, 2 - 4, 2 - 11, 1 - 7, 5 - 7, 3 - 7, 7 - 9, 1 - 11, 5 - 11, 9 - 11, 3 - 11.

Polytope composed of eight tetrahedrons

1 - 2 - 9 - 11, 2 - 9 - 3 - 11, 2 - 5 - 3 - 11, 1 - 2 - 5 - 11, 1 - 2 - 5 - 7, 2 - 5 - 3 - 7, 2 - 9 - 3 - 7, 1 - 2 - 9 - 7.

To these three - dimensional figures you need to add an octahedron without a center

1 - 5 - 3 - 7 - 9 - 11.

Thus, the total number of three - dimensional figures is equal to 9 (f_3=9). Each flat face of any of these nine polyhedrons is simultaneously one of the flat faces of some other of these nine polyhedrons. Therefore, the total number of two - dimensional faces in the polytope in Figure 3 is 16 (f_2=16):

1 - 9 - 11, 9 - 3 - 11, 5 - 3 - 11, 1 - 5 - 11, 5 - 3 - 7, 1 - 5 - 7, 1 - 7 - 9, 7 - 3 - 9, 2 - 9 - 11, 2 - 3 - 11, 1 - 2 - 11, 5 - 2 - 11, 1 - 2 - 7, 2 - 5 - 7, 2 - 3 - 7, 2 - 11 - 9, 1 - 2 - 5, 2 - 5 - 3, 2 - 9 - 3, 1 - 2 - 9.

Define dimension polytope in Figure 3 on the Euler - Poincaré equation (Poincaré, 1895)

$$\sum_{i=0}^{n-1}(-1)^i f_i(P) = 1-(-1)^i, \qquad (1)$$

f_i is the number of the elements with the dimension i at polytope P; n is dimension of the polytope P.

Substituting the obtained values $f_i(i=0,1,2,3)$ in the equation (1) you can see that the Euler – Poincaré equation is satisfied in this case with a value of n equal to 4. This proves that the dimension of an octahedron with center is 4.

In a linear homogeneous chain of metal atoms, tetrahedrons penetrate into each other (Figure 2). This greatly complicates the geometric analysis of the compound. Here one could use the same method of forming a convex figure, corresponding to a chain, as in the analysis of individual molecules. However, the construction of an additional system of edges, taking into account the possible growth of the chain, greatly complicates the convex figure. In this case, the dimension of the whole figure, the corresponding chain, will increase dramatically with the growth of the chain. This approach seems to be irrational. It would be logical to present these chains in the form of a convex figure consisting of a sequence of convex polytopes adjacent to each other in the geometric analysis of the chains of metal atoms.

Theorem 1. Linear homo - element metal chains with ligands are a sequence of convex polytopes of dimension 4 adjacent to each other along whole sections of octahedrons passing through the center and the vertices of the octahedrons.

Proof. Taking into account that metal chains with two chain links exist (Cotton & Walton, 1982), consider two adjacent chain links in Figure 2. To create a closed shape from two octahedrons interconnected, it is necessary to fill the free space between the octahedrons. This can be done by connecting

Figure 4. Convex figure built on a metal chain of two metal atoms with ligands

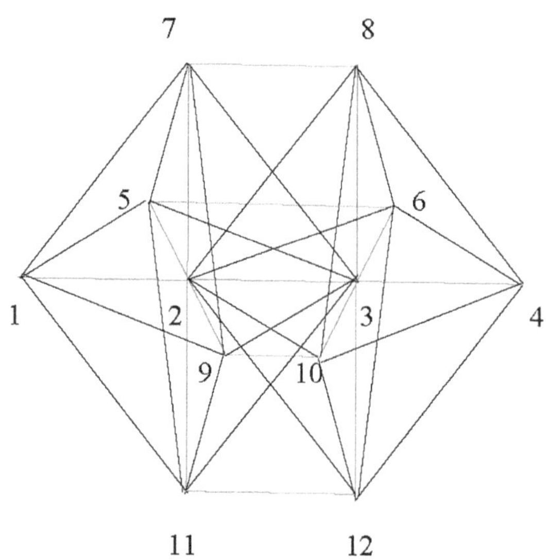

the vertices of two octahedrons, which were not yet connected to each other (Figure 4). These compounds

(7 - 8, 5 - 6, 9 - 10, 11 - 12)

in Figure 4 are indicated in green.

In Figure 4, the region in which the octahedrons enter each other

5 - 7 - 8 - 6 - 10 - 12 - 9 - 11 - 2 - 3

should be distinguished. As well as two regions in which only of the octahedron is present

1 - 5 - 7 - 9 - 11 - 2, 4 - 8 - 6 - 10 - 12 - 3

Figure 5.

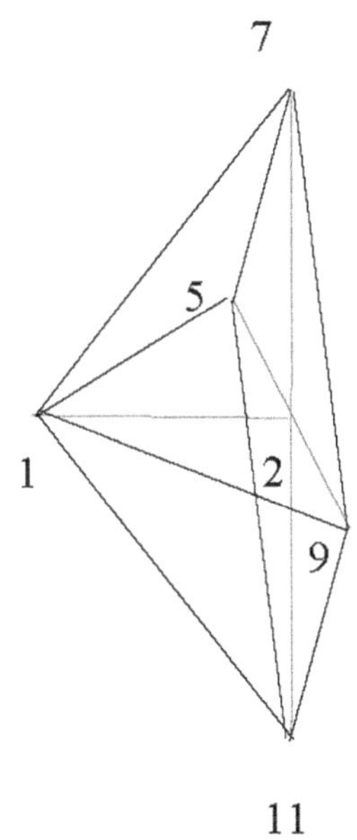

should be distinguished. An area in which only one of the octahedron is present, for example,

1 - 5 - 7 - 9 - 11 - 2

(Figure 5) has 6 vertices (f_0=6), 13 edges (f_1=13)

1 - 5, 1 - 7, 1 - 2, 1 - 9, 1 - 11, 5 - 7, 7 - 9, 9 - 11, 5 - 11, 2 - 5, 2 - 7, 2 - 9, 2 - 11,

12 flat two - dimensional faces (f_2=12)

1 - 7 - 5, 1 - 5 - 2, 1 - 7 - 2, 1 - 9 - 7, 1 - 2 - 9, 1 - 9 - 11, 1 - 2 - 11, 1 - 5 - 11, 2 - 5 - 7, 2 - 7 - 9, 2 - 11 - 9, 5 - 2 - 11,

5 three - dimensional areas (f_3=5)

1 - 5 - 2 - 7, 1 - 2 - 7 - 9, 1 - 2 - 9 - 11, 1 - 5 - 2 - 11, 1 - 7 - 9 - 5 - 11 - 2.

Substituting the obtained values f_i(i=0,1,2,3) in the equation (1), you can see that the Euler - Poincaré equation is satisfied in this case with a value of n equal to 4

$$6 - 13 + 12 - 5 = 0.$$

This proves that each of these regions has dimension 4. It should be noted that the region

1 - 7 - 9 - 5 - 11 - 2

does not contain edge 1 - 2, therefore it has dimension 3. As soon as edge 1 - 2 is added, region has dimension 4. The base of this figure is constructed from four triangles that are faces of other three - dimensional figures in this polytope.

The region

5 - 7 - 8 - 6 - 10 - 12 - 9 - 11 - 2 - 3

in which the octahedrons penetrate into each other is quite complicated. However, here we must bear in mind that the edges of the octahedrons in this region, which are not valence bonds, were necessary for the creation of convex figures (octahedrons) when considering these figures separately.

When the general region, which includes octahedrons, is considered, their

Figure 6. Simplified region penetrated by adjacent octahedrons in chains

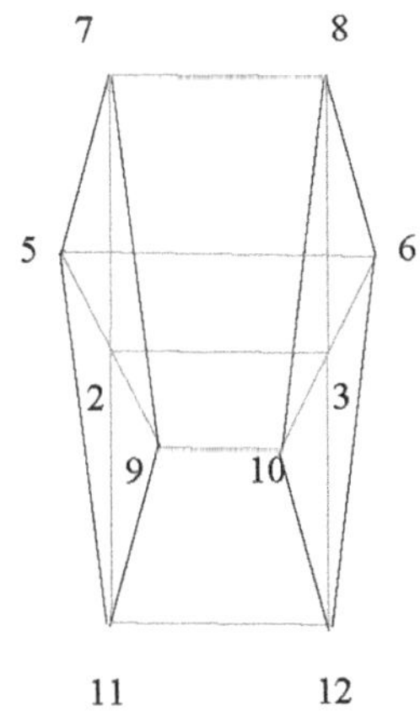

purpose is lost, due to the introduction of other edges that create a unifying closed convex figure. Therefore, in this region one should leave only the edges corresponding to the valence bonds and the edges forming the boundary of the region. Other edges

7 - 3, 8 - 2, 5 - 3, 2 - 6, 2 - 10, 3 - 9

must be removed. In this case, the region takes the form shown in Figure 6.

The polytope on Figure 6 has 10 vertices (f_0=10), 21 edges (f_1=21)

7 - 8, 6 - 5, 2 - 3, 9 - 10, 11 - 12, 5 - 7, 7 - 9, 9 - 11, 11 - 5, 2 - 5, 2 - 7, 2 - 9, 2 - 11, 8 - 6, 6 - 12, 12 - 10, 10 - 8, 3 - 8, 3 - 6, 3 - 12, 3 - 10,

16 two - dimensional faces (f_2=16)

5 - 7 - 8 - 6, 5 - 2 - 6 - 3, 7 - 8 - 2 - 3, 7 - 8 - 9 - 10, 9 - 10 - 11 - 12, 2 - 3 - 11 - 12, 2 - 3 - 9 - 10, 2 - 3 - 5 - 6, 2 - 5 - 7, 2 - 7 - 9, 2 - 9 - 11, 2 - 5 - 11, 3 - 6 - 8, 3 - 6 - 12, 3 - 10 - 12, 3 - 8 - 10,

5 tree - dimensional faces (f_3=5)

2 - 5 - 7 - 3 - 8 - 6, 2 - 7 - 9 - 3 - 8 - 10, 2 - 9 - 11- 3 - 10 - 12, 2 - 5 - 11 - 3 - 6 - 12, 5 - 7 - 8 - 6 - 9 - 10 - 11 - 12 - 2 - 3

(not contain edge 2 - 3) .

Figure 7. Image of a chain with three metal atoms in a higher dimension space

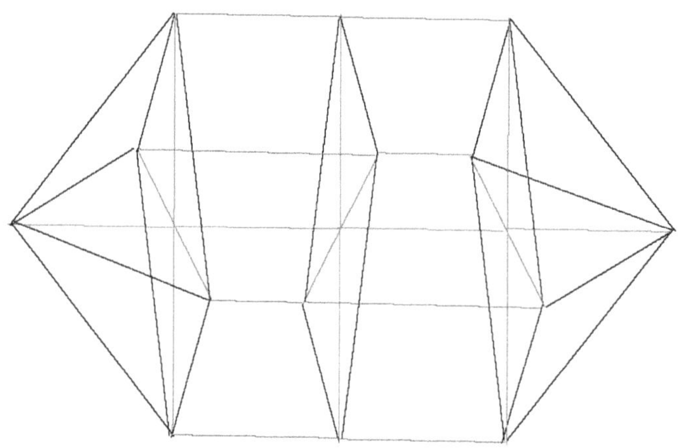

Substituting the obtained values $f_i(i=0,1,2,3)$ in the equation (1) you can see that the Euler - Poincaré equation is satisfied in this case with a value of n equal to 4

$$10 - 21 + 16 - 5 = 0.$$

This proves that polytope on Figure 6 has dimension 4.

Let us return to the chain containing three metal atoms (Figure 2). Its image using spaces of higher dimension, according to the developed description, has the form shown in Figure 7.

It is a chain of convex closed polytopes of dimension 4 adjoining each other along vertical sections of octahedrons passing through the center of the octahedrons and their vertices. It is obvious that the chain of polytopes can be continued. It will include a chain of polytopes in Figure 6, in alternating positions. At the ends of the chain are the polytopes in Figure 5, also in reversed positions. This is the spatial implementation of a polymetallic linear chain using a higher dimension space. Obviously, other linear metal chains with similar attachments of ligands will have the same appearance.

For example, metal chains $[Mn_3(CO)_{14}]^-$ (Bau, Kirtley, Sorrel, & Winarko, 1974), in which the atoms of the metal are manganese atoms, and all ligands

Figure 8. One of the possible of hydride $H_2Os_3(CO)_{12}$ structures

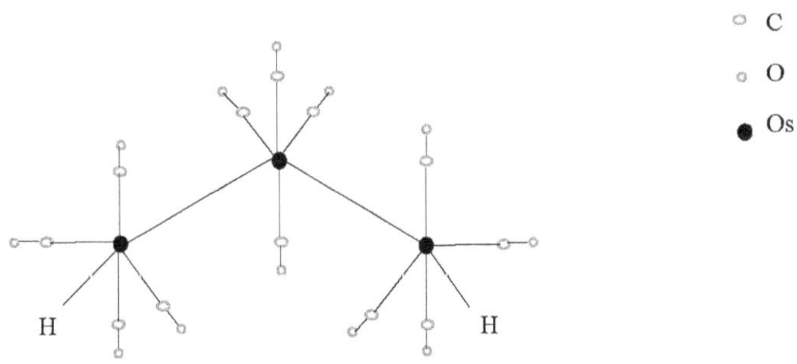

are functional groups CO.

CLUSTER CONNECTIONS WITH A BACKBONE IN THE FORM OF NONLINEAR HOMOELEMENT METALLOCHAIN

The existence of cluster compounds with a skeleton in the form of nonlinear (bent) homo - element metal chains was experimentally established. An

Figure 9. The spatial image of hydride $H_2Os_3(CO)_{12}$

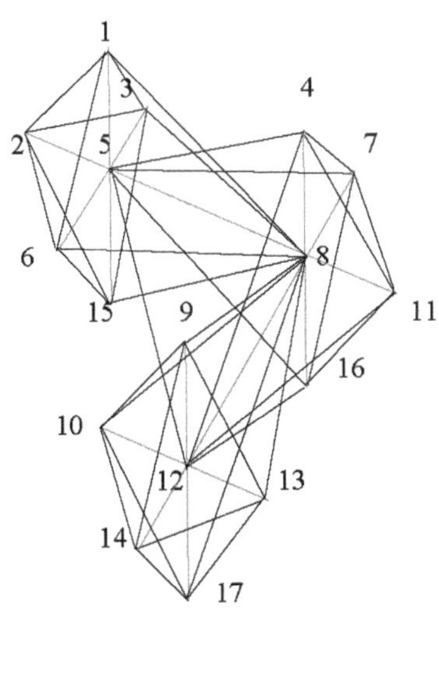

example of such a compound can be hydride $H_2Os_3(CO)_{12}$ (Aleksandrov, Zolnikova, Kritskaya, & Pods, 1980). A diagram of it is presented in Figure 8.

Theorem 2. The spatial figure correspondent to hydride $H_2Os_3(CO)_{12}$ is a polytope consists of two parts, each of which is a polytope of dimension 4. Both polytopes of dimension 4 are adjacent to each other along a flat section that separates them and are symmetric with respect to this section.

Proof. The octahedral coordination of elements manifests itself around each of the three atoms of the osmium and the octahedrons (in the center of each octahedron, an osmium atom is located) penetrate each other. The spatial image of this compound is shown in Figure 9.

On Figure 9 the valence bonds are marked in red, it is taken into account that the metal - metal bond length is substantially greater than the bond length of the metal with the ligands. The edges of the octahedrons are marked in black. At the vertices 5, 8, 12 (centers of octahedrons), atoms of the osmium are located; hydrogen atoms are located at the vertices 2, 14; at vertices

3, 4, 6, 7, 9, 10, 11, 13, 14, 15, 16, 17

functional groups CO are located. Each octahedron with a center has a dimension of 4. However, the common connection of octahedrons is a non - convex and closed figure. To bring the image of this compound to the form of a figure closed and convex, one should fill the space between the octahedrons with additional figures. The result of such a construction will give a complex figure for analysis. However, here we must recall the procedure applied in the previous section. It is necessary to connect the vertices cut off from each other by the edges and remove the edges of the octahedrons, which have lost their significance for the creation of a convex closed figure around an Os atom and do not solve this problem in a combined figure. At same time, all the valence bonds are preserved and the edges delineating the outer contour of the figure are preserved. So, let is connect vertices of octahedrons 1, 4, 9 edges with each other (blue color), connect vertices of octahedrons 15, 16, 17 edges with each other (blue color). Connect vertex 3 with vertex 7, connect vertex 6 with vertex 10, connect vertex 11 with vertex 13 (all edges are black). Remove the edges

5 - 7, 1 - 8, 3 - 8, 12 - 11, 13 - 8, 5 - 16, 6 - 8, 15 - 8, 12 - 16.

Figure 10. The top of the polytope corresponding to the compound $H_2Os_3(CO)_{12}$

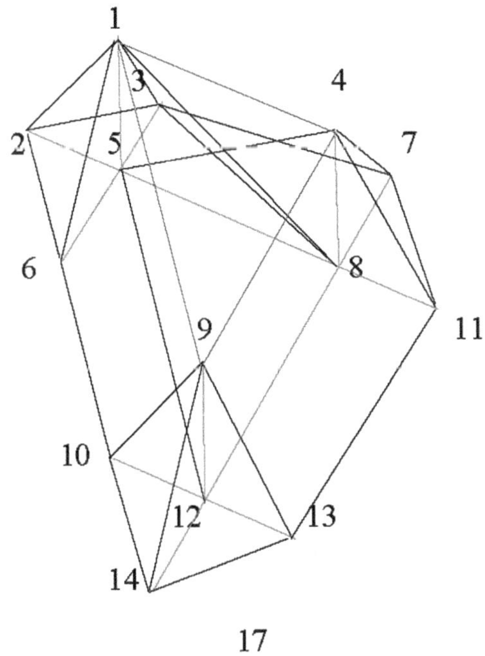

The number of remote edges turned out to be equal to the number of newly entered edges. Then all valence compounds will be inside the new

Figure 11. The bottom of the polytope corresponding to the compound $H_2Os_3(CO)_{12}$

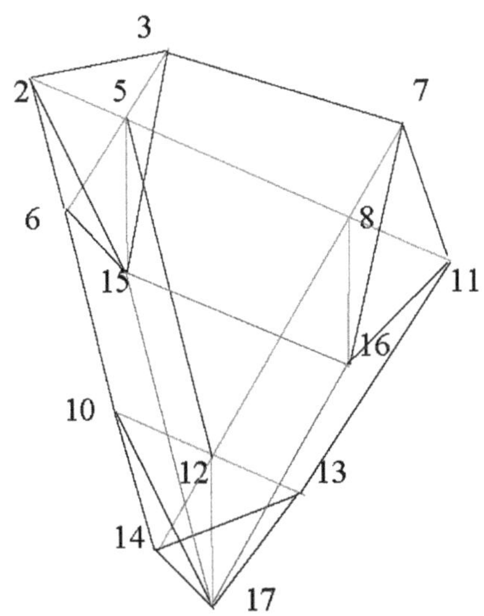

closed convex figure. Moreover, the figure turned out to be a folded of two convex closed polytopes, applied to each other along a two - dimensional (horizontal) flat section

2 - 3 - 7 - 11 - 13 - 14 - 10 - 6 - 2

made up of several sections. The image of the upper polytope (above the horizontal section) is shown in Figure 10.

The image of the lower polytope (below the horizontal section) is shown in Figure 11.

One define the dimension of the polytope in Figure 10.The polytope on Figure 10 has 14 vertices (f_0=14), 32 edges (f_1=32)

1 - 2, 1 - 3,1 - 4, 1 - 5, 1 - 6, 1 - 9, 2 - 5, 2 - 6, 3 - 5, 3 - 7, 4 - 9, 4 - 7, 4 - 11, 4 - 8, 5 - 8, 5 - 6, 6 - 10, 7 - 8, 7 - 11, 8 - 11, 8 - 12, 9 - 10, 9 - 12, 9 - 13, 9 - 14, 10 - 12, 10 - 14, 11 - 13, 12 - 13, 12 - 14, 13 - 14, 5 - 12,

28 flat two - dimension sections (f_2=28)

1 - 4 - 9, 1 - 2 - 6, 1 - 2 - 3, 1 - 2 - 5, 1 - 5 - 3, 1 - 4 - 5 - 8, 1 - 5 - 6, 1 - 3 - 4 - 7, 1 - 5 - 9 - 12, 1 - 6 - 9 - 10, 2 - 3 - 5, 2 - 5 - 6, 3 - 5 - 7 - 8, 4 - 7 - 8, 4 - 7 - 11, 4 - 8 - 11, 4 - 9 - 11 - 13, 4 - 8 - 9 - 12, 5 - 6 - 10 - 12, 5 - 8 - 12, 7 - 8 - 11, 8 - 11 - 12 - 13, 9 - 10 - 12, 9 - 12 - 13, 9 - 10 - 14, 9 - 13 - 14, 10 - 12 - 14, 12 - 13 - 14,

10 tree – dimension area (f_3=10)

1 - 2 - 3 - 5, 1 - 2 - 5 - 6, 1 - 3 - 4 - 7 - 5 - 8, 1 - 4 - 9 - 5 - 8 - 12, 1 - 9 - 5 - 12 - 6 - 10, 4 - 7 - 8 - 11, 4 - 8 - 11 - 9 - 12 - 13, 9 - 12 - 13 - 14, 1 - 2 - 3 - 4 - 5 - 6 - 7 - 8 - 9 - 10 - 11 - 12 - 13 - 14

(without edges 1 - 5, 4 - 8, 9 - 12).

Substituting the obtained values $f_i(i=0,1,2,3)$ in the equation (1) you can see that the Euler - Poincaré equation is satisfied in this case with a value of n equal to 4

$$14 - 32 + 28 - 10 = 0.$$

This proves that polytope on Figure 10 has dimension 4.

One define the dimension of the polytope in Figure 11. The polytope on Figure 11 has 14 vertices (f_0=14), 32 edges (f_1=32)

15 - 2, 15 - 3,15 - 16, 15 - 5, 15 - 6, 15 - 17, 2 - 5, 2 - 6, 3 - 5, 3 - 7, 16 - 17, 16 - 7, 16 - 11, 16 - 8, 5 - 8, 5 - 6, 6 - 10, 7 - 8, 7 - 11, 8 - 11, 8 - 12, 17 - 10, 17 - 12, 17 - 13, 17 - 14, 10 - 12, 10 - 14, 11 - 13, 12 - 13, 12 - 14, 13 - 14, 5 - 12,

28 flat two - dimension sections (f_2=28)

15 - 16 - 17, 15 - 2 - 6, 15 - 2 - 3, 15 - 2 - 5, 15 - 5 - 3, 15 - 16 - 5 - 8, 15 - 5 - 6, 15 - 3 - 16 - 7, 15 - 5 - 17 - 12, 15 - 6 - 17 - 10, 2 - 3 - 5, 2 - 5 - 6, 3 - 5 - 7 - 8, 16 - 7 - 8, 16 - 7 - 11, 16 - 8 - 11, 16 - 17 - 11 - 13, 16 - 8 - 17 - 12, 5 - 6 - 10 - 12, 5 - 8 - 12, 7 - 8 - 11, 8 - 11 - 12 - 13, 17 - 10 - 12, 17 - 12 - 13, 17 - 10 - 14, 17 - 13 - 14, 10 - 12 - 14, 12 - 13 - 14,

10 tree - dimension area (f_3=10)

15 - 2 - 3 - 5, 15 - 2 - 5 - 6, 15 - 3 - 16 - 7 - 5 - 8, 15 - 16 - 17 - 5 - 8 - 12, 15 - 17 - 5 - 12 - 6 - 10, 16 - 7 - 8 - 11, 16 - 8 - 11 - 17 - 12 - 13, 17 - 12 - 13 - 14, 15 - 2 - 3 - 16 - 5 - 6 - 7 - 8 - 17 - 10 - 11 - 12 - 13 - 14

(without edges 15 - 5, 16 - 8, 17 - 12).

Substituting the obtained values f_i(i=0,1,2,3) in the equation (1) you can see that the Euler - Poincaré equation is satisfied in this case with a value of n equal to 4

Figure 12. The scheme of compound $Os_3(\mu\text{-}Br)_2(CO)_{10}$

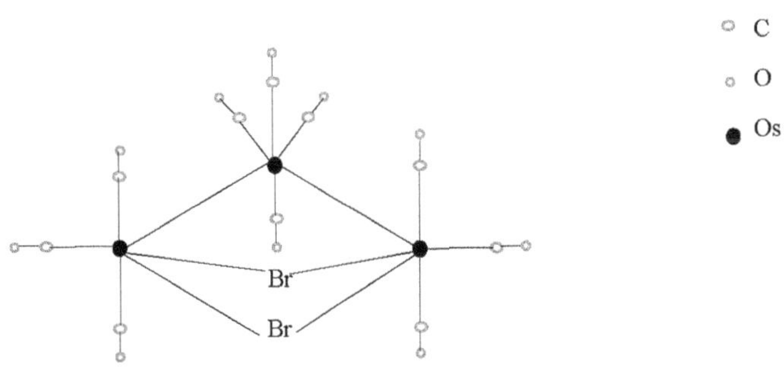

$14 - 32 + 28 - 10 = 0.$

This proves that polytope on Figure 11 has dimension 4.

It can be noted that the upper and lower parts of the polytope corresponding to the compound are symmetrical to each other with respect to the flat section separating them.

A feature of homo - element metal chains is the possibility of bridging

Figure 13. Spatial image of compound $Os_3(\mu\text{-}Br)_2(CO)_{10}$

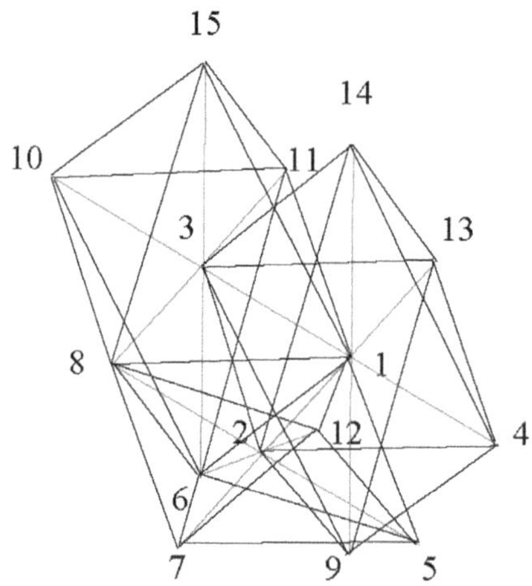

compounds between metal atoms. An example of such a cluster can be a chemical compound $Os_3(\mu\text{-}Br)_2(CO)_{10}$ (Aleksandrov, Zolnikova, Kritskaya, & Struchkov, 1980). The scheme of this compound is shown in Figure 12.

Theorem 3. The spatial figure correspondent to hydride $Os_3(\mu\text{-}Br)_2(CO)_{10}$

It is a polytope consists of two parts, each of which is a polytope of dimension 4. Both polytopes of dimension 4 are adjacent to each other along a flat section that separates.

Proof. The octahedral coordination of elements around osmium atoms is also visible here. Due to bridging compounds, bromine atoms in this case simultaneously three octahedrons penetrate each other (Figure 13).

At the same time, one of the octahedrons has a slope with respect to two other octahedrons. As before, the edges corresponding to the valence bonds are marked in red, and the edges of the octahedrons proper are marked in black. At the vertices 1, 2, 3 (the centers of the octahedrons), atoms of osmium are located. At the vertices 6, 8 are located bromine atoms. In the remaining vertices

4, 5, 7, 9, 10, 11, 12, 13, 14, 15

there are functional groups CO.

Each octahedron with a center has a dimension of 4 there are, however, the common connection of octahedrons is a non - convex and closed figure. To bring the image of this compound to the form of a figure closed and convex, one should fill the space between the octahedrons with additional figures. The result of such a construction will give a complex figure for analysis. However, here one must recall the procedure applied in the previous section. It is necessary to connect the vertices cut off from each other by the edges and remove the edges of the octahedrons, which have lost their significance for the creation of a convex closed figure around an Os atom and do not solve this problem in a combined figure. At same time, all the valence bonds are preserved and the edges delineating the outer contour of the figure are preserved. So, let is connect vertices of octahedrons 6, 9 edge with each other (blue color), connect vertices of octahedrons 15, 14, 12 edges with each other (blue color). Connect vertex 11 with vertex 13, connect vertex 9 with

Figure 14. The top of the polytope corresponding to the compound $Os_3(\mu\text{-}Br)_2(CO)_{10}$

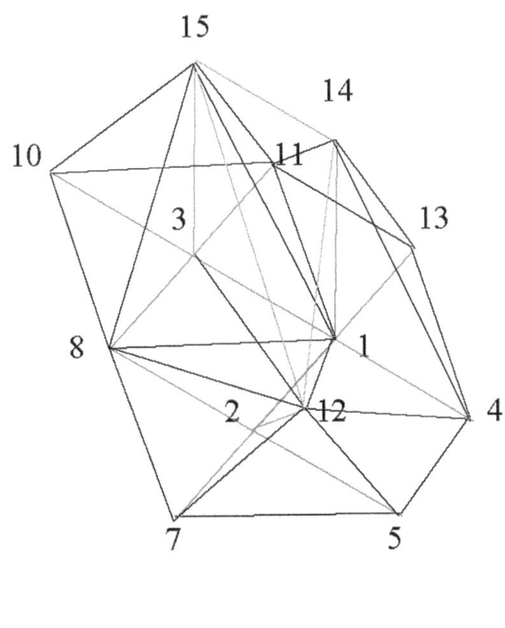

vertex 7, connect vertex 9 with vertex 5, connect vertices 5 with vertices 4, connect vertices 11 with vertices 14 (all edges are black). Remove the edges

Figure 15. The bottom of the polytope corresponding to the compound $Os_3(\mu\text{-}Br)_2(CO)_{10}$

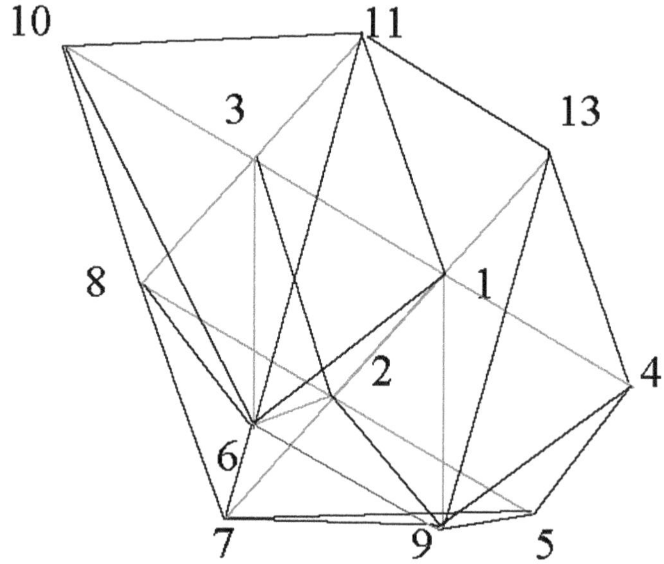

2 - 14, 3 - 2, 2 - 4, 3 - 13, 14 - 3, 5 - 1, 3 - 9, 5 - 6, 1 - 8.

The number of remote edges turned out to be equal to the number of newly entered edges (9). Then all valence compounds will be inside the new closed convex figure. Moreover, the figure turned out to be a folded of two convex closed polytopes, applied to each other along a two - dimensional (horizontal) flat section 4 - 5 - 7 - 8 - 10 - 11 - 13 made up of several sections. The image of the upper polytope (above the horizontal section) is shown in Figure 14.

The image of the lower polytope (below the horizontal section) is shown in Figure 15.

One define the dimension of the polytope in Figure 14. The polytope on Figure 14 has 13 vertices (f_0=13), 35 edges (f_1=35)

1 - 4, 1 - 2, 1 - 3, 1 - 13, 1 - 12, 1 - 14, 1 - 15, 1 - 11, 2 - 5, 2 - 7, 2 - 8, 2 - 12, 3 - 8, 3 - 10, 3 - 11, 3 - 12, 3 - 15, 4 - 13, 4 - 14, 4 - 12, 4 - 5, 5 - 7, 5 - 12,

7 - 12, 7 - 8, 8 - 10, 8 - 12, 8 - 15, 10 - 11, 10 - 15, 11 - 13, 11 - 14, 11 - 15, 13 - 14, 14 - 15,

37 flat two - dimension sections (f_2=37)

1 - 11 - 14, 1 - 3 - 15, 1 - 11 - 15, 1 - 11 - 13, 1 - 13 - 4, 1 - 4 - 12, 1 - 12 - 14, 1 - 3 - 11, 1 - 2 - 12, 1 - 12 - 15, 1 - 12 - 3, 2 - 8 - 12, 2 - 12 - 5, 2 - 7 - 5, 2 - 7 - 12, 2 - 8 - 7, 3 - 12 - 8, 3 - 12 - 15, 3 - 15 - 8, 3 - 11 - 15, 3 - 10 - 8, 3 - 15 - 10, 3 - 11 - 15, 4 - 13 - 14, 4 - 12 - 14, 4 - 5 - 12, 5 - 12 - 7, 8 - 12 - 15, 8 - 10 - 15, 8 - 7 - 12, 10 - 11 - 15, 11 - 14 - 15, 11 - 13 - 14, 12 - 14 - 15, 1 - 2 - 3 - 8, 1 - 2 - 4 - 5, 1 - 3 - 14 - 15,

15 tree - dimension area (f_3=15)

1 - 11 - 14 - 15, 1 - 12 - 14 - 15, 1 - 12 - 3 - 15, 1 - 11 - 3 - 15, 1 - 4 - 13 - 14, 2 - 7 - 8 - 12, 3 - 8 - 12 - 15, 3 - 8 - 10 - 15, 1 - 4 - 12 - 14, 1 - 13 - 11 - 14, 1 - 2 - 4 - 5 - 12, 1 - 2 - 3 - 8 - 12, 1 - 3 - 14 - 15 - 11, 4 - 5 - 7 - 8 - 10 - 15 - 1 4 - 13 - 1 - 2 - 3 - 12 - 11

(without edges 3 - 15, 1 - 14, 2 - 12).

Substituting the obtained values f_i(i=0,1,2,3) in the equation (1) you can see that the Euler - Poincaré equation is satisfied in this case with a value of n equal to 4

$$13 - 35 + 37 - 15 = 0.$$

This proves that polytope on Figure 14 has dimension 4.

One define the dimension of the polytope in Figure 15. The polytope on Figure 15 has 12 vertices (f_0=12), 30 edges (f_1=30)

1 - 9, 1 - 2, 1 - 13, 1 - 6, 1 - 4, 1 - 3, 2 - 9, 2 - 6, 2 - 7, 2 - 8, 2 - 5, 3 - 6, 3 - 8, 3 - 10, 3 - 11, 4 - 5, 4 - 9, 4 - 13, 5 - 9, 5 - 7, 6 - 7, 6 - 8, 6 - 10, 6 - 11, 7 - 9, 8 - 10, 9 - 13, 10 - 11, 11 - 13, 1 - 11,

31 flat two - dimension sections (f_2=31)

1 - 6 - 3, 1 - 6 - 2, 1 - 3 - 11, 1 - 6 - 9, 1 - 9 - 13, 1 - 9 - 2, 1 - 4 - 9, 1 - 6 - 11, 2 - 6 - 8, 2 - 9 - 5, 2 - 9 - 6, 2 - 6 - 7, 1 - 13 - 11, 2 - 5 - 7, 3 - 6 - 8, 3 - 8 - 10, 3 - 10 - 11, 3 - 6 - 11, 3 - 6 - 10, 4 - 9 - 13, 4 - 5 - 9, 5 - 9 - 7, 6 - 7 - 9, 6 - 7

- 8, 6 - 8 - 10, 6 - 10 - 11, 1 - 2 - 8 - 3, 1 - 2 - 4 - 5, 1 - 9 - 6 - 3, 9 - 6 - 11 - 13,

13 tree - dimension area (f_3=13)

1 - 4 - 9 - 13, 2 - 1 - 9 - 6, 3 - 6 - 11 - 10, 3 - 6 - 8 - 10, 1 - 3 - 6 - 11, 2 - 6 - 8 - 7, 9 - 6 - 2 - 7, 9 - 2 - 5 - 7, 4 - 1 - 13 - 9, 1 - 2 - 5 - 4 - 9, 1 - 2 - 3 - 8 - 6, 1 - 6 - 9 - 11 - 13, 4 - 5 - 9 - 7 - 8 - 10 - 11 - 13 - 1 - 2 - 3 - 6

(without edges 1 - 9, 3 - 6, 2 - 6).

Substituting the obtained values f_i(i=0,1,2,3) in the equation (1) you can see that the Euler - Poincaré equation is satisfied in this case with a value of n equal to 4

Figure 16. Schemes of hetero - element metal chains of transition metal atoms

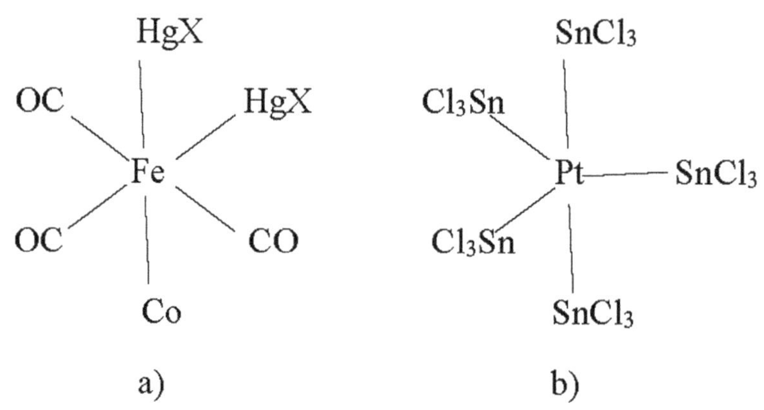

$12 - 30 + 31 - 13 = 0.$

This proves that polytope on Figure 15 has dimension 4.

HETERO - ELEMENT METALL CHAINS

Figure 17. Schemes of compound $Pt(SnCl_3)_5$

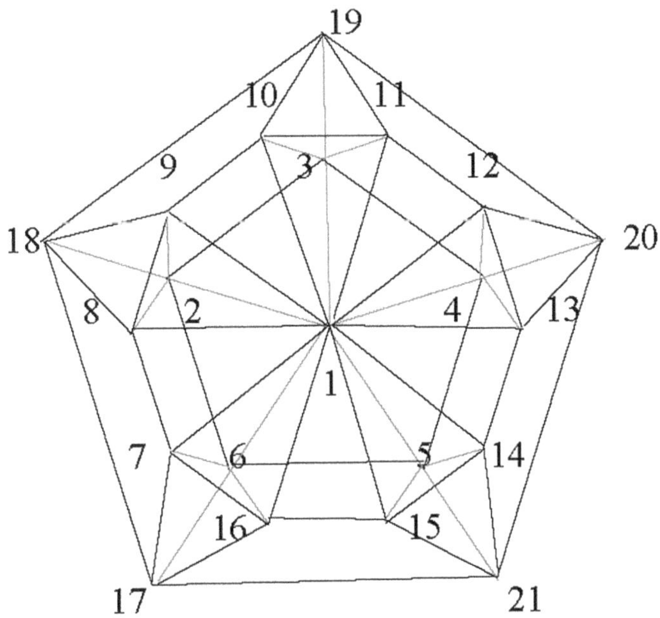

Figure 16 a, b shows typical examples of nonlinear hetero - element metal chains constructed from transition metal atoms (Cramer et al., 1965; Chiswell, & Venanzi, 1966).

From Figure 16 it follows that these compounds have, respectively, the shape of an octahedron with a center and a tetrahedron with a center in which the metal atom is located, and ligands are located at the vertices of the octahedron and tetrahedron. Moreover, the ligands also include transition elements that differ from the metal in the center of the figure. The functional dimension of these compounds in both cases is 4.

Metal chains can contain both transition elements and non - transition elements. Figure 17 shows the schemes of compound $Pt(SnCl_3)_5$. It is a limiting case of the replacement of all ligands at the central atom of the transition metal by metal - containing ligands with non - transition metal (Cramer et al., 1965).

Each tin atom in this compound exhibits tetrahedral coordination, i.e. the compounds of the platinum atom with the functional group are geometrically a tetrahedron with the center, therefore it has the dimension 4. In Figure 17, a platinum atom is located at vertex 1, tin atoms are located at vertices 2, 3, 4, 5, 6, and at vertices

7, 8, 9, 10, 11, 12, 13, 14, 15, 16, 17, 18, 19, 20, 21

located chlorine atoms. Valence bonds in Figure 17 are marked in red, and all other edges are marked in black. The space between the tetrahedrons with the center in this compound is also a polytope. Let us turn to one of these five polytopes, for example, the polytope

1 - 2 - 18 - 9 - 3 - 10 - 19.

It has seven vertices (f_0=7), 13 edges (f_1=13)

18 - 9, 9 - 2, 18 - 2, 2 - 1, 9 - 1, 1 - 3, 1 - 10, 10 - 3, 10 - 19, 19 - 3, 18 - 19, 9 - 10, 2 - 3,

9 flat faces (f_2=9)

2 - 9 - 1, 2 - 9 - 18, 18 - 19 - 9 - 10, 2 - 3 - 9 - 10, 19 - 10 - 3, 10 - 3 - 1, 2 - 3 - 1, 9 - 10 - 1, 9 - 2 - 10 - 3,

3 three - dimensional figures (f_3=3)

1 - 2 - 3 - 9 - 10, 2 - 9 - 18 - 3 - 10 - 19, 1 - 2 - 3 - 4 - 5 - 6 - 7 - 8 - 9 - 10 - 11 - 12 - 13 - 14 - 15 - 16 - 17 - 18 - 19 - 20 - 21

Figure 18. The tetrahedron with center

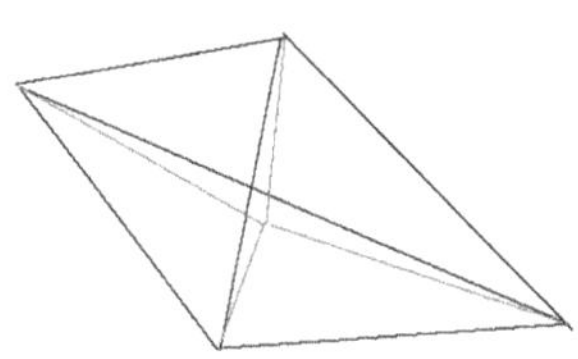

(without flat section 9 - 10 - 2 - 3).

Substituting the obtained values $f_i(i=0,1,2,3)$ in the equation (1) you can see that the Euler - Poincaré equation is satisfied in this case with a value of n equal to 4

$$7 - 13 + 9 - 3 = 0.$$

This proves that polytope 1 - 2 - 18 - 9 - 3 - 10 - 19 has dimension 4.

Thus, all five polytopes between tetrahedrons with center have dimension 4. It should be noted that the image of tetrahedrons with center in Figure 17 is schematized to give symmetry to Figure 17. A more accurate image of the tetrahedron with center (Figure 18) avoids the overlapping of edges on each other occurring in Figure 17 and confirm the convexity of the polytope located between the tetrahedrons with the center.

Consequently, the cluster $Pt(SnCl_3)_5$ is a curious example of a closed cycle of polytopes of dimension 4 adjoining each other along two - dimensional faces of a tetrahedron centered inside tetrahedrons, although there is no closed loop of metal atoms in this compound. Clusters with closed cycles of metal atoms will be discussed in the next chapter.

CONCLUSION

Geometrically investigated the structure of clusters, the core of which represent the metal chains of both identical and different elements. Chains are linear or curved. For the first time it was shown that the dimension of the structures of these clusters is more than three. The octahedral coordination of ligands around metal atoms in chains leads to the penetration of octahedrons into each other in both linear and non - linear metal chains. The dimension of octahedron with a metal atom in the center is 4. To create a model of these chains in a higher dimension space, a new geometric approach has been developed, which allows us to construct a convex, closed polytopes of these chains. It consists in removing part of the octahedron edges necessary for constructing the octahedron and adding the same number of new edges necessary to build a closed polytope chain, while maintaining the number of metal atoms and ligands and their valence bonds. As a result, it was found that metal chain polytopes consist of polytopes of higher dimension, adjacent to each other along flat sections. This allowed us to simplify the geometry of the clusters and to construct their structure. Including it was possible to detect the existence of clusters with a closed cycle of polytopes in the absence of closed cycles of metal atoms.

REFERENCES

Aleksandrov, G.G., Zolnikova, G.P., Kritskaya, I.I., & Struchkov, Yu.T. (1980). Molecular structure $Os_3(\mu\text{-}Br)_2(CO)_{10}$ of the reaction product $Os_3(CO)_2$ with (3 -bromomethylnaphthalene). *Coordination Chemistry*, *6*(4), 626 - 628.

Bau, R., Kirtley, S. W., Sorrell, T. N., & Winarko, S. (1974). Preparation and structure determination of the manganese carbonyl - anion. Staggered vs eclipsed carbonyl groups and linear vs bent M – M – M - [metal – metal – metal] and M – H – M [metal – hydrogen – metal] bonds. *Journal of the American Chemical Society*, *96*(4), 988–993. doi:10.1021/ja00811a006

Bender, R., Braunstein, P., Dusausoy, Y., & Protas, J. (1979). Synthesis and Crystal Structure of a Chiral Complex a Non – linear Heterotrimetal Chain $PtMn_2(2\text{-}PPh_2)_2(CO)_9$. *Journal of Organometallic Chemistry*, *172*, 51. doi:10.1016/S0022-328X(00)92321-0

Chiswell, B., & Venanzi, L. M. (1966). A bidentate gold ligand. *Journal of the Chemical Society*, 901. doi:10.1039/j19660000901

Cotton, F. L., & Walton, R. A. (1982). *Multiple bonds between metal atoms*. New York: Wiley.

Cramer, R. D., Lindsey, R. V., Prewitt, C. T., & Stolberg, U. G. (1965). Article. *Journal of the American Chemical Society*, *7*, 503.

Evans, J., Okrasunski, S. J., Pribula, A. J., & Norton, J. K. (1976). Synthesis of di- and trinuclear methyl osmium complexes via cis-hydridomethyltetracarbonylosmium. *Journal of the American Chemical Society*, *98*(13), 4000–4001. doi:10.1021/ja00429a045

Gochin, M., & Moss, Y. R. (1980). Article. *Journal of Organometallic Chemistry*, *192*, 409.

Gubin, S. P. (2019). *Chemistry clusters. Basics of classification and structure*. Moscow: Science.

Poincare, A. (1895). Analysis situs. *J. de é Ecole Polytechnique, 1*, 1 – 121.

Smart, L., Cook, N., & Woodward, P. (1977). Article. *Journal of the Chemical Society, Dalton Transactions: Inorganic Chemistry*, 4312.

Zhizhin, G. V. (2016). The structure, topological and functional dimension of biomolecules. *J. Chemoinformatics and Chemical Engineering*, *5*(2), 44–58.

Zhizhin, G. V. (2018). *Chemical Compound Structures and the Higher Dimension of Molecules: Emerging Research and Opportunities*. Hershey, PA: IGI Global. doi:10.4018/978-1-5225-4108-0

Zhizhin, G. V. (2019a). *The Geometry of Higher – Dimensional Polytopes*. Hershey, PA: IGI Global. doi:10.4018/978-1-5225-6968-8

Zhizhin, G. V. (2019b). *Attractors and Higher Dimensions in Population and Molecular Biology: Emerging Research and Opportunities*. Hershey, PA: IGI Global. doi:10.4018/978-1-5225-9651-6

KEY TERMS AND DEFINITIONS

Dimension of the Space: The member of independent parameters needed to describe the change in position of an object in space.

Hetero-Element Metal Chains: Clusters in which the core contains atoms of different elements.

Homo-Element Metal Chains: Clusters in which the core contains atoms of the same element.

Linear Metal Chains: Metal chains in which metal atoms are located in a straight line.

Metal Chains: Cluster compounds that have a skeleton in the form of metal chains, that is, polymetallic chains formed by metal-metal localized covalent bonds.

Nanocluster: A nanometric set of connected atoms, stable either in isolation state or in building unit of condensed matter.

Nonlinear (Curved) Metal Chains: Clusters in which metal-metal valent bonds form an angle different from 180 degrees between them.

Polytope: Polyhedron in the space of higher dimension.

Chapter 5
Closed Metal Cycles in Clusters With Ligands

ABSTRACT

This chapter considers closed three-membered metal cycles of one or several chemical elements surrounded by ligands connected to them. It has been proven that the widespread opinion in the literature about the formation of ligands by atoms in some cases of the semi-correct polyhedron of the anti-cube-octahedron is wrong. Geometrical analysis of the interpenetration of the coordinates of ligand atoms around each of the metal atoms of a closed chain showed that this leads to a different class of special three-dimensional irregular polyhedrons for different clusters. In all cases of homo-element and hetero-element closed metal chains, the cycle itself, located in a certain plane, creates a cross section of the cluster, dividing the cluster into two parts. Each of the parts of a cluster has dimension 4.

INTRODUCTION

Cluster compounds containing a skeleton in the form of metal cycles are a widespread type of cluster metal compounds (Gubin, 2019). Homo - cyclic structures with metal atoms from 3 to 8, bicyclic formations are known. Particularly, systems consisting of several condensed metal cycles can be considered. Cluster compounds that have one hundred in the form of three -membered metal cycles are obtained for most metals and have ligands of various types. In cluster compounds with a large number of metal atoms, the

DOI: 10.4018/978-1-7998-3784-8.ch005

main part of the faces of the metal frame polyhedrons is triangles. Therefore, three - membered metal rings to some extent can simulate the properties of more complex clusters.

HOMO: ELEMENT METAL CYCLES WITH LIGANDS

The absence of bridging ligands and high symmetry make three - nuclear carbonyls $Ru_3(CO)_{12}$, $Os_3(CO)_{12}$ convenient support compounds for structural and theoretical studies of three - membered homo - element metal cycles. In molecules, each metal atom is associated with four functional groups (Figure 1).

Figure 1. Shema of three - nuclear carbonyls Ru and Os

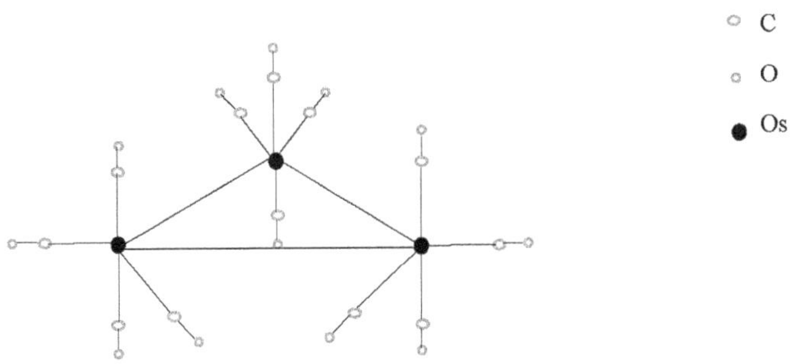

It is believed that 12 ligands are arranged so that they form an anti – cube – octahedron as a ligand polyhedron (Mason & Rae, 1968; Benfield & Jonson, 1981; Gubin, 2019). However, evidence of this assumption has not yet been provided. The proof of this assertion could be a concrete construction of an anti – cube – octahedron with a three - link metal cycle enclosed in it, connection by valence bonds of the metal cycle atoms to the vertices of the anti – cube - octahedron. After this, it is required to determine the partition of the anti – cube - octahedron with the constructed valence bonds into elementary three - dimensional cells and the verification of the implementation of the Euler – Poincaré (Poincoré, 1895) equation for the constructed polytope. No such evidence was carried out. In this chapter, this question will be considered as part of the proof of the following theorem:

Theorem 1. The geometric model of three nuclear carbonyls Ru and Os consists of two polytopes of dimension 4, touching each other in a two - dimensional section, containing a three nuclear metal cycle.

Proof. Let us assume that carbonyl ligands form a ligand polyhedron in the form of an anti - cube - octahedron. Then the three nuclear metal cycle contained in an anti - cube - octahedron must be connected by vertices of an

Figure 2. The anti - cube - octahedron with the three nuclear metal cycle

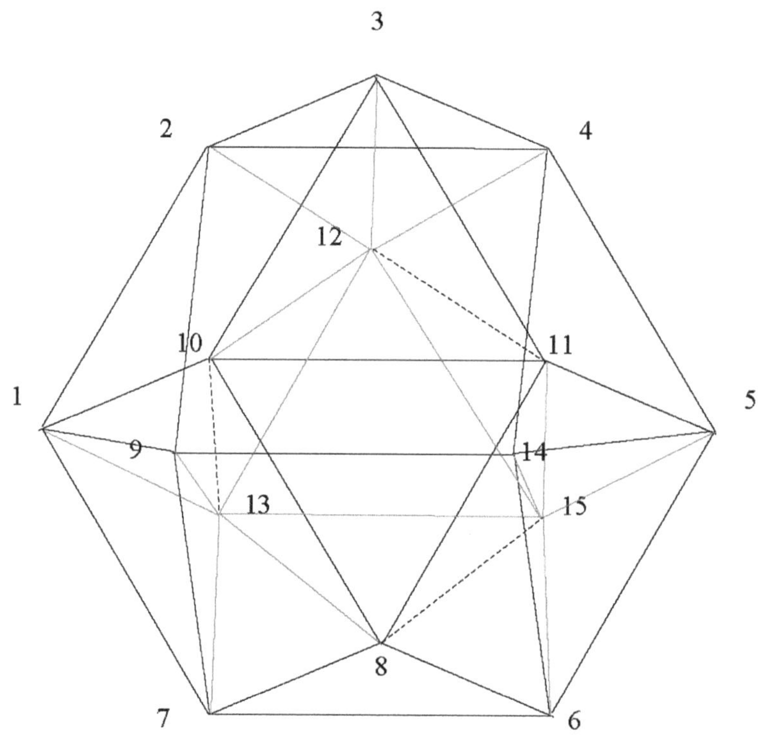

anti - cube - octahedron by valence bonds. Each vertex of the metal cycle must be connected with the four nearest vertices of the anti - cube - octahedron. In an arbitrary general form will look like that shown in Figure 2.

At the vertices 12, 13, 14, belonging to the metal cycle, there are metal atoms, and at the 12 vertices of the anti - cube - octahedron functional groups CO are present. The valence bonds of the metal atoms with each other and the functional groups are indicated by red edges. The edges of the anti - cube - octahedron is denoted by solid black lines. Valence bonds and edges of anti - cube - octahedron already create convex three - dimensional bodies

1) 2 - 3 - 4 - 12,
2) 5 - 6 - 15 - 14,
3) 1 - 9 - 13 - 7,
4) 9 - 12 - 13 - 14 - 15,
5) 1 - 2 - 12 - 9 - 13,
6) 3 - 4 - 12 - 11 - 5 - 15,
7) 4 - 12 - 5 - 15 - 14,
8) 2 - 4 - 9 - 14 - 12,
9) 7 - 9 - 14 - 6 - 13 - 15,
10) anti - cube - octahedron.

However, body particles do not yet create a partition of the space inside the anti - cube - octahedron into elementary three - dimensional cells, i.e. do not create the structure of this space. To create it, you need to add more edges between the 15 vertices of the system. You need to do this carefully enough so as not to come to possible contradictions. To do this, choose in the system the smallest possible number of places for holding these edges so that each added edge leads to the creation of the maximum number of three-dimensional bodies. Such places can be symmetrically located edges (indicated by dotted lines)

13 - 10, 12 - 11, 11 - 15.

These edges lead to the creation of a whole series of three - dimensional bodies:

1-7 - 8 - 10 - 13, 12) 13 - 10 - 11 - 15 - 12 - 8, 13) 10 - 3 - 11 - 12, 14) 7 - 8 - 13 - 15 - 6, 15) 9 - 13 - 12 - 14 - 15, 16) 1 - 3 - 10 - 2 - 13 - 12, 17) 3 - 10 - 12 - 11, 18) 5 - 6 - 8 - 11 - 15. 11)

Now the whole space of the anti - cube - octahedron is divided into three - dimensional polyhedrons. For further it is necessary use the Euler - Poincare` equation (Poincaré, 1895)

$$\sum_{i=0}^{n-1}(-1)^i f_i(P) = 1-(-1)^i, \tag{1}$$

f_i is the number of the elements with the dimension i at polytope P; n is dimension of the polytope P.

Two numbers are already there f_0=15, f_3=18 . It is necessary to count the number of edges and the number of flat two - dimensional elements. It is possible to determine that Figure 2 has 45 edges (f_1=45)

1 - 2, 1 - 9, 1 - 10, 1 - 13, 1 - 7, 2 - 3, 2 - 4, 2 - 12, 2 - 9, 3 - 10, 3 - 12, 3 - 11, 3 - 4, 4 - 12, 4 - 14, 4 - 5, 5 - 11, 5 - 14, 5 - 15, 5 - 6, 6 - 14, 6 - 15, 6 - 8, 6 - 7, 7 - 8, 7 - 13, 7 - 9, 8 - 15, 8 - 10, 8 - 13, 8 - 11, 9 - 12, 9 - 14, 9 - 13, 10 - 12, 10 - 11, 10 - 13, 11 - 2, 11 - 15, 12 - 13, 12 - 14, 12 – 15, 13 - 15, 14 - 15,

31 two - dimensional elements (f_2=31) (two - dimensional elements that are a section of three -dimensional figures are not taken into account)

1 - 2 - 3 - 10, 1 - 2 - 9, 1 - 9 - 13, 1 - 10 - 13, 1 - 9 - 7, 2 - 3 - 4, 2 - 4 - 12, 2 - 4 - 9 - 14, 3 - 10 - 12, 3 - 10 - 11, 3 - 4 - 12, 3 - 4 - 5 - 11, 3 - 12 - 11, 4 - 14 - 12, 4 - 5 - 14, 5 - 14 - 15, 5 - 15 - 6, 5 - 14 - 6, 6 - 14 - 15, 5 - 6 - 8 - 11, 6 - 8 - 15, 7 - 8 - 6, 7 - 8 - 13, 7 - 9 - 13, 8 - 13 - 10, 8 - 11 - 15, 8 - 10 - 11, 1 - 7 - 8 - 10, 1 - 2 - 12 - 13, 9 - 12 - 13, 11 - 15 - 12.

Substituting the obtained values f_i(i=0,1,2,3) in the equation (1) you can see that the Euler - Poincaré equation is not satisfied in this case

$$15 - 45 + 31 - 18 = -17 < 0.$$

This proves that Figure 2 is not polytope.

The reason for this lies in the fact that not all the flat faces of the 45 faces are the faces of two adjacent polyhedrons. It can be determined from Figure 2 that the faces

3 - 10 - 12, 3 - 12 - 11

are not at the same time the face of two neighboring polyhedrons (they are only a part of the face of neighboring polyhedrons). This is a necessary condition for the existence of a polytope in this case, dimension 4. With increasing polytope dimension, each flat face must simultaneously be the face of an even larger number of three - dimensional faces (Zhizhin, 2019).

Attempts to draw additional edges in Figure 2, in order to create additional three - dimensional bodies with faces

3 - 10 - 12, 3 - 12 - 11,

lead to the emergence of new flat faces, which need to be closed by new polyhedrons, and this process is difficult to complete. Make it fails. It should be recognized that the assumption made at the beginning about the anti - cube - octahedron as a ligand polytope of compounds $Ru_3(CO)_{12}$, $Os_3(CO)_{12}$ is not correct, since it is not provable.

To create a geometric image of a cluster with the skeleton of a three - bar metal cycle, one should turn to the coordination of ligands around each atom of the cycle. This coordination is octahedral. In the case of a closed cycle, coordination around each atom overlaps each other. Here it is necessary to apply the method developed in the previous chapter when considering

Figure 3. A section of a cluster containing a closed three - bar metal cycle

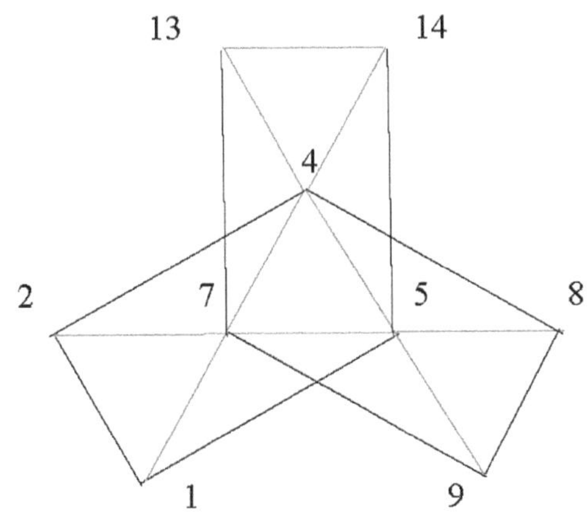

metal chains. This will be a logical continuation of the method during the transition to closed cycles. Consider the plane in which the metal cycle is located (Figure 3).

In this plane, there are three intersecting cross sections of the octahedron coordination of ligands around each metal atom. In the figures in Figure 3, these sections are

1 - 2 - 4 - 5, 4 - 7 - 8 - 9, 5 - 7 - 13 - 14.

Valence bonds are indicated by red edges. At the vertices 4, 5, 7 are located the metal atoms. In other vertices are located functional groups. Passing from the octahedron section into space, one should draw segments perpendicular to the section plane at vertices 4, 5, 7. These segments should intersect the section plane and are located above the section plane for the bond length and below the section plane for the bond length too. From the free vertices of these segments, the edges of the octahedrons connecting these vertices with the vertices of each corresponding octahedron should be drawn. The overall picture of the intersection of three octahedrons in the projection on the plane will be quite complicated. In addition, to create a geometric image of the entire cluster in the form of a convex closed figure, it is necessary to fill the space between the octahedrons. Here, as well as in the previous chapter, it is necessary to remove part of the edges of octahedrons that have ceased to bear the function of creating a convex figure in the cluster image, and add

Figure 4. Spatial image of the cluster with a closed three - bar metal cycle

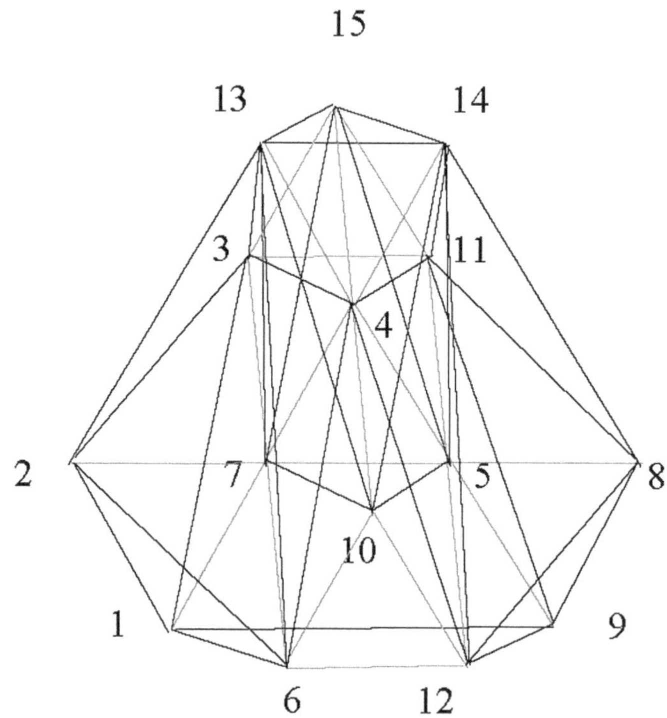

the edges necessary to create a convex figure in the cluster image. The final cluster figure is shown in Figure 4.

The vertices of the octahedrons above the sectional plane (Figure 3) are marked with numbers 3, 11, 15. The vertices of the octahedrons below the section plane (Figure 3) are designated with numbers 6, 10, 12. In accordance with what was said earlier, in Figure 4 eight edges of

6 - 5, 7 -9, 3 - 5, 7 - 11, 8 - 4, 2 - 4, 12 - 7, 1 - 5

from all edges of octahedrons with marked vertices are removed. At the same time, eight other edges are added

2 - 13, 8 - 14, 6 - 10, 10 - 12, 6 - 12, 3 - 15, 15 - 11, 3 - 11.

The edges connecting the indicated vertices of three parallel octahedrons have blue color. The edges corresponding to the valence bonds are marked in red, the other edges have a black color. The construction in Figure 4 can be divided into two polytopes: one polytope, located above the section in

Figure 5. The polytope upper part of a cluster with a three - bar metal atom cycle

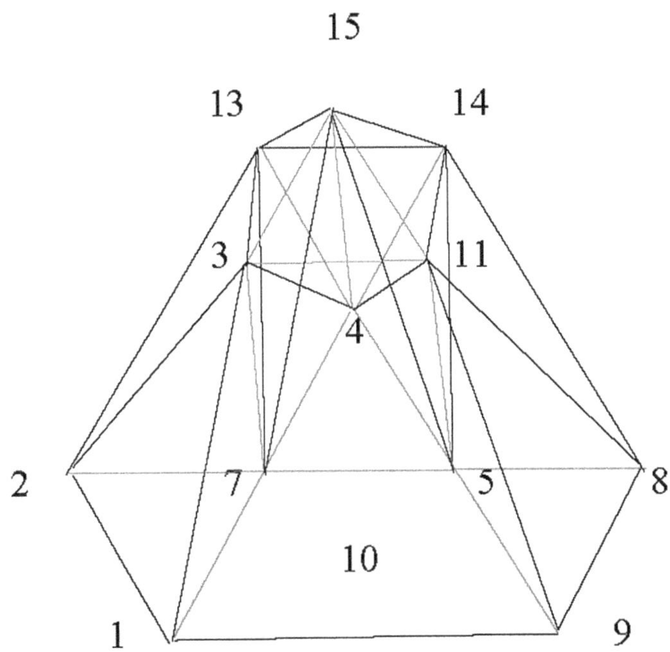

Figure 6. The polytope lower part of a cluster with a three - bar metal atom cycle

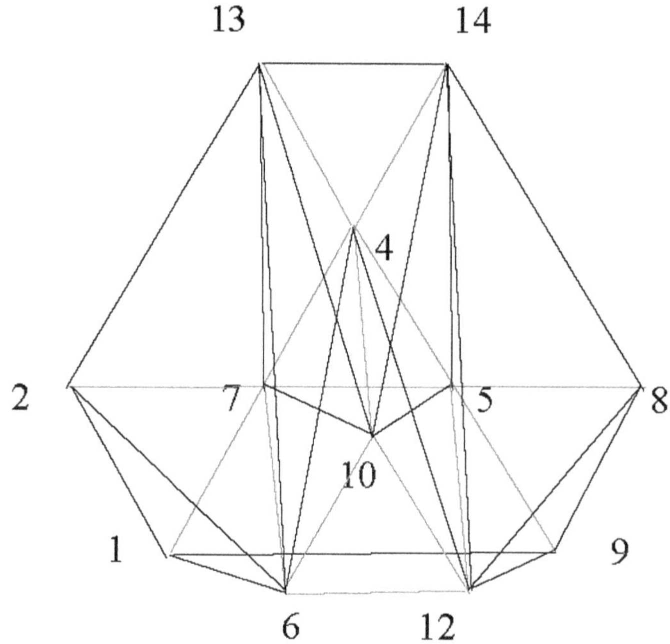

which the closed cycle of metal atoms is located (Figure 5); another polytope is located under this section (Figure 6).

Figure 7. A section dividing a cluster with a three - bar closed cycle of metal atoms into two polytopes

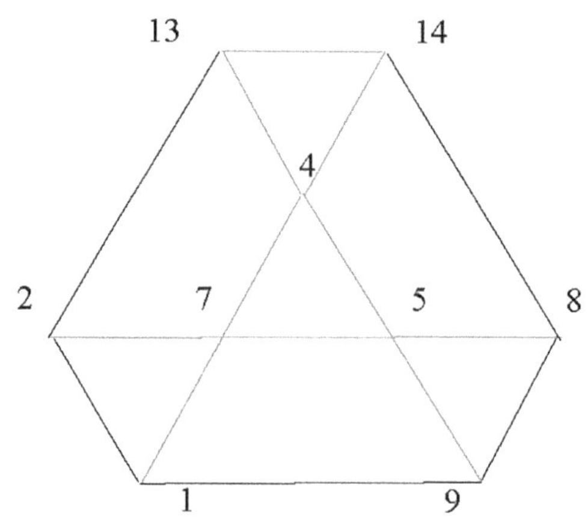

The section that separates both polytopes and at the same time belongs to both polytopes, taking into account the added and removed edges, has the form shown in Figure 7.

One defined the dimension of the upper part of the cluster shown in Figure 5. This part has 12 vertices (f_0=12), 35 edges (f_1=35)

1 - 2, 1 - 7, 1 - 3, 1 - 9, 2 - 3, 2 - 7, 2 - 13, 3 - 7, 3 - 4, 3 - 11, 3 - 13, 3 - 15, 4 - 7, 4 - 13, 4 - 15, 4 - 14, 4 - 11, 4 - 5, 5 - 11, 5 - 8, 5 - 9, 5 - 7, 5 - 14, 5 - 15, 7 - 13, 7 - 15, 8 - 11, 8 - 14, 8 - 9, 9 - 11, 11 - 14, 11 - 15, 13 - 14, 13 - 15, 14 - 15,

41 two - dimensional elements (f_2=41)

1 - 2 - 7, 1 - 7 - 5 - 9, 1 - 7 - 3, 1 - 2 - 3, 1 - 3 - 11 - 9, 2 - 3 - 7, 2 - 3 - 13, 2 - 13 - 7, 3 - 13 - 15, 3 - 4 - 11, 3 - 4 - 15, 3 - 4 - 7, 3 - 7 - 15, 3 - 13 - 14 - 11, 3 - 13 - 4, 4 - 7 - 5, 4 - 5 - 11, 4 - 7 - 13, 4 - 5 - 14, 4 - 11 - 14, 3 - 13 - 7, 4 - 13 - 15, 4 - 11 - 15, 4 - 13 - 14, 4 - 15 - 14, 4 - 15 - 5, 4 - 7 - 15, 5 - 9 - 11, 5 - 15 - 11, 5 - 11 - 14, 5 - 14 - 8, 5 - 11 - 8, 5 - 15 - 14, 5 - 8 - 9, 8 - 11 - 14, 8 - 9 - 11, 11 - 15 - 14, 13 - 14 - 15, 3 - 7 - 4 - 15, 5 - 11 - 4 - 15, 3 - 7 - 5 - 11,

18 three - dimensional elements (f_3=18)

1 - 2 - 7 - 3, 1 - 3 - 7 - 11 - 5 - 9, 2 - 3 - 7 - 13, 3 - 7 - 4 - 13, 3 - 7 - 4 - 15, 3 - 13 - 15 - 4, 3 - 13 - 15 - 14 - 11, 3 - 13 - 14 - 11 - 7 - 5, 3 - 4 - 11 - 14 - 13, 3 - 4 - 7 - 5 - 11, 4 - 7 - 5 - 15, 4 - 13 - 14 - 15, 4 - 5 - 11 - 14, 4 - 15 - 11 - 14, 5 - 15 - 11 - 14, 5 - 14 - 11 - 8, 5 - 11 - 8 - 14, 5 - 11 - 8 - 9,

external three - dimensional surface of Figure 5.

Substituting the obtained values $f_i(i=0,1,2,3)$ in the equation (1) you can see that the Euler - Poincaré equation is satisfied in his case for $n = 4$

$$12 - 35 + 41 - 18 = 0.$$

This proves that Figure 5 is polytope with dimension 4.

The lower part of the cluster (Figure 6) is symmetrical to the upper part of the cluster (Figure 5) relative to the cross section separating them. The perpendiculars to the section

3-7, 5-11, 4-15

in the upper part of the cluster are equal in length to the perpendiculars to the section in the lower part of the cluster passing through the same points in the section, respectively, 7, 4, 5. Each edge in the upper part of the cluster has an edge in the lower part of the cluster. Therefore, the dimension of the polytope of the lower part of the cluster is equal to the dimension of the polytope of the upper part of the cluster, i.e. it is equal to 4. This can also be seen by directly counting the number of elements of different dimensions at the bottom of the cluster.

Following. The cluster ligand polyhedron is not an anti - cube - octahedron.

Figure 8. Top and bottom views of ligand polyhedron

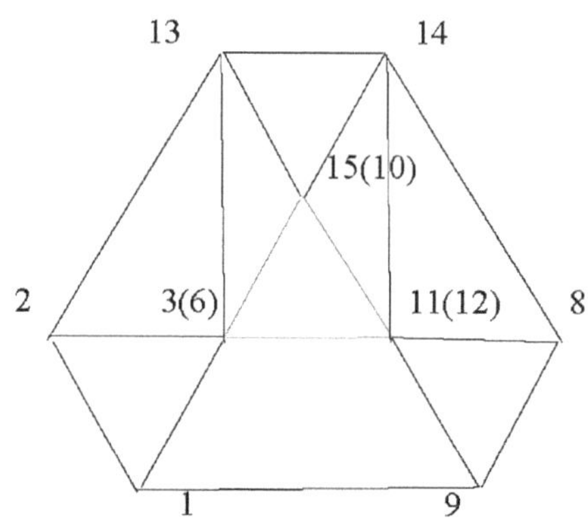

Figure 9. The spatial image of the ligand polyhedron

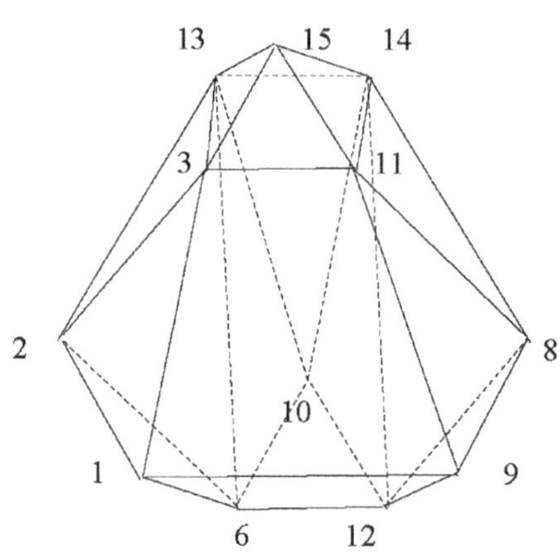

The top view of the cluster (from a point above the triangle 3 - 15 - 11) coincides with the view of the cluster from the bottom (from a point under triangle 6 - 10 - 12), as follows from Figure 8.

Only the numbers of the points of the blue triangle change (the numbers for the lower part of the cluster are indicated in brackets). The spatial image of the ligand polyhedron is shown in Figure 9.

As follows from Figures 8, 9 the ligand polyhedron contains 12 vertices (f_0=12), 28 edges (f_1=28), 18 flat faces (2 trapeziums, 16 triangles) (f_2=18).

Substituting the obtained values f_i(i=0,1,2) in the equation (1) you can see that the Euler – Poincaré equation is satisfied in his case for $n = 3$

$$12 - 28 + 18 = 2.$$

This proves that Figure 9 is polyhedron with dimension 3.

It is clearly seen that the ligand polyhedron is not an anti - cube - octahedron, as was assumed earlier.

The ligand polyhedron cross section divides the ligand polyhedron into two parts, which are also three - dimensional surfaces. If the section is a hexagon, then in each half there are 9 vertices (f_0=9), 17 edges (f_1=17), 10 two - dimensional faces (f_2=10) . Substituting the obtained values f_i(i=0,1,2) in the equation (1) you can see that the Euler – Poincaré equation is satisfied in his case for $n = 3$

$$9 - 17 + 10 = 2.$$

Substituting the obtained values f_i(i=0,1,2) in the equation (1) you can see that the Euler – Poincaré equation is satisfied in his case for $n = 3$

$$12 - 28 + 18 = 2.$$

If the cross section of the ligand polyhedron is the section in Figure 7 (it was used to analyze the upper and lower parts of the cluster when determining their dimensions), then each of the halves also represents a three - dimensional surface. In this case, for each of the halves, the number of vertices increases by 3, the number of edges increases by 9, the number of flat faces increases by 6. The total change in the right side of the Euler-Poincaré equation is 3 – 9 + 6 = 0. Thus, the right side of equation (1) does not change. The surfaces remain three - dimensional.

HOMO: ELEMENT METAL CYCLES WITH THE BRIDGING LIGANDS

The carbonyls Ru and Os discussed above are essentially the only structurally characterized compounds in which the metal – containing fragments are connected to each other only by metal - metal bonds. In many cases, there are a number of bridging ligands in the cycle. A typical example of such a compound is hydride $H_3Mn_3(CO)_{12}$ (Kirtley et al., 1973). Here, along the metal – metal bonds, three bridging hydrogen atoms are located in the cycle plane. The scheme of such a compound is shown in Figure 10.

Figure 10. Shema of a hydride $H_3Mn_3(CO)_{12}$

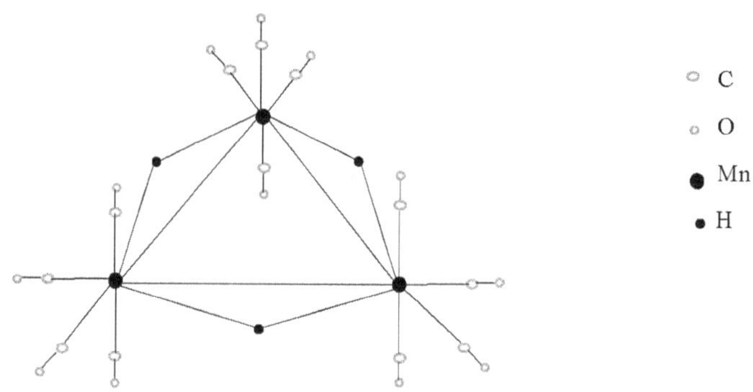

In this compound, the coordination of a double hexagonal pyramid is observed around each metal atom. Passing into each other the coordination creates the spatial structure. It also, as in the previous case, in a three - link metal cycle without bridge connections consists of two parts that are symmetrical with respect to the flat section separating them. The upper part of this structure is shown in the Figure 11.

The cross section separating both parts of the compound's spatial structure is a hexagon. However, compared to the section in Figure 7, this section contains three more vertices. Hydrogen atoms are located in these additional vertices 2, 5, 11. At the vertices 13, 14, 15, there are manganese atoms. In the remaining vertices of both the upper and lower parts there are functional groups of CO. The edges of the structure corresponding to the valence bonds are indicated in red. The remaining edges of the pyramids are marked in black. To determine the dimension of this compound, one will use the method developed in this book. As a first step, it is necessary to transform

Figure 11. Spatial structure of compound $H_3Mn_3(CO)_{12}$

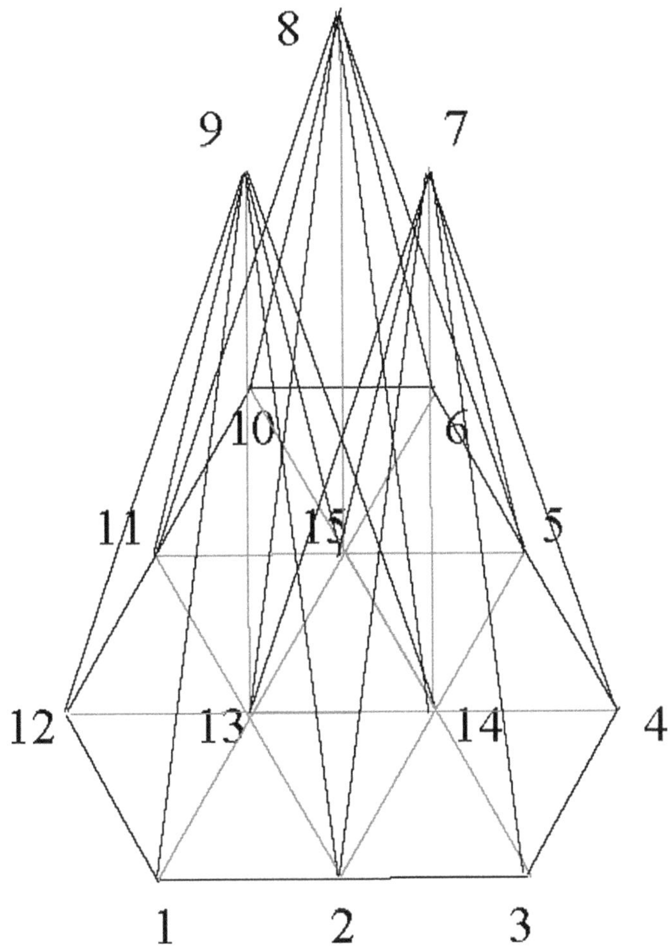

this structure into a closed convex body. To do this, add the edges connecting the free vertices of the three hexagonal pyramids. These are edges

9 - 8, 8 - 7, 9 - 7.

These edges are marked in blue (Figure 12).

From the hexagonal pyramids, it is necessary to remove the edges that have lost their significance in the creation of pyramids, like convex bodies. In Figure 12, the edges

Figure 12. The upper part of the structure of the compound $H_3Mn_3(CO)_{12}$, as a convex body

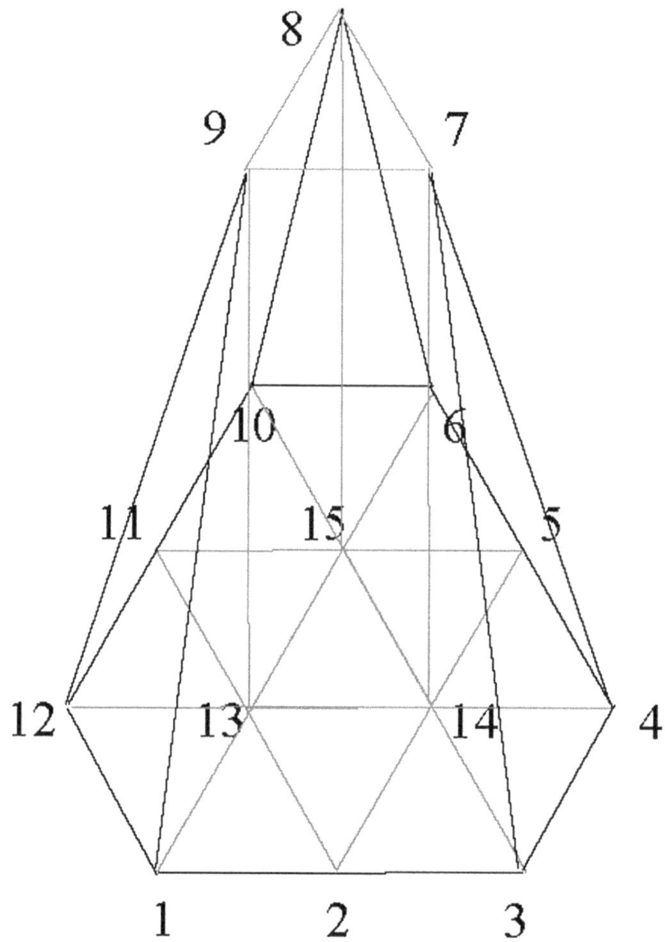

11 - 9, 11 - 8, 5 - 7, 5 - 8, 2 - 9, 2 - 7, 13 - 7, 13 - 8, 15 - 9, 15 - 7, 14 - 8, 14 - 9

are removed. As a result, in Figure 12 a convex figure is obtained. As a second step in determining the dimension of a compound, it is necessary to apply the Euler - Poincaré formula to this figure (Poincaré, 1895). To do this, you first need to determine which three - dimensional figures are included in the structure in Figure 12. These figures are tetrahedrons with edges, which are the short edges of the base, and the heights of the pyramids. These are three tetrahedrons

1 - 12 - 13 - 9, 3 - 4 - 14 - 7, 6 - 15 - 10 - 8,

prism

13 - 14 - 15 - 7 - 8 - 9,

and three figures located between the tetrahedrons and the prism. It is required to prove that these figures are three - dimensional. Consider one of them

1 - 2 - 3 - 7 - 9 - 13 - 14

(Figure 13).

Figure 13. The figure 1 - 2 - 3 - 7 - 9 - 13 - 14 in the structure of the compound $H_3Mn_3(CO)_{12}$

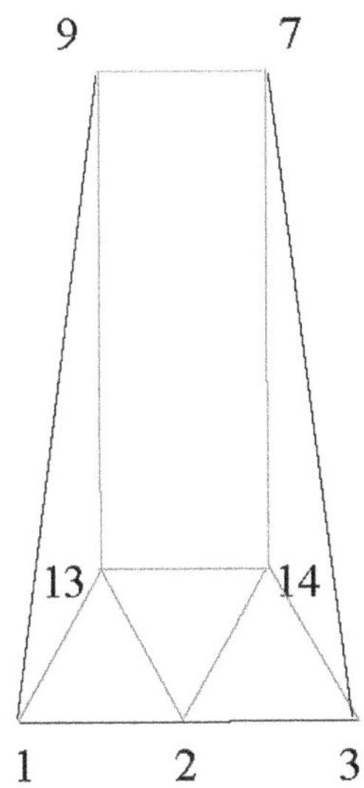

The polyhedron in Figure 13 has 7 vertices (f_0=7), 12 edges (f_1=12)

1 - 2, 2 - 3, 1 - 13, 2 - 13, 2 - 14, 3 - 14, 3 - 7, 1 - 9, 7 - 9, 7 - 14, 9 - 13, 13 - 14,

7 two - dimensional faces (f_2=7)

1 - 2 - 13, 2 - 13 - 14, 2 - 3 - 14, 3 - 7 - 14, 1 - 13 - 9, 7 - 9 - 13 - 14, 1 - 2 - 3 - 7 - 9.

Substituting the obtained values $f_i(i=0,1,2)$ in the equation (1) you can see that the Euler - Poincaré equation is satisfied in his case for $n = 3$

$7 - 12 + 7 = 2.$

This proves that Figure 13 is polyhedron with dimension 3.

In addition to this figure, the polytope in Figure 12 has two more similar figures

10 - 11 - 12 - 9 - 8, 4 - 5 - 6 - 7 - 8.

Another three - dimensional polyhedron in Figure 12 is the outer surface of the structure in this figure. It includes all vertices and all edges except for the three heights of the hexagonal pyramids

13 - 9, 14 - 7, 15 - 8.

Thus, this surface includes 15 vertices (f_0=15), edges (f_1=33)

1 - 2, 1 - 13, 1 - 12, 1 - 9, 2 - 13, 2 - 14, 2 - 3, 3 - 14, 3 - 4, 3 - 7, 4 - 5, 4 - 14, 4 - 7, 14 - 5, 5 - 6, 5 - 15, 6 - 15, 6 - 10, 6 - 8, 7 - 9, 7 - 8, 9 - 8, 8 - 10, 9 - 12, 10 - 11, 10 - 15, 11 - 15, 11 - 13, 11 - 12, 12 - 13, 13 - 14, 13 - 15, 14 - 15,

20 two - dimensional faces (f_2=20)

1 - 12 - 13, 1 - 13 - 2, 1 - 12 - 9, 1 - 9 - 7 - 3 - 2, 2 - 14 - 3, 2 - 13 - 14, 3 - 14 - 4, 3 - 7 - 4, 4 - 5 - 6 - 7 - 8, 4 - 5 - 14, 5 - 14 - 15, 5 - 15 - 6, 6 - 8 - 10, 6 - 15 - 10, 10 - 8 - 9 - 12, 10 - 15 - 11, 11 - 15 - 13, 11 - 13 - 12, 13 - 14 - 15, 7 - 8 - 9.

Substituting the obtained values $f_i(i=0,1,2)$ in the equation (1) you can see that the Euler - Poincaré equation is satisfied in his case for $n = 3$

$15 - 33 + 20 = 2.$

This proves that the outer surface of Figure 12 is polyhedron with dimension 3.

Thus, the polytope on Figure 12 has 8 polyhedrons with dimension 3 (f_3=8). The number of edges in Figure 12 on three edges more of the number of edges in its outer surface (three heights of the hexagonal pyramids 9 - 13, 7 - 14, 8 - 15). Therefore, the number of the edges in Figure 12 is equal to 36 (f_1=36). Adding the heights of hexagonal pyramids into the surface in Figure 12 leads to an increase in the number of two - dimensional faces compared with the number of such faces on the surface. Added 9 two - dimensional faces

1 - 13 - 9, 12 - 13 - 9, 3 - 14 - 7, 4 - 14 - 7, 13 - 14 - 7 - 9, 6 - 15 - 8, 10 - 15 - 8, 14 - 15 - 7 - 8, 13 - 15 - 8 - 9.

Figure 14. A ligand polyhedron of a cluster with a three - membered metal chain with bridging ligands

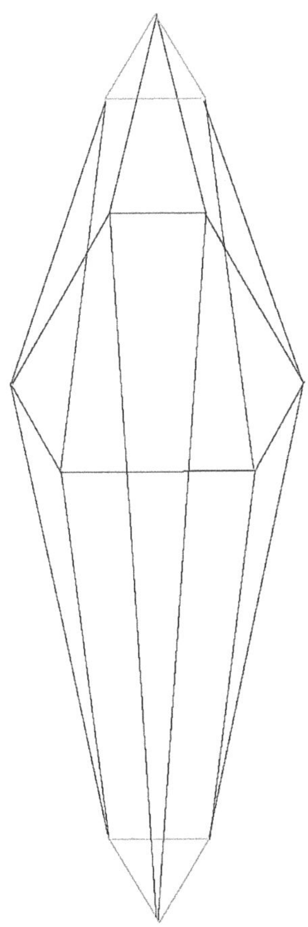

Therefore, the number of two - dimensional faces in the polytope in Figure 12 is 29 (f_2=29).

Substituting the obtained values f_i(i=0,1,2,3) in the equation (1) you can see that the Euler - Poincaré equation is satisfied in his case for $n = 4$

$$15 - 36 + 29 - 8 = 0.$$

This proves that Figure 12 is polytope with dimension 4.

The polytope on Figure 12 is only the upper part of the polytope, corresponding to a cluster with a three - membered metal chain with bridging ligands. Due to symmetry, the lower part of this polytope also has dimension 4. The surface of the polytope corresponding to the cluster is a ligand polyhedron. It has the form shown in Figure 14.

The ligand polyhedron consists of eight triangles (two equilateral and six isosceles) and six quadrangles in the form of trapezoids. Topologically, it is close to an anti - cube - octahedron and, therefore, it undoubtedly has dimension 3. It differs from an anti - cube - octahedron in that the two - dimensional figures that form it are mostly incorrect.

HETERO: ELEMENT METAL CYCLES WITH THE BRIDGING LIGANDS

Other atoms can be used as atoms in the bridged compounds of the three - membered metal chain, i.e. metals other than chain metals. In this case, the cluster becomes hetero element. Moreover, the ligand metal atom can be associated with all the metal atoms of the three - membered metal chain. It is obvious that in this case the ligand metal atom can no longer lie in the plane of the three - membered metal chain. An example of such a cluster is cluster $FeCo_2(CO)_9(\mu\text{-}\Theta)(\Theta = S, Se, Te)$ (Raithby, 1980). The scheme of this cluster is shown in Figure 15.

In this compound, octahedral coordination is manifested around each atom of the metal chain. Moreover, the three vertices of the octahedrons are combined in one atom of the ligand. The spatial structure of this cluster is shown in Figure 16.

In Figure 16, at the vertices 8, 9, 10 are located the atoms of the metal of the chain, at the vertex of 4 there is an atom of the ligand, common to the three octahedrons of the chain, in the other vertices the functional groups of

Figure 15. Scheme of the cluster $FeCo_2(CO)_9(\mu\text{-}\Theta)(\Theta=S,Se,Te)$.

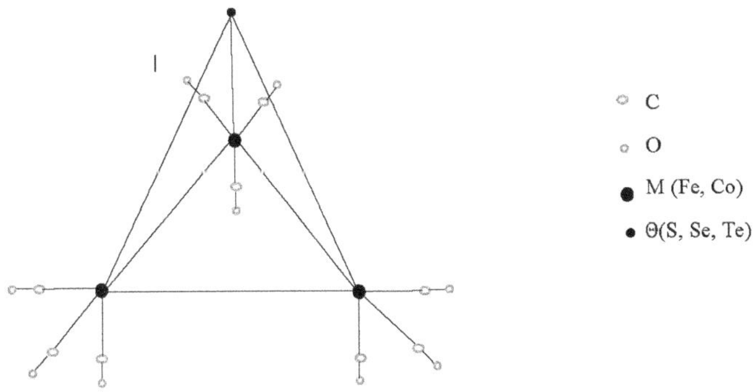

Figure 16. Spatial structure of the cluster $FeCo_2(CO)_9(\mu\text{-}\Theta)(\Theta=S,Se,Te)$.

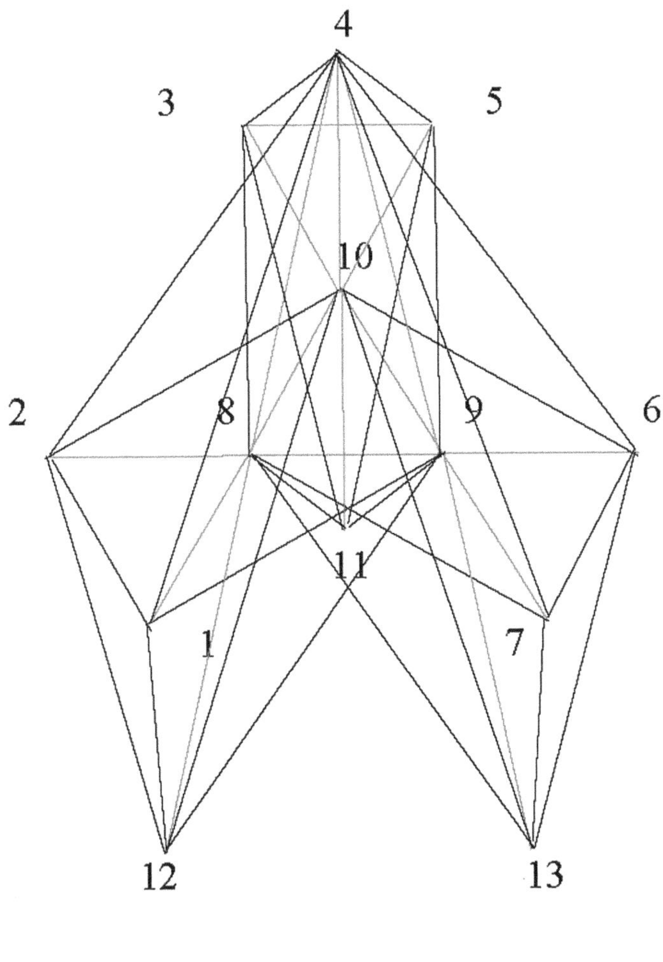

CO are located. The edges corresponding to the valence bonds are marked in red. The edges of the octahedrons are marked in black. To determine the dimension of this compound, one will use the method developed in this book. As a first step, it is necessary to transform this structure into a closed convex body. To do this, add the edges connecting the free vertices of the three octahedrons. These are edges

1 - 7, 2 - 3, 5 - 6, 12 - 11, 11 - 13, 12 - 13.

These edges are marked in blue (Figure 17).

Figure 17. The structure of the compound $FeCo_2(CO)_9(\mu\text{-}\Theta)(\Theta=S,Se,Te)$, as a convex body

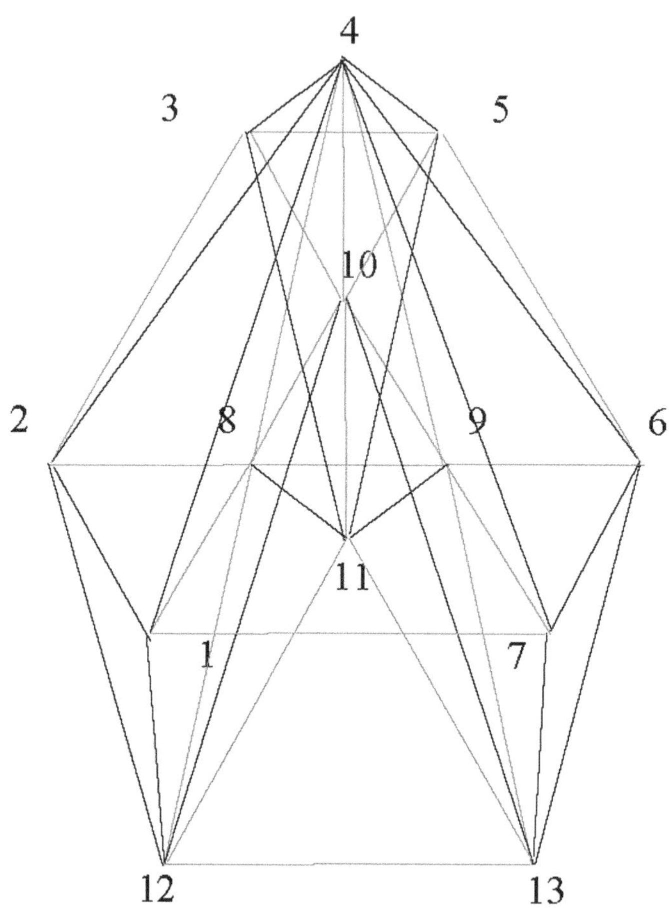

From the octahedrons, it is necessary to remove the edges that have lost their significance in the creation of octahedrons, like convex bodies. In Figure 17, the edges

2 - 10, 1 - 9, 8 - 7, 10 - 6, 3 - 8, 5 - 9, 12 - 10, 13 - 10, 11 - 9, 11 - 8

are removed. As a result, in Figure 17 a convex figure is obtained.

The construction in Figure 17 can be divided into two polytopes: one polytope, located above the section in which the closed cycle of metal atoms is located (Figure 18); another polytope is located under this section (Figure 19).

Figure 18. The polytope upper part of a cluster $FeCo_2(CO)_9(\mu\text{-}\Theta)(\Theta=S,Se,Te)$.

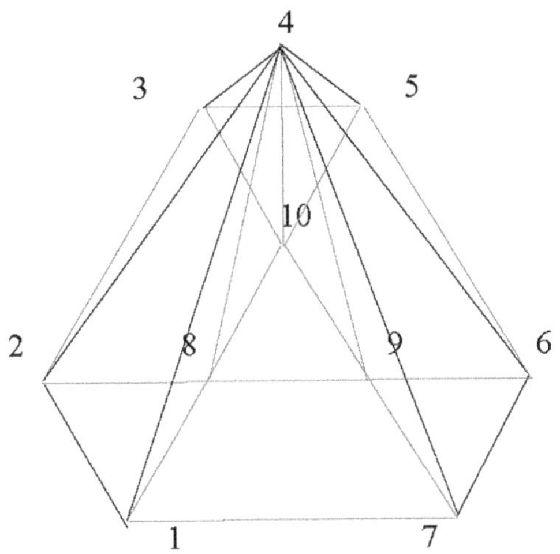

One can define the dimension of the upper part of the cluster shown in Figure 18. This part has 10 vertices (f_0=10), 24 edges (f_1=24)

1 - 4, 1 - 2, 1 - 8, 1 - 7, 2 - 8, 2 - 4, 2 - 3, 3 - 4, 3 - 5, 3 - 10, 4 - 8, 4 - 10, 4 - 9, 4 - 7, 4 - 6, 4 - 5, 5 - 6, 5 - 10, 6 - 7, 6 - 9, 7 - 9, 8 - 9, 8 - 10, 9 - 10,

22 two - dimensional elements (f_2=22)

1 - 2 - 8, 1 - 8 - 4, 1 - 8 - 9 - 7, 1 - 4 - 7, 2 - 8 - 10 - 3, 2 - 3 - 4, 2 - 8 - 4, 3 - 4 - 5, 3 - 4 - 10, 3 - 5 - 10, 4 - 5 - 10, 4 - 5 - 6, 4 - 9 - 6, 4 - 10 - 9, 4 - 8 - 10, 4 - 8 - 9, 4 - 6 - 7, 4 - 9 - 7, 8 - 7 - 10, 5 - 10 - 9 - 6, 6 - 9 - 7,

Figure 19. The polytope lower part of a cluster $FeCo_2(CO)_9(\mu\text{-}\Theta)(\Theta=S,Se,Te)$.

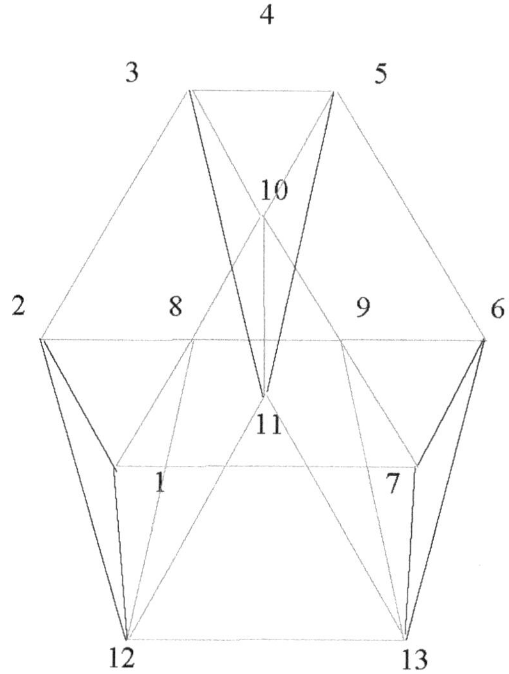

8 three - dimensional elements (f_3=8)

1 - 2 - 8 - 4, 1 - 7 - 8 - 9 - 4, 2 - 8 - 3 - 10 - 4, 3 - 4 - 5 - 10, 5 - 6 - 9 - 10 - 4, 7 - 6 - 9 - 4, 8 - 9 - 10 - 4,

external surface of the upper part of the cluster.

Substituting the obtained values $f_i(i=0,1,2,3)$ in the equation (1) you can see that the Euler - Poincaré equation is satisfied in his case for $n = 4$

$$10 - 24 + 22 - 8 = 0.$$

This proves that Figure 18 is polytope with dimension 4.

One can define the dimension of the lower part of the cluster shown in Figure 19. This part has 12 vertices (f_0=12), 26 edges (f_1=26)

1 - 2, 1 - 8, 1 - 7, 1 - 12, 2 - 3, 2 - 8, 2 - 12, 3 - 5, 3 - 10, 3 - 11, 5 - 10, 5 - 11, 5 - 6, 6 - 9, 6 - 7, 6 - 13, 7 - 9, 7 - 13, 8 - 10, 8 - 9, 9 - 13, 9 - 10, 10 - 11, 11 - 12, 11 - 13, 12 - 13,

21 two - dimensional elements (f_2=21)

1 - 2 - 8, 1 - 8 - 9 - 7, 1 - 7 - 12 - 13, 1 - 2 - 12, 1 - 8 - 12, 2 - 3 - 10 - 8, 3 - 5 - 10, 3 - 10 - 11, 3 - 5 - 11, 2 - 3 - 11 - 12, 5 - 10 - 11, 5 - 10 - 9 - 6, 5 - 6 - 11 - 13, 6 - 7 - 13, 6 - 9 - 7, 7 - 9 - 13, 8 - 10 - 9, 8 - 9 - 12 - 13, 8 - 10 - 11 - 12, 9 - 10 - 11 - 13, 11 - 12 - 13,

8 three - dimensional elements (f_3=8)

1 - 2 - 8 - 12, 1 - 7 - 8 - 9 - 12 - 13, 2 - 8 - 12 - 3 - 10 - 11, 3 - 5 - 10 - 11, 5 - 10 - 9 - 6 - 11 - 13, 9 - 6 - 7 - 13, 8 - 10 - 9 - 11 - 12 - 13,

external surface of the lower part of the cluster.

Substituting the obtained values f_i(i=0,1,2,3) in the equation (1) you can see that the Euler - Poincaré equation is satisfied in his case for $n = 4$

$$12 - 26 + 21 - 8 = 0.$$

Figure 20. A ligand polyhedron of a hetero - element cluster $FeCo_2(CO)_9(\mu\text{-}\Theta)$ (Θ=S,Se,Te) with a three - membered metal chain with bridging ligand

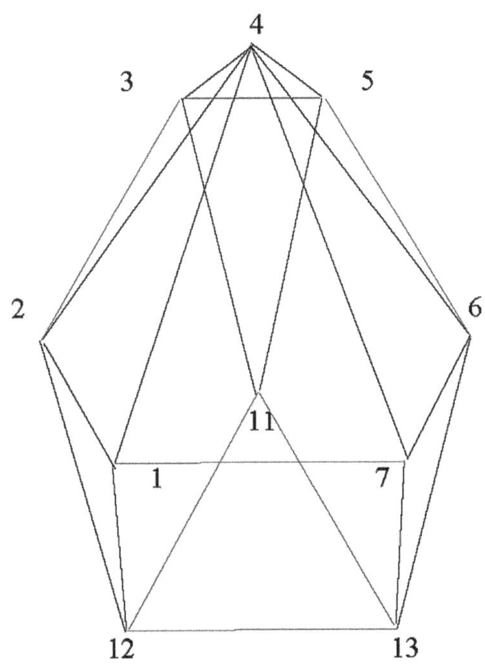

This proves that Figure 19 is polytope with dimension 4.

The surface of the polytope consists from two parts corresponding to the clusterFe$Co_2(CO)_9(\mu$-$\Theta)(\Theta$=S,Se,Te) is a ligand polyhedron. It has the form shown in Figure 20.

It differs from an anti - cube - octahedron.

CONCLUSION

There are considered with closed three - membered metal cycles of one or several chemical elements surrounded by ligands connected to them. It has been proven that the widespread opinion in the literature about the formation of ligands by atoms in some cases of the semi -correct polyhedron of the anti - cube - octahedron is wrong. In alleged cases, ligands do not form this polyhedron. Geometrical analysis of the interpenetration of the coordinates of ligand atoms around each of the metal atoms of a closed chain showed that this leads to a different class of special three - dimensional irregular polyhedrons for different clusters. A very effective method for the geometric analysis of the distribution of atoms in clusters of such a structure turned out to be a developed method for removing and adding edges. The edges of the geometric coordination of ligands around individual atoms of a metal chain that have lost their purpose in creating a convex body around an individual metal atom during the transition of a metal into a chain are removed. The edges needed to create a convex body around the chain are added. In this case, all edges corresponding to the valence bonds of atoms and, of course, all atoms are retained. This leads to the creation of a clear, easier to understand, cluster structure with preservation of their chemical nature. In all cases of homo element and hetero element closed metal chains, the cycle itself, located in a certain plane creates a cross section of the cluster, dividing the cluster into two parts. Each of the parts of a cluster has dimension 4. The increase in dimension is connected geometrically with the fact that there are several atoms in this two - dimensional section. Thus, the structure of clusters with a skeleton in the form of a flat metal cycle is two polytope of dimension 4, separated by a section containing this metal cycle.

If the bridging ligands break the symmetry of the chemical compound, then both parts of the cluster structure are not symmetrical with respect to the cross section containing the metal cycle. If the bridging ligands do not violate the symmetry of the chemical compound, then both parts of the cluster structure are symmetric with respect to the cross section containing the metal cycle.

The geometry of clusters with a more complex spatial arrangement of metal atoms in the cluster core will be considered in the next chapter.

REFERENCES

Benfield, R. E., & Johnson, B. F. G. (1981). Structures and fluxional behaviour of transition metal cluster carbonyls. *Transition Metal Chemistry (Weinheim)*, *6*(3), 131–144. doi:10.1007/BF00624332

Gubin, S. P. (2019). *Chemistry clusters. Basics of classification and structure*. Moscow: Editorial URSS.

Kirtley, S. W., Olsen, I. P., & Bau, R. (1973). Location of the hydrogen atoms in dodecarbonyltrihydromanganese. Crystal structure determination. *Journal of the American Chemical Society*, *95*(14), 4532–4536. doi:10.1021/ja00795a012

Mason, R., & Rae, A. I. M. (1968). The crystal structure of ruthenium carbonyl, Ru3(CO)12. *Journal of the Chemical Society, Section A: Inorganic, Physical, Theoretical*, 778. doi:10.1039/j19680000778

Poincaré, A. (1895). Analysis situs. *J. de é Ecole Polytechnique, 1,* 1 – 121.

Raithby, P. R. (1980). *Transition metal clusters* (B. F. G. Jonson, Ed.). Chichester: Wiley.

Zhizhin, G. V. (2019). *The Geometry of Higher – Dimensional Polytopes*. Hershey, PA: IGI Global. doi:10.4018/978-1-5225-6968-8

KEY TERMS AND DEFINITIONS

Dimension of the Space: The member of independent parameters needed to describe the change in position of an object in space.

Hetero-Element Metal Chains: Clusters in which the core contains atoms of different elements.

Homo-Element Metal Chains: Clusters in which the core contains atoms of the same element.

Linear Metal Chains: Metal chains in which metal atoms are located in a straight line.

Metal Chains: Cluster compounds that have a skeleton in the form of metal chains, that is, polymetallic chains formed by metal-metal localized covalent bonds.

Nanocluster: A nanometric set of connected atoms, stable either in isolation state or in building unit of condensed matter.

Nonlinear (Curved) Metal Chains: Clusters in which metal-metal valent bonds form an angle different from 180 degrees between them.

Polytope: Polyhedron in the space of higher dimension.

Chapter 6
Metallic Clusters With Ligands and Polyhedral Core

ABSTRACT

The geometry of clusters with ligands and a polyhedral frame is considered by the methods of studying the geometry of higher-dimensional polytopes, developed in the author's monograph. It is shown that these methods allow us to establish important details of cluster geometry, which elude analysis based on the representations of three-dimensional geometry. It is established that the well-known Kuban cluster is a 4-cross-polytope, which allows different variants of the Kuban cluster. A cluster of gold with a tetrahedral backbone is a 5-cross-polytope. The cluster tetra anion of cobalt is a polytope of dimension 5 of a new type. Different types of ligands limit the cobalt skeleton from above and below.

INTRODUCTION

A significant part of the known cluster are molecules that have a skeleton in the form of a metal polyhedron. In other words, the shortest distances between metal atoms in such molecules form a convex closed polyhedron bounded by flat faces. As a rule, these polyhedrons are considered three - dimensional. Even if in the simplest cases this is so, then taking into account the ligands attached to the skeleton, such a molecule will have a dimension greater three. In addition, the metal core may have a higher dimension, which further increases the dimension of the cluster. Therefore to analyze

DOI: 10.4018/978-1-7998-3784-8.ch006

the structure of clusters with a metal backbone in the form of a polyhedron, it is necessary to use the geometry of high - dimensional polytopes, not the abstract geometry of spaces of higher dimension, but the geometry of high - dimensional polytopes based on the analysis of the geometry of chemical compounds (Zhizhin, 2018, 2019a, b). If the space surrounding the cluster is considered three - dimensional, the outer surface of the cluster can be constructed from two - dimensional faces, although the cluster itself has a higher dimension. The existence of a higher dimension space inside a three - dimensional space does not contradict the Riemann geometry (Riemann, 1854), since according to the Riemann geometry, space is finite (in Euclidean geometry the space is infinite).

Inside the cluster can meet a variety of convex bodies with dimensions less than the dimension of the cluster, including regular, semi - correct and irregular three - dimensional polyhedrons. Analysis of the structure of clusters, taking into account its dimensions and the dimensions of its components, is necessary for an adequate modern description of cluster geometry.

CLUSTER COMPOUNDS HAVING A SKELETON IN THE FORM OF A METAL TETREHEDRON

Among a wide variety of clusters, a special place is occupied by compounds based on the skeleton in the form of the simplest metal - polyhedron — the tetrahedron. Tetrahedral cluster compounds are widely distributed (Cubin, 2019; Garner, 1980). The special significance of tetrahedral clusters lies in

Figure 1. Homo - metallic tetrahedral iridium cluster $Ir_4(CO)_{12}$

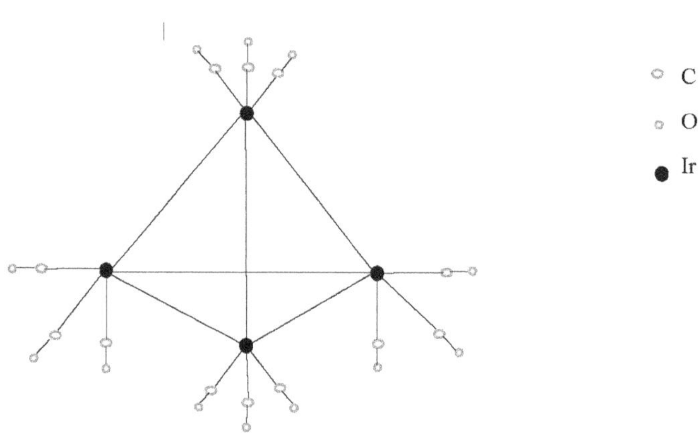

the fact that compounds based on them are found in nature in the form of multi - core redox homo - metallic and hetero - metal enzymes (Lichtensins, 1979; Neuton & Otsuka, 1980). Let us consider in more detail the structural features of cluster compounds with tetrahedral metal cores. A classic example of compounds of this type is the tetrahedral cluster $Ir_4(CO)_{12}$ (Gubin, 2019). The scheme of this compound is shown in Figure 1.

Theorem 1. The homo - metallic tetrahedral iridium cluster $Ir_4(CO)_{12}$ has dimension 4.

Proof. The spatial structure of the cluster $Ir_4(CO)_{12}$ is shown in Figure 2.

Figure 2. The spatial structure of the cluster $Ir_4(CO)_{12}$

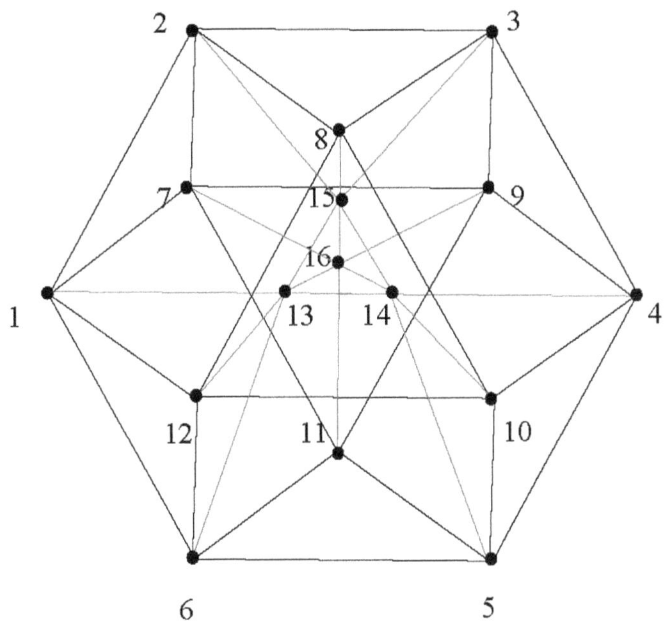

As follows from Figures 1, 2 the spatial structure of the cluster $Ir_4(CO)_{12}$ contains 16 vertices. In vertices 1÷2 is arrangement of the functional groups CO, in vertices 13÷6 is arrangement of the atoms Ir.

For further it is necessary to use the Euler - Poincaré equation (Poincaré, 1895)

$$\sum_{i=0}^{n-1}(-1)^i f_i(P) = 1-(-1)^i, \qquad (1)$$

f_i is the number of the elements with the dimension i at polytope P; n is dimension of the polytope P.

In this case f_0=16. The numbers of edges on Figure 2 is 42 (f_1=42)

1 - 13, 1 - 12, 1 - 6, 1 - 7, 1 - 2, 2 - 7, 2 - 15, 2 - 8, 2 - 3, 3 - 8, 3 - 15, 3 - 9, 3 - 4, 4 - 9, 4 - 10, 4 - 14, 4 - 5, 5 - 10, 5 - 14, 5 - 11, 5 - 6, 6 - 11, 6 - 13, 6 - 12, 12 - 10, 12 - 13, 12 - 8, 7 - 11, 7 - 16, 7 - 9, 9 - 16, 9 - 11, 8 - 15, 8 - 10, 10 - 14, 11 - 16, 13 - 15, 13 - 16, 13 - 14, 14 - 16, 14 - 15, 15 - 16.

The number of triangles on Figure 2 is 24

2 - 3 - 15, 2 - 3 - 8, 2 - 8 - 15, 3 - 8 - 15, 3 - 15 - 16, 13 - 15 - 14, 15 - 16 - 14, 13 - 16 - 14, 14 - 4 - 10, 14 - 10 - 5, 14 - 4 - 5, 5 - 4 - 10, 1 - 13 - 6, 1 - 13 - 12, 1 - 12 - 6, 6 - 13 - 12, 7 - 9 - 16, 7 - 16 - 11, 7 - 9 - 11, 16 - 9 - 11, 1 - 2 - 7, 3 - 9 - 4, 6 - 5 - 11, 12 - 8 - 10.

The flat tetragonal faces on Figure 2 is 18

1 - 2 - 8 - 12, 8 - 3 - 4 - 10, 12 - 10 - 5 - 6, 2 - 3 - 7 - 9, 1 - 7 - 6 - 11, 11 - 9 - 4 - 5, 2 - 15 - 7 - 16, 15 - 16 - 3 - 9, 9 - 4 - 16 - 14, 1 - 7 - 16 - 13, 11 - 16 - 14 - 5, 6 - 11 - 13 - 16, 2 - 15 - 13 - 1, 12 - 8 - 15 - 13, 3 - 4 - 15 - 14, 8 - 10 - 15 - 14, 6 - 5 - 13 - 14, 12 - 10 - 13 - 14.

Thus, the common number of flat faces on Figure 2 is 42, (f_2=42).

The tetrahedrons on Figure 2 is 5

1 - 13 - 12 - 6, 5 - 4 - 10 - 14, 2 - 3 - 8 - 15, 13 - 14 - 15 - 16, 7 - 9 - 11 - 16.

The trigonal prisms on Figure 2 is 4

1 - 2 - 7 - 13 - 16 - 15, 14 - 15 - 16 - 3 - 9 - 4, 13 - 16 - 14 - 6 - 11 - 5, 13 - 16 - 14 - 12 - 8 - 10.

The pyramids on Figure 2 is 6

6 - 12 - 13 - 14 - 10 - 5, 1 - 2 - 8 - 12 - 15 - 13, 8 - 3 - 4 - 10 - 15 - 14, 1 - 7 - 6 - 11 - 13 - 16, 9 - 4 - 5 - 11 - 16 - 14, 2 - 3 - 9 - 7 - 15 - 16.

On Figure 2 is one cube - octahedron

1 - 2 - 3 - 4 - 5 - 6 - 7 - 8 - 9 - 10 - 11 - 12

(ligand polyhedron). Thus, the common number of the three - dimensional faces on Figure 2 is 16 ($f_3^{'}$=16).

Substituting the obtained values f_i(i=0,1,2,3) in the equation (1) you can see that the Euler - Poincaré equation is satisfied in his case for $n = 4$

$$16 - 42 + 42 - 16 = 0.$$

This proves that Figure 2 is polytope with dimension 4.

FOUR-CORE CLUSTER CONNECTION HAVING A FRAME IN THE FORM "CUBAN CLUSTER"

Among the cluster compounds containing four metal atoms, there is a rather representative group of molecules having a skeleton in the form of a "Cuban cluster" (Garner, 1980; Gubin, 2019), at the vertices of which metal atoms and non - metal atoms alternate. Schemes of such compounds are presented in Figure 3 a, b.

Figure 3. Structural types of cores of cluster molecules containing a group M_4E_4: a) A metal tetrahedron resting on a triangular face. b) A metal tetrahedron resting on an edge.

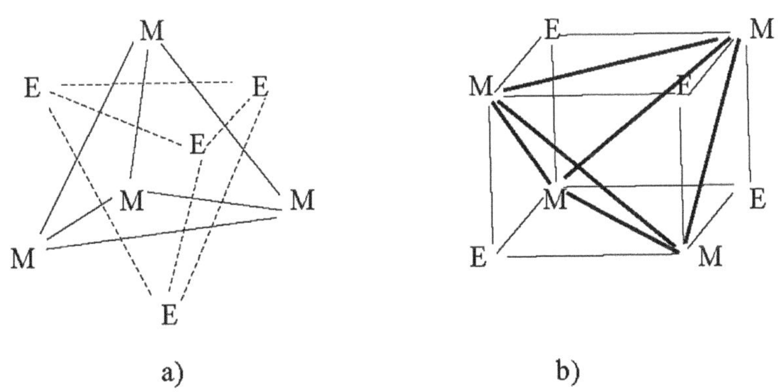

On Figure 3 M is atom of metal, E is atom non - metal (O, S, N, P, As, Sb, Cl, Br, I).

Theorem 2. "Cuban cluster" is 4 - cross - polytope.

Proof. The spatial image of clusters containing a group M_4E_4 is shown in Figure 4 a, b.

Figure 4. The spatial image of clusters containing a group M_4E_4: a) A metal tetrahedron resting on a triangular face. b) A metal tetrahedron resting on an edge.

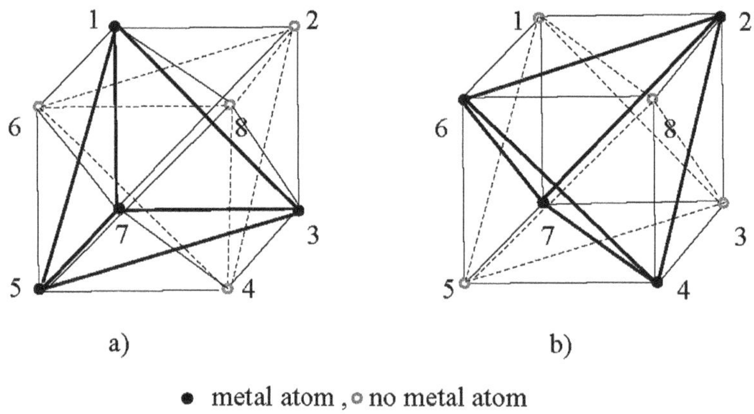

As follows from Figures 4 a, b, the metal atoms in variant a) are located at the vertices 1, 3, 5, 7, and in variant b) the metal atoms are located at the vertices 2, 4, 6, 7. In theory, non - metal atoms are located in the remaining free vertices of the cube. In both cases the number of vertices is 8 (f_0=8), the number of edges is 24 (f_1=24)

1 - 2, 1 - 3, 1 - 8, 1 - 7, 1 - 5, 1 - 6, 2 - 3, 2 - 4, 2 - 7, 2 - 8, 2 - 6, 3 - 4, 3 - 5, 3 - 7, 3 - 8, 4 - 5, 4 - 6, 4 - 7, 4 - 8, 5 - 6, 5 - 8, 5 - 7, 6 - 7, 6 - 8,

the number of two - dimension faces is 32 (f_2=32),

1 - 2 - 3, 1 - 3 - 8, 1 - 7 - 2, 1 - 5 - 7, 1 - 6 - 5, 1 - 6 - 7, 1 - 2 - 6, 1 - 6 - 8, 1 - 8 - 2, 1 - 5 - 8, 2 - 6 - 8, 2 - 8 - 3, 2 - 4 - 3, 2 - 8 - 4, 2 - 7 - 3, 2 - 7 - 4, 3 - 8 - 4, 3 - 7 - 4, 3 - 5 - 4, 4 - 5 - 7, 4 - 5 - 6, 4 - 6 - 7, 5 - 6 - 7, 5 - 6 - 8, 1 - 3 - 7, 2 - 6 - 4, 3 - 1 - 5, 3 - 5 - 7, 4 - 2 - 6, 4 - 6 - 8, 5 - 4 - 8, 5 - 8 - 3,

The number of three - dimension faces is 16 (f_3=16),

1 - 7 - 5 - 6, 1 - 2 - 8 - 6, 1 - 2 - 3 - 7, 1 - 2 - 3 - 8, 2 - 3 - 7 - 4, 2 - 3 - 8 - 4, 3 - 4 - 5 - 7, 4 - 5 - 6 - 7, 1 - 2 - 6 - 7, 1 - 3 - 7 - 5, 1 - 3 - 8 - 5, 1 - 5 - 6 - 8, 2 - 6 - 8 - 4, 2 - 6 - 7 - 4, 4 - 5 - 8 - 6, 4 - 5 - 8 - 3.

Substituting the obtained values $f_i(i=0,1,2,3)$ in the equation (1) you can see that the Euler - Poincaré equation is satisfied in his case for $n = 4$

$$8 - 24 + 34 - 16 = 0.$$

This proves that Figures 4 a, b are polytopes with dimension 4.

Obviously, it is easy to see that there are four more variants of the arrangement of the tetrahedron of metal atoms in a cube by changing the arrangement of the diagonal and the triangle with the metal atoms in the lower base of the cube. In all the cases listed above, the arrangement of the atoms of the constituents they are characterized by the fact that each vertex has a certain (opposite) vertex with which this vertex is not connected by an edge. Moreover, this vertex is connected with all other vertices by an edge. This property of polytopes, which are called cross - polytopes. For the first time the existence of cross - polytopes of dimension 4 was announced in 1880 (Stringham, 1880). The properties of cross - polytopes with any dimension and n are studied in detail in the work (Zhizhin, 2019a). Thus, “Cuban cluster” is 4 - cross - polytope.

It should be noted that the ligand polyhedron cannot be separated separately in the “Cuban cluster”, since the ligand tetrahedron and the metal tetrahedron penetrate each other.

CLUSTER WITH A GOLD TETRAHEDRON FRAME

Clusters with the participation of tetrahedron of gold atoms show a known position about the high lability of compounds of these metals (Gubin, 2019). Consider the structure of one of these clusters $Au_4I_2R_4$, R=PPh_3 is a functional group. The scheme of this compound is shown in Figure 5.

To obtain a spatial model of this compound in the form of a convex polytope, it is necessary to add several edges that have a purely geometric value. In the previous text, to create a spatial model of molecules, at the first step, the coordination of atoms around a single metal atom was taken into account and then the extra edges of this coordination were removed, losing their importance in the combination of several metal atoms. In clear

Figure 5. The scheme of cluster $Au_4I_2R_4$, $R=PPh_3$

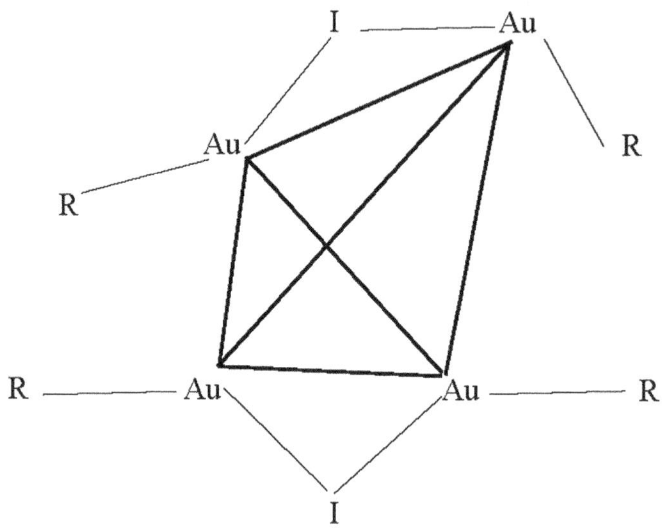

Figure 6. The spatial structure of cluster $Au_4I_2R_4$, $R=PPh_3$

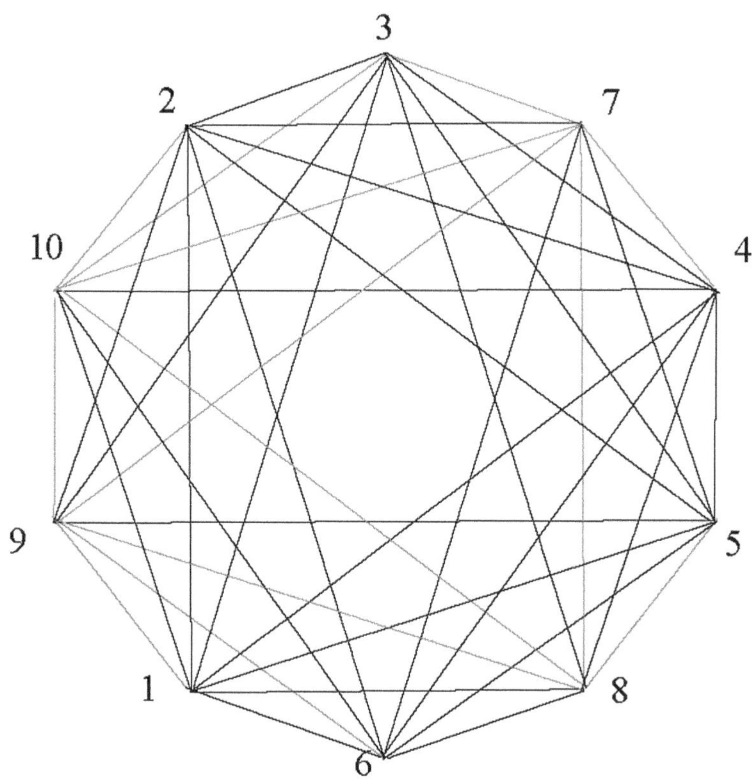

cases, you can immediately go to the second stage of adding edges to create a spatial image of the compound, like a convex body. The spatial image of this compound is shown in Figure 6.

In Figure 6, the valence bonds are marked with red edges, the added edges are marked with black. Here one had to add a significant number of edges in order to achieve a convex body of the minimum dimension in this case. From Figure 6 it can be seen that the resulting convex figure is a cross - polytope, since each vertex of the figure has an opposite vertex with which there is no edge connection. With other vertices, except the opposite, there are such edge connections. In this case, to calculate the numbers of elements of different dimensions *i*, you can use the general formula for *n* - cross - polytopes (Zhizhin, 2013, 2014, 2019a)

$$f_i(n) = 2^{1+i} C_n^{n-1-i} . \qquad (2)$$

The correctness of this formula can be checked by subtracting numbers of various dimensions for 4 - cross - polytopes with the help of this formula and comparing with the corresponding values of these numbers obtained by direct recalculation of image 4 - cross - polytopes in the previous section.

In accordance with formula (2), the number of vertices ($i = 0$) in a polytope of dimension *n* is equal to $2C_n^{n-1} = 2n$. Since the number of vertices in Figure 6 is 10, the dimension of the cross - polytope in Figure 6 is equal to 5. According to formula (2), the number of edges, two -dimensional, three - dimensional and four - dimensional elements in a 5 - cross - polytope are equal

$$f_1(5)=40, f_2(5)=80, f_3(5)=80, f_4(5)=32.$$

Substituting these numbers in the equation (1) of Euler - Poincaré, we confirm that it holds for $n = 5$

$$10 - 40 + 80 - 80 + 32 = 2.$$

Thus, cluster $Au_4I_2R_4$, R=PPh_3 is cross - polytope with dimension 5.

It should be noted that ligands in this case form a convex figure. This can be proved by counting elements of various dimensions on the set of ligand vertices. According to Figure 6, the number of ligand vertices is 6 (f_0=6), the number of edges between the ligand vertices is 14 (f_1=14)

1 - 2, 1 - 3, 1 - 4, 1 - 5, 1 - 6, 2 - 3, 2 - 4, 2 - 5, 2 - 6, 3 - 4, 3 - 5, 4 - 5, 4 - 6, 5 - 6,

the number of flat elements formed by the edges is 16 (f_2=16),

1 - 2 - 3, 1 - 2 - 4, 1 - 2 - 5, 1 - 2 - 6, 1 - 3 - 4, 1 - 3 - 5, 1 - 4 - 5, 1 - 4 - 6, 1 - 5 - 6, 2 - 3 - 4, 2 - 3 - 5, 2 - 4 - 5, 2 - 4 - 6, 2 - 5 - 6, 3 - 4 - 5, 4 - 5 - 6,

the number of three - dimensional elements is 8 (f_3=8)

1 - 2 - 3 - 4, 1 - 2 - 3 - 5, 1 - 2 - 4 - 6, 1 - 2 - 5 - 6, 1 - 3 - 4 - 5, 1 - 4 - 5 - 6, 2 - 3 - 4 - 5, 2 - 4 - 5 - 6.

Substituting these numbers in the equation (1) of Euler - Poincaré you can see that it holds for $n = 4$

$$6 - 14 + 16 - 8 = 0.$$

This proves that ligand polytope in this case has dimension equal 4. In this case, the ligand polytope 1 - 2 - 3 - 4 - 5 - 6 and metallic polyhedron 7 - 8 - 9 - 10 penetrate each other.

CLASTERS WITH A GOLD OCTAHEDRON FRAME

Theoretically, a large number of isomeric non - planar metal cores can be constructed from six metal atoms, many of which can be realized in the form of cluster molecules. However, octahedron is most common among 6 - nuclear cluster compounds. As an example, consider the structure of the gold phosphine cluster $[Au_6(PR_3)_6]^{2+}$, R=$C_6H_4CH_3$-p (Gubin, 2019). The scheme of this compound is shown in Figure 7.

The spatial image of a cluster of phosphine gold is presented in Figure 8.

It represents the gold octahedron embedded in the octahedron of ligands. In vertices 7, 8, 9, 10, 11, 12 is arrangement of the functional groups PR_3, in vertices 1, 2, 3, 4, 5, 6 is arrangement of the atoms of gold Au. In Figure 8, the valence bonds are marked with red edges. Additional edges forming the outer convex shape of the connection are indicated in black.

Let one proved, that the dimension of this compound is 4.

Figure 7. Schema of the gold phosphine cluster $[Au_6(PR_3)_6]^{2+}$, $R=C_6H_4CH_3$-p

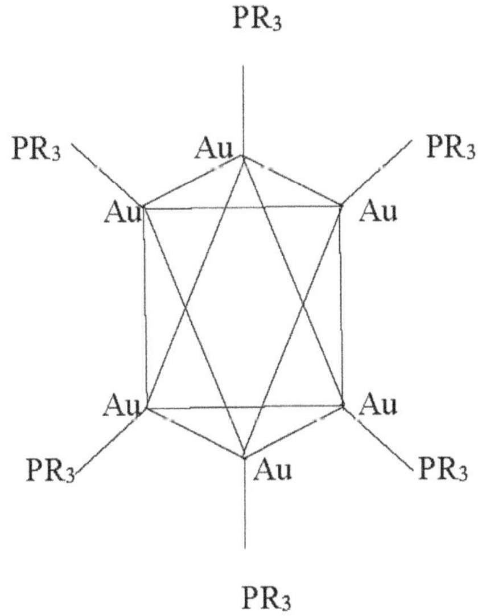

Figure 8. The spatial image of a cluster of phosphine gold $[Au_6(PR_3)_6]^{2+}$, $R=C_6H_4CH_3$-p

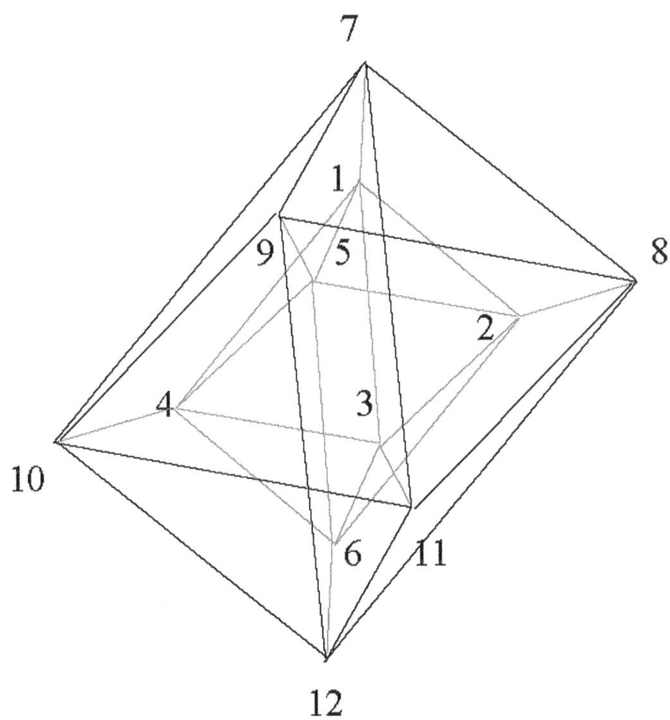

In this case the number of vertices is 12 (f_0=12). The numbers of edges on Figure 8 is 30 (f_1=30)

1 - 2, 1 - 3, 1 - 4, 1 - 5, 6 - 5, 6 - 2, 6 - 3, 6 - 4, 4 - 5, 5 - 2, 2 - 3, 3 - 4, 1 - 7, 2 - 8, 3 - 11, 4 - 10, 5 - 9, 9 - 10, 10 - 11, 11 - 8, 8 - 9, 8 - 7, 11 - 7, 10 - 7, 9 - 7, 8 - 12, 9 - 12, 10 - 12, 11 - 12, 6 - 12.

The number of triangles on Figure 8 is 16

1 - 2 - 3, 1 - 3 - 4, 1 - 4 - 5, 1 - 2 - 5, 6 - 3 - 2, 6 - 5 - 2, 6 - 4 - 5, 6 - 4 - 3, 7 - 9 - 8, 7 - 9 - 10, 7 - 10 - 11, 7 - 11 - 8, 11 - 8 - 12, 11 - 10 - 12, 10 - 9 - 12, 9 - 8 - 12.

The number of tetragon flat faces on Figure 8 is 12

3 - 2 - 8 - 11, 3 - 4 - 11 - 10, 4 - 5 - 9 - 10, 2 - 5 - 9 - 8, 1 - 2 - 7 - 8, 1 - 3 - 7 - 11, 1 - 4 - 7 - 10, 1 - 5 - 7 - 9, 3 - 6 - 11 - 12, 2 - 6 - 12 - 8, 10 - 4 - 6 - 12, 9 - 5 - 6 - 12.

Thus, the common number of flat faces on Figure 8 is 28 (f_2=28).
The number of octahedrons on Figure 8 is 2

1 - 2 - 3 - 4 - 5 - 6 (gold octahedron), 7 - 8 - 9 - 10 - 11 - 12 (ligand octahedron).

The trigonal prisms on Figure 8 is 8

1 - 2 - 3 - 7 - 8 - 11, 1 - 2 - 5 - 7 - 8 - 9, 3 - 4 - 10 - 11 - 12, 4 - 5 - 6 - 9 - 10 - 12, 4 - 1 - 3 - 7 - 10 - 11, 1 - 4 - 5 - 7 - 9 - 10, 6 - 3 - 2 - 8 - 12 - 11, 5 - 2 - 6 - 8 - 9 - 12.

Thus, the common number of the three - dimensional faces on Figure 8 is 10 (f_3=10).
Substituting the obtained values f_i(i=0,1,2,3) in the equation (1) you can see that the Euler - Poincaré equation is satisfied in his case for $n = 4$

$$12 - 30 + 28 - 10 = 0.$$

This proves that Figure 8 is polytope with dimension 4.

COMPOUNDS HAVING A METAL OCTAHEDRON FRAME

A cluster of phosphine gold, discussed in the previous section, is a special case of numerous compounds in which the octahedron participates as a structural unit. If a complex functional group is opened in gold phosphine, the dimension of the compound in this case will become much higher. Clusters with a structural unit octahedron exist with different metals in the core: Pt, Pd, Ni, Cu, Ti, Fe, Ru and other metals (Gubin, 2019). They are distinguished by sufficient structural complexity, leading to the formation of polytopes of a higher dimension of new types. So far, none of them have been analyzed in the space of higher dimension. As an illustration of such an analysis, we consider the structure of the cobalt tetra - anion cluster $[Co_6(\mu\text{-}Co)_8(CO)_6]^{4-}$. The scheme of this compound is shown in Figure 9 (Johnson, & Benfield, 1981).

Figure 9. Scheme of cobalt tetra - anion cluster $[Co_6(\mu\text{-}Co)_8(CO)_6]^{4-}$

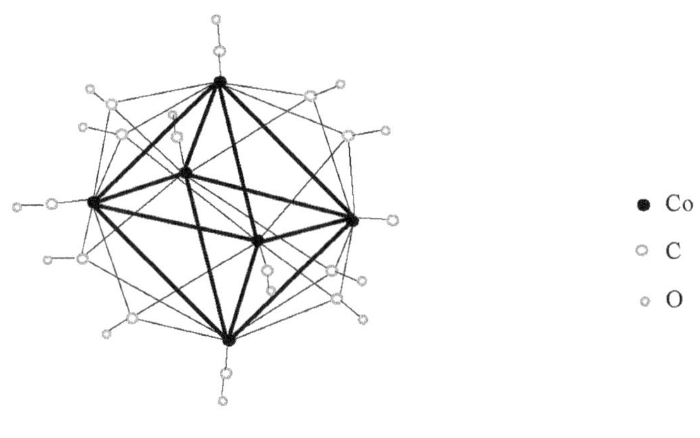

The desire to create a convex model of this compound leads to the appearance of various three -dimensional polyhedrons and four - dimensional polytopes entering into each other.

Theorem 3. The dimension of cobalt tetra - anion cluster $[Co_6(\mu\text{-}Co)_8(CO)_6]^{4-}$ equal 5.

Proof. To build a spatial model of cobalt tetra - anion cluster $[Co_6(\mu\text{-}Co)_8(CO)_6]^{4-}$, let us turn to its scheme in Figure 9. From this figure, it can be seen that the ligands (μ - CO) that bind two metal atoms form a cube. In addition, each vertex of the metal core in the form of an octahedron is formed as a result of constructing a pyramid on each face of the ligand cube. The free vertices of these pyramids give the vertices of the octahedron of atoms cobalt. The six faces of the cube thus give six vertices of the octahedron. Consider first the figure that forms as a result of this construction. The spatial image of this figure is shown in Figure 10.

Figure 10. The spatial image of compound $Co_6(\mu\text{-}Co)_8$

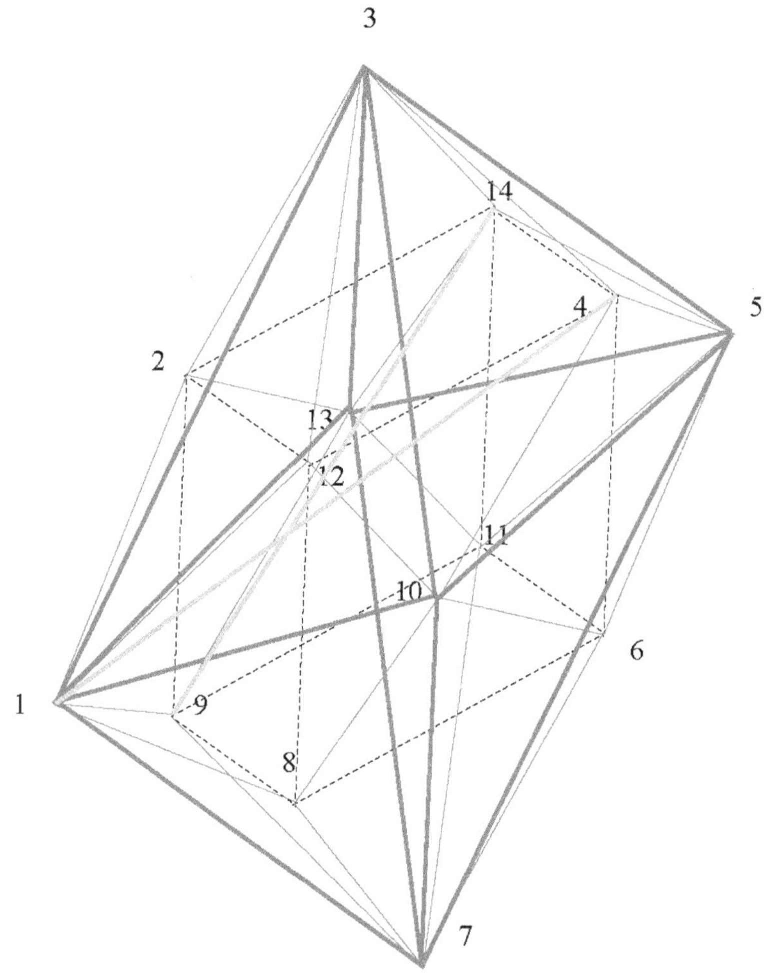

The valence bonds are indicated in this figure with red edges. The edges of the ligand cube are marked with dotted black lines. The edges of the pyramids, with the exception of the grounds, denoted by solid lines in black. Thus, the ligands (μ - CO) are located at the vertices

2, 4, 6, 8, 9, 11, 12, 14.

Metal atoms are located at the vertices

1, 3, 5, 7, 10, 13.

To create a closed convex figure in the figure, two edges are added blue 1 - 4, 9 - 14. Determine the dimension of this compound. There are 14 vertices here (f_0=14). The numbers of edges on Figure 10 is 50 (f_1=50)

1 - 2, 1 - 3, 1 - 13, 1 - 12, 1 - 4, 1 - 10, 1 - 9, 1 - 8, 1 - 7, 2 - 3, 2 - 13, 2 - 14, 2 - 12, 2 - 9, 3 – 4, 3 - 5, 3 - 14, 3 - 13, 3 - 12, 3 - 10, 4 - 5, 4 - 6, 4 - 10, 4 - 12, 4 - 14, 5 - 14, 5 - 11, 5 - 10, 5 - 6, 5 - 7, 5 - 13, 6 - 7, 6 - 10, 6 - 11, 6 - 8, 7 - 10, 7 - 11, 7 - 13, 7 - 8, 7 - 9, 8 - 9, 8 - 10, 8 - 12, 9 - 11, 9 - 13, 9 - 14, 10 - 12, 11 - 13, 11 - 14, 13 - 14.

The number of triangles on Figure 10 is 58

1 - 2 - 3, 1 - 3 - 13, 1 - 3 - 12, 1 - 2 - 13, 1 - 2 - 12, 1 - 2 - 9, 1 - 9 - 8, 1 - 9 - 7, 1 - 12 - 8, 1 - 13 - 7, 1 - 10 - 7, 1 - 10 - 8, 1 - 10 - 12, 1 - 10 - 3, 1 - 9 - 13, 1 - 8 - 7, 2 - 12 - 3, 2 - 13 - 3, 2 - 14 - 3, 2 - 14 - 13, 2 - 9 - 13, 3 - 12 - 4, 3 - 13 - 14, 3 - 13 - 5, 3 - 4 - 5, 3 - 14 - 5, 3 - 14 - 4, 3 - 10 - 12, 3 - 10 - 4, 3 - 10 - 5, 4 - 5 - 14, 4 - 10 - 12, 4 - 5 - 10, 4 - 6 - 10, 4 - 5 - 6, 5 - 13 - 11, 5 - 13 - 14, 5 - 14 - 11, 5 - 11 - 6, 5 - 11 - 7, 5 - 10 - 6, 5 - 7 - 6, 5 - 10 - 7, 6 - 10 - 7, 6 - 11 - 7, 6 - 8 - 7, 6 - 8 - 10, 7 - 11 - 13, 7 - 8 - 10, 7 - 8 - 9, 7 - 9 - 11, 7 - 9 - 13, 8 - 10 - 12, 9 - 11 - 13, 9 - 13 - 14, 9 - 14 - 11, 1 - 12 - 4, 1 - 10 - 4.

The number of tetragon flat faces on Figure 10 is 6

2 - 14 - 4 - 13, 14 - 4 - 11 - 6, 11 - 6 - 8 - 9, 9 - 8 - 2 - 13, 4 - 6 - 8 - 12, 2 - 14 - 11 - 9.

Thus, the common number of flat faces on Figure 10 is 64 (f_2=64).
The number of tetrahedrons on Figure 10 is 19

1 - 2 - 3 - 12, 1 - 12 - 9 - 7, 1 - 12 - 8 - 9, 1 - 2 - 3 - 13, 1 - 2 - 12 - 9, 1 - 3 - 11 - 13, 1 - 9 - 8 - 7, 2 - 12 - 14 - 3, 3 - 4 - 5 - 14, 3 - 13 - 4 - 11, 13 - 8 - 11

- 1, 4 - 5 - 6 - 11, 5 - 14 - 10 - 12, 5 - 11 - 7 - 6, 5 - 10 - 7 - 6, 5 - 12 - 7 - 10, 6 - 11 - 7 - 8, 7 - 9 - 10 - 12, 1 - 8 - 11 - 7.

The number of pyramids on Figure 10 is 6

2 - 12 - 4 - 6 - 3, 4 - 14 - 6 - 11 - 5, 6 - 8 - 9 - 11 - 7, 9 - 8 - 12 - 2 - 1, 4 - 6 - 8 - 12 - 10, 2 - 14 - 11 - 9 - 13.

On Figure 10 is 1 cube

2 - 14 - 4 - 12 - 6 - 8 - 9 - 11,

and is 1 octahedron

Figure 11. The image a closed body S = 2 - 13 - 11 - 10 - 6 - 12 - 3 - 5 - 1 - 7

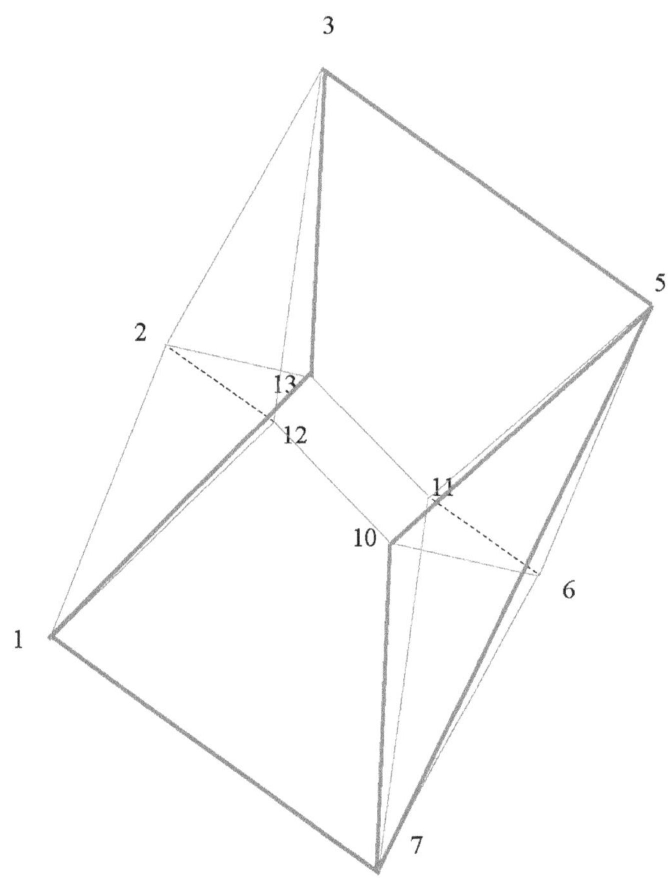

1 - 3 - 5 - 7 - 10 - 13.

In addition, in Figure 10 there is also a closed body with 12 vertices

2 - 13 - 11 - 10 - 6 - 12 - 3 - 5 - 1 - 7.

His image is presented separately in Figure 11.

Determine the dimension of the body S. For this body is f_0=10 . The number of edges on Figure 11 is 22 (f_1=22)

1 - 2, 1 - 13, 1 - 12, 1 - 10, 1 - 7, 2 - 3, 2 - 13, 2 - 12, 3 - 5, 3 - 10, 3 - 13, 3 - 12, 5 - 13, 5 - 11, 5 - 10, 6 - 7, 6 - 10, 6 - 11, 7 - 10, 7 - 11, 10 - 12, 11 - 13.

The number of triangles on Figure 11 is 8

2 - 3 - 12, 2 - 3 - 13, 1 - 2 - 13, 6 - 5 - 10, 6 - 5 - 11, 1 - 2 - 12, 7 - 6 - 10, 7 - 6 - 11.

The number of tetragon flat faces on Figure 11 is 4

13 - 3 - 5 - 11, 12 - 3 - 5 - 10, 1 - 7 - 13 - 11, 1 - 12 - 10 - 7.

Thus, the common number of flat faces on Figure 11 is 12 (f_2=12).

Substituting the obtained values f_i(i=0,1,2) in the equation (1) you can see that the Euler - Poincaré equation is satisfied in his case for $n = 3$

$$10 - 22 + 12 = 2.$$

This proves that Figure 11 is polyhedron with dimension 3. Consequently, the total number of three - dimensional figures included in the polytope in Figure 10 is 19 + 9 = 28, i.e. for this figure f_3=28.

Substituting the obtained for Figure 10 values f_i(i=0,1,2,3) in the equation (1) you can see that the Euler - Poincaré equation is satisfied in his case for $n = 4$

$$14 - 50 + 64 - 28 = 0.$$

This proves that Figure 10 is polytope with dimension 4. Now back to the review of cobalt tetra - anion cluster $[Co_6(\mu\text{-}Co)_8(CO)_6]^{4-}$ in Figure 9.

Figure 10 shows the spatial image of only part of this compound. To create the spatial image of this compound, it is necessary to release the edges into the outer part from the vertices of the octahedron in Figure 10 and build a larger octahedron

15 - 16 - 17 - 18 - 19 - 20

on the free vertices of these edges. This are red edges

1 - 18, 3 - 15, 5 - 16, 7 - 17, 13 - 19, 10 – 20

on Figure 12.

Figure 12. Spatial image of cobalt tetra - anion cluster $[Co_6(\mu\text{-}Co)_8(CO)_6]^{4-}$

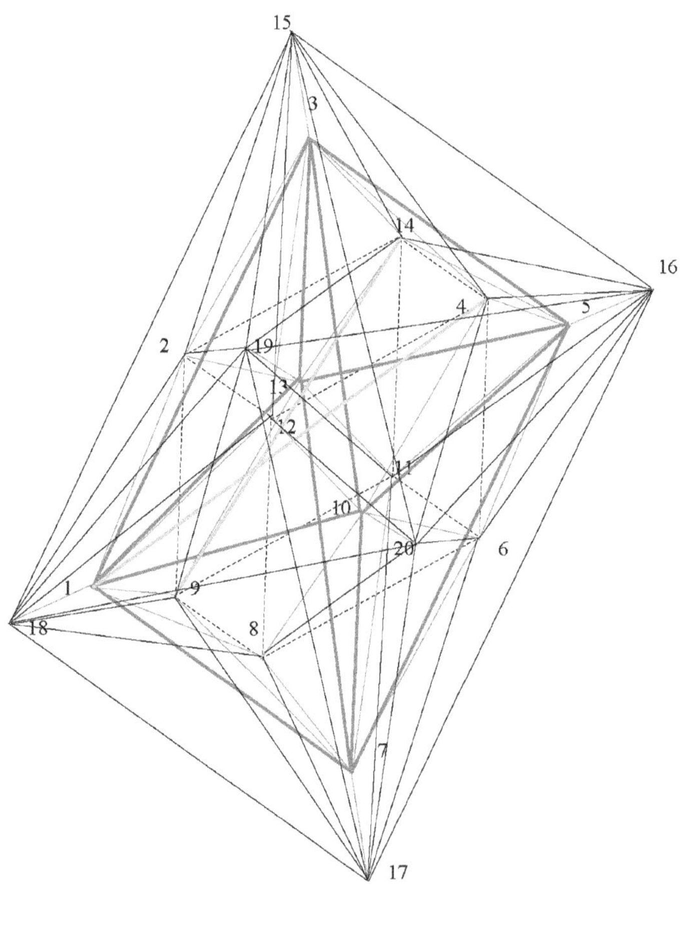

Thus, the number of vertices on Figure 12 equal 20 (f_0=20). The construction of a large octahedron taking into account the edges connecting two octahedrons adds 16 edges to the body in Figure 10. A three - dimensional figure is formed between each face of a large octahedron and the corresponding face of a small octahedron. For example, a three - dimensional figure is formed between the face of 1 - 10 - 7 small octahedron and the face 18 - 17 - 20 of the large octahedron (see Figure 12)

1 - 10 - 7 - 18 - 17 - 20.

The number of such three - dimensional figures is equal to the number of faces of the octahedron - 8. In addition, to fill the space between the large octahedron and other convex bodies that make up the polytope in Figure 10, it is necessary to connect the vertices of the large octahedron with the vertices of the cube

2 - 14 - 4 - 12 - 6 - 8 - 9 - 11.

This adds to the body in Figure 10 another 24 edges

2 - 15, 12 - 15, 14 - 15, 4 - 15, 14 - 16, 4 - 16, 6 - 16, 11 - 16, 6 - 17, 8 - 17, 9 - 17, 11 - 17, 8 - 18, 9 - 18, 2 - 18, 12 - 18, 2 - 19, 14 - 19, 9 - 19, 11 - 19, 4 - 20, 12 - 20, 8 - 20, 6 - 20.

Thus, a common number of the edges on Figure 12 is 50 + 16 + 24 = 90 (f_1=90).

The construction of a large octahedron taking into account the edges connecting two octahedrons adds 20 flat faces to body on Figure 10 (see previous section). Connection the vertices of the large octahedron with the vertices of the cube adds 48 flat faces to Figure 10. For example, the face of the cube 2 - 14 - 4 - 12 when building a large pyramid on it gives an additional 8 flat faces

2 - 12 - 15, 4 - 12 - 15, 4 - 14 - 15, 2 - 14 - 15, 12 - 3 - 15, 2 - 3 - 15, 4 - 3 - 15, 14 - 3 - 15.

Since the faces at cube 6 one gets an additional 48 flat faces. In this way, the total number of flat edges in Figure 12 is 64 + 20 + 48 = 132, (f_2=132).

The construction of a large octahedron taking into account the edges connecting two octahedrons adds 9 three - dimensional faces to body on Figure 10 (see previous section). Connection the vertices of the large octahedron with the vertices of the cube adds 30 three - dimensional faces to Figure 10. For example, the face of the cube 2 - 14 - 4 - 12 when building a large pyramid on it gives an additional 5 three - dimensional faces

2 - 12 - 15 - 3, 4 - 12 - 15 - 3, 4 - 14 - 15 - 3, 2 - 14 - 15 - 3, 12 - 2 - 15 - 4 - 14.

Since the faces at cube 6 one get an additional 30 three - dimensional faces. In addition, 4 tetrahedrons are additionally formed between the cube and the large octahedron, each of which has an edge of one of the edges of a rectangular cross section of 15 - 16 - 17 - 18 large octahedron. For example, a tetrahedron 18 - 17 - 8 - 9 is formed on the edge 17 - 18. Thus, the total number of three - dimensional areas in the body in figure 12 is 28 + 9 + 30 + 4 = 71 (f_3=71).

Connection the vertices of the large octahedron with the vertices of the cube adds 2 four - dimensional faces to Figure 10, one of which has an edge 18 - 17 and second has edges 15 - 16 of a rectangular cross section of 15 - 16 - 17 - 18 large octahedron. For example, on the edge 17 - 18 formed area

18 - 1 - 9 - 8 - 7 - 17

(see Figure 12). This area has 6 vertices (f_0=6). It has 14 edges (f_1=14)

1 - 9, 1 - 8, 1 - 7, 1 - 18, 7 - 8, 7 - 9, 7 - 17, 8 - 9, 8 - 18, 8 - 17, 9 - 18, 9 - 17, 9 - 8, 18 - 17.

It has 14 two - dimensional faces (f_2=14)

1 - 9 - 18, 1 - 9 - 8, 1 - 8 - 7, 1 - 7 - 17 - 18, 7 - 9 - 17, 7 - 8 - 9, 7 - 9 - 17, 8 - 17 - 18, 8 - 9 - 17, 8 - 9 - 18, 9 - 18 - 17.

It has 6 three - dimensional faces (f_3=6)

1 - 9 - 8 - 18, 1 - 9 - 8 - 7, 18 - 9 - 8 - 17, 1 - 18 - 9 - 7 - 17, 9 - 8 - 7 - 17, 1 - 8 - 18 - 7 - 17.

Substituting the obtained for Figure 10 values f_i(i=0,1,2,3) in the equation (1) you can see that the Euler - Poincaré equation is satisfied in his case for $n = 4$

$14 - 50 + 64 - 28 = 0.$

This proves that body

18 - 1 - 9 - 8 - 7 - 17

is polytope with dimension 4. It is also proved that the body

15 - 16 - 5 - 4 - 14 - 3

has the dimension 4. It can be shown that, due to the orientation of the cube with respect to the large octahedron, similar four - dimensional polytopes with the edges of the large octahedron 18 - 15 and 17 - 16 do not form.

Connection the vertices of the large octahedron with the vertices of the cube adds yet 6 four - dimensional faces to Figure 10, associated with the formation of structures of the pyramid in the pyramid on the faces of the cube. Consider for example two pyramids on the face of a cube 2 - 12 - 14 - 4. This body has 6 vertices (f_0=6), 13 edges (f_1=13)

2 - 3, 2 - 15, 12 - 3, 12 - 15, 4 - 3, 14 - 3, 4 - 15, 14 - 15, 3 - 15, 2 - 12, 2 - 14, 4 - 12, 4 - 14,

13 two - dimensional faces (f_2=13)

2 - 15 - 12, 2 - 3 - 12, 4 - 3 - 14, 4 - 15 - 14, 3 - 4 - 12, 15 - 4 - 12, 2 - 14 - 3, 2 - 14 - 15, 12 - 3 - 15, 2 - 3 - 15, 4 - 3 - 15, 14 - 3 - 15, 2 - 14 - 12 - 4,

and 6 three - dimensional faces (f_3=6)

2 - 14 - 4 - 12 - 3, 2 - 14 - 4 - 12 - 15, 4 - 3 - 15 - 14, 12 - 4 - 3 - 15, 2 - 12 - 3 - 15, 2 - 3 - 14 - 15.

Substituting the obtained for two pyramids values f_i(i=0,1,2,3) in the equation (1) you can see that the Euler - Poincaré equation is satisfied in his case for $n = 4$

$14 - 50 + 64 - 28 = 0.$

This proves that construction from two pyramids is polytope with dimension 4.

The four - dimensional structure of two pyramids on the remaining five faces of the cube is proved in a similar way. In addition, the polytope in Figure 12 has two polytopes formed by inscribing a cube into an octahedron (large and small). Each of these polytopes has dimension 4 (this has already been proved). Finally, the polytope in Figure 12 has a polytope octahedron in an octahedron. This polytope also has dimension 4 (see previous section). Therefore, the total number of polytopes of dimension 4 that make up a polytope in Figure 12 is 11 (f_4=11). Substituting the obtained for Figure 12 values f_i(i=0,1,2,3,4) in the equation (1) you can see that the Euler - Poincaré equation is satisfied in his case for $n = 5$

$20 - 90 + 132 - 71 + 11 = 2.$

This proves that dimension of cobalt tetra - anion cluster $[Co_6(\mu\text{-}Co)_8(CO)_6]^{4-}$ equal 5.

CONCLUSION

The geometry of clusters with a polyhedral metal core with ligands was studied. The attachment of ligands to the skeleton of metal atoms forming a three - dimensional polyhedron should naturally lead to the formation of a cluster of the higher dimension. This is shown in this chapter on the example of the widespread metal cores in the form of a tetrahedron and octahedron. As a result, the cluster assumes the form of both well - known polytopes of a higher dimension, and polytopes of a higher dimension of a new type. In particular, it has been established that the well - known Kuban clusters are nothing more than the well - known cross - polytopes of dimension 4. The bridged metal compounds by ligands further increases their dimension. For example, the cluster, as it turned out, has dimension 5, forming 5 - cross - polytope. The most complex structures of polytopes of higher dimension have clusters with ligands at the octahedral metal core. The simplest spatial model of a metallic octahedron with ligands is a metallic octahedron embedded in a ligand octahedron and has dimension 4. It should be noted that in many cases, ligands penetrate into the metallic core, so ideas about the surroundings of the metallic skeleton ligands are not realized. Inside clusters with a polyhedral metallic skeleton and ligands, various geometric shapes of various dimensions are formed. The analysis of the geometry of a tetra - anion cobalt cluster can serve as an effective example of the application of geometric analysis of clusters of higher dimensions. It is proved that the dimension of this cluster is 5. The structure of this cluster includes octahedrons, a cube, a set of twin pyramids, and different three – dimensional body, 11 four - dimensional bodies. In this cluster, the metal core was located between the two shells of the ligands. Binding ligands form a small ligand polyhedron. Terminal ligands form a large ligand polyhedron. Both ligand polyhedrons limit the metal core "bottom" and "top". It should be noted that the geometrical analysis using the methods of analysis of polytopes of higher dimension (Zhizhin, 2019 a) allows us to establish the important features of cluster geometry that elude the analysis from the standpoint of three - dimensional geometry.

REFERENCES

Gubin, S. P. (2019). *Chemistry clusters. Basics of classification and structure*. Moscow: Editorial URSS.

Johnson, B. F. G., & Benfield, R. E. (1981). *Topics in inorganic and organometallic stereochemistry* (G. Geoffroy, Ed.). New York: Wiley.

Lichtensins, G. I. (1979). *Multicore oxidative - reducing metal - enzymes*. Moscow: Science.

Neuton, W. E., & Otsuka, N. Y. (1980). *Molybdenum chemistry of biological significance*. New York: Plenum press. doi:10.1007/978-1-4615-9149-8

Poincaré, A. (1895). Analysis situs. *J. de é Ecole Polytechnique, 1,* 1 - 121.

Riemann, B. (1854). *On the Hypotheses Which Lie at the Foundations of Geometry*. Gottingen: Gottingen Observatory.

Stringham, W. I. (1880). Regular figures in n - dimensional space. *American Journal of Mathematics*, *3*(1), 1–14. doi:10.2307/2369441

Zhizhin, G. V. (2013). Relations for the numbers of faces of various dimensions in the tower of n - dimensional convex regular polytopes. In *Proceedings of the 9th All-Russian School "Mathematical Studies in the Natural Sciences."* Apatity: Geological Institute, Russian Academy of Sciences.

Zhizhin, G. V. (2014). *World - 4D*. St. - Petersburg: Polytechnic Service.

Zhizhin, G. V. (2018). *Chemical Compound Structures and the Higher Dimensional of Molecules: Emerging Research and Opportunities*. Hershey, PA: IGI Global. doi:10.4018/978-1-5225-4108-0

Zhizhin, G. V. (2019a). *The Geometry of Higher - Dimensional Polytopes*. Hershey, PA: IGI Global. doi:10.4018/978-1-5225-6968-8

Zhizhin, G. V. (2019b). *Attractors and Higher Dimensions in Population and Molecular Biology: Emerging Research and Opportunities*. Hershey, PA: IGI Global. doi:10.4018/978-1-5225-9651-6

KEY TERMS AND DEFINITIONS

Binding (or Bridge) Ligands: Non-metal atoms or functional groups bonded by a chemical bond to several atoms of the metal backbone of the cluster simultaneously.

Dimension of the Space: The member of independent parameters needed to describe the change in position of an object in space.

Nanocluster: A nanometric set of connected atoms, stable either in isolation state or in building unit of condensed matter.

Polytope: Polyhedron in the space of higher dimension.

Terminal Ligands: Non-metal atoms or functional groups bonded by chemical bonding to only one of the atoms in the metal core of the cluster.

Chapter 7
Incident Conservation Law

ABSTRACT

This chapter first establishes the existence of integral equality in relation to the issue of the transmission of information by elements of lower and higher dimensions in the polytopes of higher dimension that describe natural objects. This integral equality is called the law of conservation of incidents. There is the incidence interpreted as the transfer of information from one material body to another. The fulfillment of the law of conservation of incidents for the n-simplex of the n-cube and the n-cross-polytope is proved in general terms. It is shown that the law of conservation of incidents is valid for both regular bodies and irregular bodies, which can be clusters of chemical compounds. The incident conservation law can serve as a mathematical basis for the recently discovered epigenetic principle of the transmission of hereditary information without changing the sequence of genes in DNA and RNA molecules.

INTRODUCTION

In previous chapters, it was convincingly shown that various objects of the nanoworld have a higher dimension. Geometrically, these objects are convex closed polytopes whose components are geometric elements with a monotonic change of the dimension from zero to the dimension of the polytope. In the monograph "The Geometry of Higher - Dimensional Polytopes" (Zhizhin, 2019a) has studied in detail the geometry of higher - dimensional polytopes, based on an analysis of the structures of chemical compounds (Zhizhin, 2018).

DOI: 10.4018/978-1-7998-3784-8.ch007

The monograph (Zhizhin, 2019a) introduced the concept of the incidence coefficients of elements of lower dimension with respect to elements of higher dimension and elements of higher dimension with respect to elements of lower dimension. The first characterizes the number of elements of a certain higher dimension to which the given element of a lower dimension belongs. The second characterizes the number of elements of a given lower dimension that are included in a particular element of a higher dimension. Here we must remember that the vertices of geometric elements of various dimensions are atoms, molecules or functional groups. Therefore, the incidence of geometric elements to a friend means contact between particles of matter, including living matter. The contact between particles of matter can be interpreted as the transfer of information on material structures, including biological structures. The monograph "Attractors and higher dimensions in population and molecular biology: emerging research and opportunities" (Zhizhin, 2019b) showed that the elements of nanostructures of living matter are objects of higher dimension. Information sharing is an inherent property of living beings, without which even primitive organisms would not be able to maintain an extremely delicate balance, which depends on their survival (Mancuso &Viola, 2013; Mancuso, 2017). The monograph (Zhizhin, 2019b) found that DNA, RNA molecules are polytopes of higher dimension, and nitrogenous bases connecting the double helix in the DNA molecule form a cross - polytopes of dimension 13. In this connection, the phenomenon of living organisms associated with DNA modification due to the binding of the methyl group CH_3 to the nitrogenous bases of DNA (methylation). This provides the memory of living organisms, i.e. transfer of inherited traits without changing the sequence of genes in DNA (Mancuso, 2017; Lindquist et al., 2016; Sanbonmatsu et al., 2016). Therefore, it is of interest to study in detail the incidence of the entire polygamy of elements of different dimensions in higher dimension polytopes, considering it as a study of the possibility of transmitting information in biological molecules. This study is devoted to this chapter. It considers all possible incidents in convex bodies, starting with the simplest bodies and gradually complicating them.

INCIDENT CONSERVATION LAW FOR N - SIMPLEX

In n - simplex (Zhizhin, 2013, 2014, 2019 a) the number of elements with dimension d is

$$f_d(n) = C_{n+1}^{d+1}. \tag{1}$$

For example, if $d = 0$, so

$$f_0(n) = C_{n+1}^{1} = \frac{(n+1)!}{n!} = n+1. \tag{2}$$

This is the number of vertices in n - simplexes. If $d = 1$, so

$$f_1(n) = C_{n+1}^{2} = \frac{(n+1)!}{(n-1)!2!} = \frac{(n+1)n}{2}. \tag{3}$$

This is the number of edges in n - simplexes. We introduce the notation: $k_{d_j^i d_u}$ is the number of elements of dimension u, which include an element of dimension j ($u > j$) with number i. Thus, $k_{d_j^i d_u}$ is the incidence factor of element i with dimension j relative to elements with dimension u. We introduce the notation also: $k_{d_j d_u^i}$ is the number of elements of dimension j, which included in element i with a dimension u ($u > j$). Thus, $k_{d_j d_u^i}$ is the incidence factor of element i with dimension u relative to elements with dimension j.

If $n = 2$, so $f_0(2)=3$, $f_1(2)=3$, see (2), (3). The factors of incidents (from smaller dimension to larger) are

$$k_{d_0^i d_1} = 2, i = 1,2,3;\ k_{d_1^i d_2} = 1, i = 1,2,3.$$

Sum up the incidence coefficients for all vertices and edges of the triangle

$$\sum_{i=1}^{3} k_{d_0^i d_1} = 6, \sum_{i=1}^{3} k_{d_1^i d_2} = 3, \sum_{i=1}^{3} k_{d_0^i d_1} + \sum_{i=1}^{3} k_{d_1^i d_2} = 9. \tag{4}$$

The factors of incidents (from larger dimension to smaller) are

$$k_{d_0 d_1^i} = 2, i = 1,2,3; k_{d_1 d_2} = 3.$$

Sum up the incidence coefficients for all elements of the triangle with dimension larger of zero

$$\sum_{i=1}^{3} k_{d_0 d_1^i} + k_{d_1 d_2} = 9. \tag{5}$$

Comparing (4) and (5) you can see that the sum of incidents in a triangle from elements with a lower dimension to elements with a higher dimension is equal to the sum of incidents from elements with a higher dimension to elements with a lower dimension. Thus, the sum of incidents retains its value when the direction of the relationship between the elements it is change.

If $n = 3$, so

$f_0(3)=4, f_1(3)=6, f_2(3)=4.$

The factors of incidents (from smaller dimension to larger) are

$$k_{d_0^i d_1} = 3, i = 1 \div 4;\; k_{d_0^i d_2} = 3, i = 1 \div 4;\; k_{d_0^i d_3} = 1, i = 1 \div 4;$$
$$k_{d_1^i d_2} = 2, i = 1 \div 6;\; k_{d_1^i d_3} = 1, i = 1 \div 6;\; k_{d_2^i d_3} = 1, i = 1 \div 4.$$

Sum up the incidence coefficients for all elements of the tetrahedron

$$\sum_{i=1}^{4} k_{d_0^i d_1} = 12, \sum_{i=1}^{4} k_{d_0^i d_2} = 12, \sum_{i=1}^{4} k_{d_0^i d_3} = 4, \sum_{i=1}^{4} k_{d_1^i d_2} = 12,$$
$$\sum_{i=1}^{6} k_{d_1^i d_3} = 6, \sum_{i=1}^{4} k_{d_2^i d_3} = 4,$$

$$\sum_{i=1}^{4} k_{d_0^i d_1} + \sum_{i=1}^{4} k_{d_0^i d_2} + \sum_{i=1}^{4} k_{d_0^i d_3} + \sum_{i=1}^{6} k_{d_1^i d_2} + \sum_{i=1}^{6} k_{d_1^i d_3} + \sum_{i=1}^{4} k_{d_2^i d_3} = 50. \tag{6}$$

The factors of incidents (from larger dimension to smaller) are

$$k_{d_0 d_1^i} = 2, i = 1 \div 6; k_{d_0 d_2^i} = 3, i = 1 \div 4; k_{d_1 d_2^i} = 3, i = 1 \div 4; k_{d_0 d_3} = 4, k_{d_1 d_3} = 6, k_{d_2 d_3} = 4.$$

Sum up the incidence coefficients for all elements of the tetrahedron

$$\sum_{i=1}^{6} k_{d_0 d_1^i} = 12, \sum_{i=1}^{4} k_{d_0 d_2^i} = 12, \sum_{i=1}^{4} k_{d_1 d_2^i} = 12,$$

$$\sum_{i=1}^{6} k_{d_0 d_1^i} + \sum_{i=1}^{4} k_{d_0 d_2^i} + \sum_{i=1}^{4} k_{d_1 d_2^i} + k_{d_0 d_3} + k_{d_1 d_3} + k_{d_2 d_3} = 50\,. \tag{7}$$

Comparing (6) and (7) you can see that the sum of incidents in a tetrahedron from elements with a lower dimension to elements with a higher dimension is equal to the sum of incidents from elements with a higher dimension to elements with a lower dimension. Thus, the sum of incidents retains its value when the direction of the relationship between the elements it is change.

Let us prove the general statement for a simplex of arbitrary finite dimension *n*.

Theorem 1. In any simplex of dimension *n*, the sum of all incidents of elements of a lower dimension with respect to all elements of a higher dimension is equal to the sum of all incidents of elements of a higher dimension with respect to all elements of a lower dimension and equals the sum of the series

$$f_0(n)(2^n - 1) + f_1(n)(2^{n-1} - 1) + f_2(n)(2^{n-2} - 1) + ... + f_{n-1}(n), f_d(n) = C_{n+1}^{d+1}, d = 0 \div (n-1).$$

Proof. The number of edges (elements of dimension 1) that have a certain vertex d_0^i its end point (given that in a simplex each vertex is connected by edges with all other vertices) is determined by the incidence factor

$$k_{d_0^i d_1} = C_n^1 = f_0(n) - 1 = n, i = 1 \div f_0(n). \tag{8}$$

The number of elements of dimension 2 that have its vertex a certain vertex d_0^i is determined by the incident factor

$$k_{d_o^i d_2} = C_n^2, i = 1 \div f_0(n)\,. \tag{9}$$

The number of elements of dimension 3 that have its vertex a certain vertex d_0^i is determined by the incident factor

$$k_{d_o^i d_3} = C_n^3, i = 1 \div f_0(n).$$

And so on.

The number of elements of dimension (n - 1) that have its vertex a certain vertex d_0^i is determined by the incident factor

$$k_{d_o^i d_{n-1}} = C_n^{n-1}, i = 1 \div f_0(n).$$

The number of elements of dimension n that have its vertex a certain vertex d_0^i is determined by the incident factor

$$k_{d_o^i d_n} = C_n^n = 1, i = 1 \div f_0(n).$$

The total number of incidents of vertices to elements of higher dimensions in a n - simplex will be

$$f_0(n)\sum_{j=1}^{n} k_{d_0^i d_j} = f_0(n)(C_n^1 + C_n^2 + ... + C_n^n) = f_0(n)(2^n - 1).$$

The number of elements of dimension 2 that have a certain edge d_1^i its edge is determined by the incidence factor

$$k_{d_1^i d_2} = C_{n-1}^1 = f_0(n) - 2 = n - 1, i = 1 \div f_1(n).$$

The number of elements of dimension 3 that have its edge an element d_1^i is determined by the incident factor

$$k_{d_1^i d_3} = C_{n-1}^2, i = 1 \div f_1(n).$$

And so on.

The number of elements of dimension (n - 1) that have its edge a certain edge d_1^i is determined by the incident factor

$$k_{d_1^i d_{n-1}} = C_{n-1}^{n-2}, i = 1 \div f_1(n).$$

The number of elements of dimension n that have its edge a certain edge d_1^i is determined by the incident factor

$$k_{d_1^i d_n} = C_{n-1}^{n-1} = 1, i = 1 \div f_1(n).$$

The total number of incidents of edges to elements of higher dimensions in n - simplex will be

$$f_1(n)\sum_{j=2}^{n} k_{d_1^i d_j} = f_1(n)(C_{n-1}^1 + C_{n-1}^2 + ... + C_{n-1}^{n-1}) = f_1(n)(2^{n-1} - 1). \quad (10)$$

In the same way, it is proved that the total number of incidences of elements of dimension 2 to elements of higher dimension in a n - simplex is determined by the expression

$$f_2(n)\sum_{j=3}^{n} k_{d_2^i d_j} = f_2(n)(C_{n-2}^1 + C_{n-2}^2 + ... + C_{n-2}^{n-2}) = f_2(n)(2^{n-2} - 1).$$

The total number of incidences of elements of dimension 3 to elements of higher dimension in a n - simplex is determined by the expression

$$f_3(n)\sum_{j=4}^{n} k_{d_3^i d_j} = f_3(n)(C_{n-3}^1 + C_{n-3}^2 + ... + C_{n-3}^{n-3}) = f_3(n)(2^{n-3} - 1).$$

And so on. Thus, the sum of all incidents of elements of a lower dimension with respect to all elements of a higher dimension in n - simplexes is equal

$$f_0(n)(2^n - 1) + f_1(n)(2^{n-1} - 1) + f_2(n)(2^{n-2} - 1) + ... + f_{n-1}(n). \quad (11)$$

To complete the proof of Theorem 1, it is necessary to show that the same expression is equal to the sum of all incidents of elements of a higher dimension with respect to all elements of a lower dimension in n - simplexes.

We pose the question: how many elements of a lower dimension belong to one element of a higher dimension? For example, how many vertices belong to a single edge $d_1^i, i = 1 \div f_1(n)$? Obviously, this number is 2. This is the value of the coefficient of incidence $k_{d_0 d_1^i} = C_2^1 = 2$. The product $C_2^1 f_1(n)$ gives the number of vertices belonging to all edges of an *n* - simplex. This number

$$2C_{n+1}^2 = 2\frac{(n+1)!}{(n-1)!2!} = (n+1)n$$

is equal to the product

$$k_{d_o^i d_1} = C_n^1 C_{n+1}^1 = n(n+1),$$

i.e. the number of edges with a peak point $d_0^i, i = 1 \div f_0(n)$ (8). The number of vertices belonging to a single element of dimension 2 (triangle) is given by the incidence coefficient $k_{d_0 d_2^i}, i = 1 \div f_2(n)$. Obviously, this number is $3 = C_3^1$. The product $C_3^1 f_2(n)$ corresponds to the number of vertices belonging to all elements of dimension 2 in the *n* - simplex. This number is

$$C_3^1 f_2(n) = \frac{3!}{2!} C_{n+1}^1 = \frac{3!}{2!} \frac{(n+1)!}{(n-2)!3!} = \frac{(n+1)!}{(n-2)!2!}.$$

Let us compare this number with the number of elements of dimension 2 with vertices (9), i.e. with the product

$$C_n^2 f_0(n) = \frac{n!}{(n-2)!2!} C_{n+1}^1 = \frac{n!}{(n-2)!2!} \frac{(n+1)!}{n!} = \frac{(n+1)!}{(n-2)!2!}.$$

It is clear that $C_n^2 f_0(n) = C_3^1 f_2(n)$. This proves that the number of vertices belonging to all elements of dimension 2 in an *n* - simplex is equal to the number of elements of dimension 2 that have their vertices d_0^i.

Let $k_{d_0 d_j^i}$ be the number of vertices belonging to some one element of dimension *j*, that is d_j^i. Obviously, this number is equal to $C_{j+1}^1 = j+1$,

$i=1 \div f_j(n)$. The product $C^1_{j+1} f_j(n)$ corresponds to the number of vertices belonging to all elements of dimension *j* in the *n* - simplex. Compare this number with the number of elements of dimension *j* having vertices d^i_0 at $i = f_0(n)$, i.e. with the product

$$C^j_n f_0(n) = (n+1)\frac{n!}{(n-j)!\,j!}.$$

On the other hand, we have

$$C^1_{j+1} f_0(n) = \frac{(j+1)!}{j!}\frac{(n+1)!}{(n-j)!(\text{j}+1)!} = \frac{n!(n+1)}{j!(n-j)!}.$$

It can be seen that these values coincide. This proves that the number of vertices belonging to all elements of dimension *j* in the *n* - simplex is equal to the number of elements of dimension *j* that have their vertices d^i_0.

Let $k_{d_1 d^i_j}$ be the number of edges belonging to some one element of dimension *j*, that is

$$d^i_j, j > 1, i = 1 \div f_j(n).$$

Obviously, this number edges belonging all elements with dimension *j* is

$$f_1(j) f_j(n) = C^2_{j+1} C^{J+1}_{n+1} = \frac{(j+1)!}{(j-1)!2!}\frac{(n+1)!}{(n-j)!(j+1)!} = \frac{(n+1)!}{(j-1)!2!(n-j)!}.$$

Compare this number with the number of elements dimension *j* having element d_1 (see (10))

$$f_1(n) k_{d^i_1 d_j} = f_1(n) C^{j-1}_{n-1} = C^2_{n+1} C^{j-1}_{n-1} = \frac{(n+1)n}{2}\frac{(n-1)!}{(n-j)!(j-1)!}.$$

It can be seen that these numbers are the same.

Let $k_{d_h d_j^i}$ be the number of elements with dimension *h* belonging to some one element of dimension *j* (*h* < *j*), that is $d_j^i, i = 1 \div f_j(n)$. Obviously, this number equal $f_h(j)$. The product $f_h(j)f_j(n)$ corresponds to the number of elements with dimension *h* belonging to all elements of dimension *j* in the *n* - simplex. This number is

$$f_h(j)f_j(n) = C_{j+1}^{h+1}C_{n+1}^{j+1} = \frac{(j+1)!}{(j-h)!(h+1)!}\frac{(n+1)!}{(n-j)!(j+1)!} = \frac{(n+1)!}{(j-h)!(h+1)!(n-j)!}.$$

Compare this number with the number of elements dimension *j* having elements with dimension *h* ($d_h^i, i = f_h(n)$)

$$f_h(n)k_{d_h^i d_j} = f_h(n)C_{n-h}^{j-h} = C_{n+1}^{h+1}C_{n-h}^{j-h} = \frac{(n+1)!}{(n-h)!(h+1)!}\frac{(n-h)!}{(n-j)!(j-h)!}.$$

It can be seen that these numbers are the same.

This proves that the number of elements of dimension *h*, belonging to all elements of dimension *j* in *n* - simplex, is equal to the number of elements of dimension *j*, which have elements of dimension *h*. Thus, the total number of incidents of elements of a smaller dimension to elements of a larger dimension is equal to the total number of incidents of elements of a higher dimension to elements of a smaller dimension. This exists in spite of the difference in the incidence coefficients of elements of a smaller dimension to elements of a higher dimension and the incidence coefficients of elements of a higher dimension to elements of a smaller dimension.

This fact can be considered the law of conservation of incidences in the *n* - simplex.

The total number of incidents in the *n* - simplex is described by equality (11).

It can be seen that the total number of incidents in the *n* - simplex depends only on the dimension *n* and increases sharply with increasing *n*. For example, it is 12 for $n = 2$, 50 for $n = 3$, 180 for $n = 4$, 602 for $n = 5$.

INCIDENT CONSERVATION LAW FOR N - CUBE

In n - cube (Zhizhin, 2013, 2014, 2019 a) the number of elements with dimension d is

$$f_d(n) = 2^{n-d} C_n^d, d = 1 \div n. \tag{12}$$

For example, if $d = 0$, so

$$f_0(n)=2^n. \tag{13}$$

This is the number of vertices in n - cube. If $d = 1$, so

$$f_1(n) = 2^{n-1} C_n^1 = 2^{n-1} n. \tag{14}$$

This is the number of edges in n - cube.

We introduce the notation: $k_{d_j^i d_u}$ is the number of elements of dimension u, which include an element of dimension j ($u > j$) with number i. Thus, $k_{d_j^i d_u}$ is the incidence factor of element i with dimension j relative to elements with dimension u. We introduce the notation also: $k_{d_j d_u^i}$ is the number of elements of dimension j, which included in element i with a dimension u ($u > j$). Thus, $k_{d_j d_u^i}$ is the incidence factor of element i with dimension u relative to elements of dimension j.

If $n = 2$, so $f_0(2)=4, f_1(2)=4$, see (13), (14). The factors of incidents (from smaller dimension to larger, Figure 1) are

$$k_{d_0^i d_1} = 2 = C_2^1, i = 1 \div 4;\ k_{d_0^i d_2} = 1, i = 1 \div 4;\ k_{d_1^i d_2} = 1, i = 1 \div 4.$$

On Figure 1 the edges are

$$d_1^1 = d_0^1 d_0^2, d_1^2 = d_0^2 d_0^3, d_1^3 = d_0^3 d_0^4, d_1^4 = d_0^4 d_0^1.$$

Sum up the incidence coefficients for all vertices and edges of the triangle

Figure 1. The 2-cube

$$\sum_{i=1}^{4} k_{d_0^i d_1} = 2 \cdot 4 = 8, \sum_{i=1}^{4} k_{d_0^i d_2} = 1 \cdot 4 = 4, \sum_{i=1}^{4} k_{d_1^i d_2} = 1 \cdot 4 = 4, \sum_{i=1}^{4} k_{d_0^i d_1} + \sum_{i=1}^{4} k_{d_0^i d_2} + \sum_{i=1}^{4} k_{d_1^i d_2} = 16. \tag{15}$$

The factors of incidents (from larger dimension to smaller, Figure 1) are

$$k_{d_0 d_1^i} = 2, i = 1 \div 4; k_{d_1 d_2} = 4, k_{d_0 d_2} = 4.$$

Sum up the incidence coefficients for all elements of the 2 - cube with dimension larger of zero

$$\sum_{i=1}^{4} k_{d_0 d_1^i} + k_{d_1 d_2} + k_{d_0 d_2} = 16. \tag{16}$$

Comparing (15) and (16) you can see that the sum of incidents in a 2 - cube from elements with a lower dimension to elements with a higher dimension is equal to the sum of incidents from elements with a higher dimension to elements with a lower dimension. Thus, the sum of incidents remains unchanged when the direction of the relationship between the elements it is change.

If $n = 3$, so

$$f_0(3)=8, f_1(3)=12, f_2(3)=6,$$

see (13), (14). The factors of incidents (from smaller dimension to larger, Figure 2) are

$$k_{d_0^i d_1} = 3, i = 1 \div 8; k_{d_0^i d_2} = 3, i = 1 \div 8; k_{d_0^i d_3} = 1, i = 1 \div 8;$$
$$k_{d_1^i d_2} = 2, i = 1 \div 12; k_{d_2^i d_3} = 1, i = 1 \div 6; k_{d_1^i d_3} = 1, i = 1 \div 12.$$

Figure 2. The 3-cube

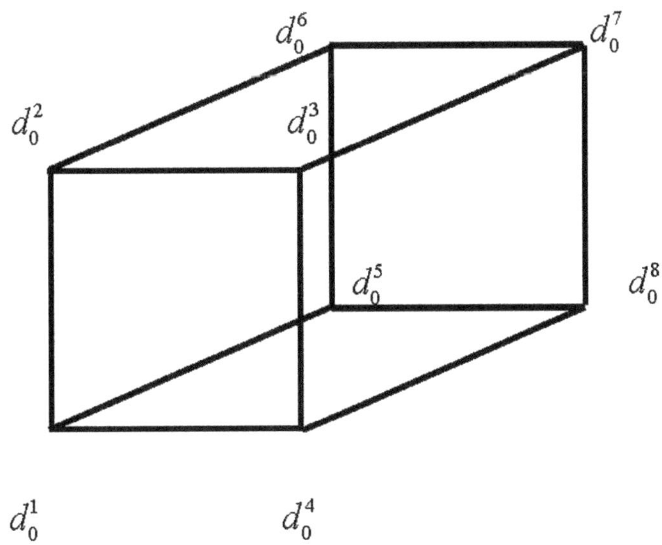

Sum up the incidence coefficients for all vertices, edges and two - dimension faces of the 3 - cube

$$\begin{gathered}\sum_{i=1}^{8} k_{d_0^i d_1} = 3 \cdot 8 = 24, \sum_{i=1}^{8} k_{d_0^i d_2} = 3 \cdot 8 = 24, \sum_{i=1}^{8} k_{d_0^i d_3} = 1 \cdot 8 = 8,\\ \sum_{i=1}^{12} k_{d_1^i d_2} = 2 \cdot 12 = 24, \sum_{i=1}^{6} k_{d_2^i d_3} = 1 \cdot 6 = 6, \sum_{i=1}^{12} k_{d_1^i d_3} = 1 \cdot 12 = 12,\\ \sum_{i=1}^{8} k_{d_0^i d_1} + \sum_{i=1}^{8} k_{d_0^i d_2} + \sum_{i=1}^{12} k_{d_1^i d_2} + \sum_{i=1}^{8} k_{d_0^i d_3} + \sum_{i=1}^{6} k_{d_2^i d_3} + \sum_{i=1}^{12} k_{d_1^i d_3} = 98.\end{gathered} \tag{17}$$

The factors of incidents (from larger dimension to smaller, Figure 2) are

$$k_{d_0 d_1^i} = 2, i = 1 \div 12; k_{d_1 d_2^i} = 4, i = 1 \div 6; k_{d_0 d_2^i} = 4, i = 1 \div 6; k_{d_0 d_3} = 8; k_{d_1 d_3} = 12; k_{d_2 d_3} = 6.$$

Sum up the incidence coefficients for all elements of the 3 - cube with dimension larger of zero

$$\sum_{i=1}^{12} k_{d_0 d_1^i} + \sum_{i=1}^{6} k_{d_0 d_2^i} + \sum_{i=1}^{6} k_{d_1 d_2^i} + k_{d_1 d_3} + k_{d_0 d_3} + k_{d_2 d_3} = 98. \quad (18)$$

Comparing (17) and (18) you can see that the sum of incidents in a 3 - cube from elements with a lower dimension to elements with a higher dimension is equal to the sum of incidents from elements with a higher dimension to elements with a lower dimension. Thus, the sum of incidents retains its value when the direction of the relationship between the elements it is changes.

If $n = 4$, so

$$f_0(4)=16, f_1(4)=32, f_2(4)=24, f_3(4)=8,$$

see (12) - (14). The factors of incidents (from smaller dimension to larger, Figure 3) are

$$k_{d_0^i d_1} = 4, i = 1 \div 16; k_{d_0^i d_2} = 6, i = 1 \div 16; k_{d_0^i d_3} = 4, i = 1 \div 16; k_{d_0^i d_4} = 1,$$
$$i = 1 \div 16; k_{d_1^i d_2} = 3, i = 1 \div 32; k_{d_1^i d_3} = 3, i = 1 \div 32; k_{d_1^i d_4} = 1, i = 1 \div 32;$$
$$k_{d_2^i d_3} = 2, i = 1 \div 24; k_{d_2^i d_4} = 1, i = 1 \div 24; k_{d_3^i d_4} = 1, i = 1 \div 8.$$

Figure 3. The 4-cube

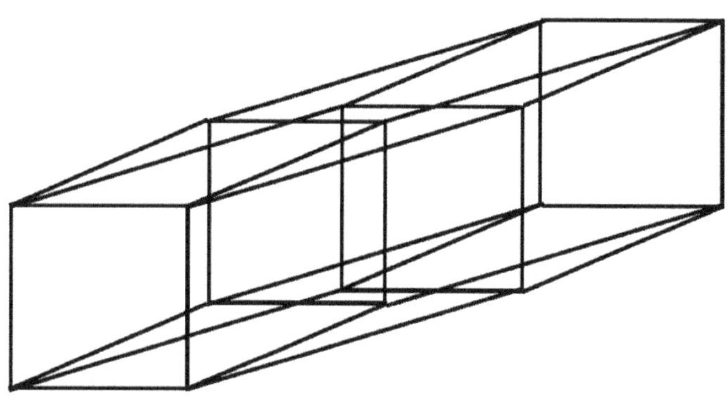

Sum up the incidence coefficients for all vertices, edges, two - dimension faces and three - dimension faces of the 4 - cube

$$\sum_{i=1}^{16} k_{d_0^i d_1} = 4 \cdot 16 = 64, \sum_{i=1}^{16} k_{d_0^i d_2} = 6 \cdot 16 = 96, \sum_{i=1}^{16} k_{d_0^i d_3} = 4 \cdot 16 = 64, \sum_{i=1}^{16} k_{d_0^i d_4} = 16,$$
$$\sum_{i=1}^{32} k_{d_1^i d_2} = 3 \cdot 32 = 96, \sum_{i=1}^{32} k_{d_1^i d_3} = 3 \cdot 32 = 96, \sum_{i=1}^{32} k_{d_1^i d_4} = 32, \sum_{i=1}^{24} k_{d_2^i d_3} = 2 \cdot 24 = 48,$$
$$\sum_{i=1}^{24} k_{d_2^i d_4} = 24, \sum_{i=1}^{8} k_{d_3^i d_4} = 1 \cdot 8 = 8. \sum_{i=1}^{16} k_{d_0^i d_1} + \sum_{i=1}^{16} k_{d_0^i d_2} + \sum_{i=1}^{16} k_{d_0^i d_3} + \sum_{i=1}^{16} k_{d_0^i d_4} + \sum_{i=1}^{32} k_{d_1^i d_2} \quad (19)$$
$$+ \sum_{i=1}^{32} k_{d_1^i d_3} + \sum_{i=1}^{32} k_{d_1^i d_4} + \sum_{i=1}^{24} k_{d_2^i d_3} + \sum_{i=1}^{24} k_{d_2^i d_4} + \sum_{i=1}^{8} k_{d_3^i d_4} = 544.$$

The factors of incidents (from larger dimension to smaller, Figure 3) are

$$k_{d_0 d_1^i} = 2, i = 1 \div 32; k_{d_0 d_2^i} = 4, i = 1 \div 24; k_{d_0 d_3^i} = 8, i = 1 \div 8; k_{d_0 d_4} = 16; k_{d_1 d_2^i} = 4,$$
$$i = 1 \div 24; k_{d_1 d_3^i} = 12, i = 1 \div 8; k_{d_1 d_4} = 32; k_{d_2 d_3^i} = 6, \text{i} = 1 \div 8; k_{d_2 d_4} = 24; k_{d_3 d_4} = 8.$$

Sum up the incidence coefficients for all elements of the 4 - cube with dimension larger of zero

$$\sum_{i=1}^{32} k_{d_0 d_1^i} {}_0 \sum_{i=1}^{24} k_{d_0 d_2^i} + \sum_{i=1}^{24} k_{d_1 d_2^i} + \sum_{i=1}^{8} k_{d_0 d_3^i} + \sum_{i=1}^{8} k_{d_1 d_3^i} + \sum_{i=1}^{8} k_{d_2 d_3^i} + k_{d_0 d_4} + k_{d_1 d_4} + k_{d_2 d_4} + k_{d_3 d_4} = 544. \quad (20)$$

Comparing (19) and (20) you can see that the sum of incidents in a 4 - cube from elements with a lower dimension to elements with a higher dimension is equal to the sum of incidents from elements with a higher dimension to elements with a lower dimension. Thus, the sum of incidents retains its value when the direction of the relationship between the elements it is changes.

It is easy to see that each incidence coefficient from a smaller dimension to a higher dimension corresponds to its inverse incidence factor from a larger dimension to a smaller dimension. For example, $k_{d_0^i d_2} = 6, k_{d_0 d_2^i} = 4$. Moreover, the number of such incidents from a smaller dimension to a higher dimension is 16, the number of corresponding inverse incidents is 24. So, the product of the incidence coefficient and the corresponding number retains its value,

6•16= 4•24= 96. It can be seen that this situation exists for all incidence rates. Therefore, the sum of incidents in the forward and reverse direction is preserved.

In order to write down the sum of incidents for a cube of any dimension in general, it is necessary to write expressions for the incidence coefficients and numbers of incidences in the forward and reverse directions, as a function of the cube size. This will also allow us to prove the next statement:

Theorem 2. In any cube of dimension n, the sum of all incidents of elements of a lower dimension with respect to all elements of a higher dimension is equal to the sum of all incidents of elements of a higher dimension with respect to all elements of a lower dimension and equals the sum of the series

$$f_0(n)(2^n - 1) + f_1(n)(2^{n-1} - 1) + f_2(n)(2^{n-2} - 1) + ... + f_{n-1}(n), f_d(n) = C_{n+1}^{d+1}, d = 0 \div (n-1).$$

Proof. The number of vertices in n - cube is equal $f_0 = 2^n$ (see (12)). The number of edges in n - cube incident to any of the vertices of a cube, by its definition, is equal to $k_{d_0^i d_1} = C_n^1, i = 1 \div f_0$.

The number of elements of two - dimension incident to any of the vertices of a cube is equal $k_{d_0^i d_2} = C_n^2, i = 1 \div f_0$. We can also record the number of elements of any dimension incident to any of the vertices of the cube up to the cube itself $k_{d_0^i d_n} = C_n^n, i = 1 \div f_0$.

Multiplying each of these incidence coefficients by the number of vertices in the n - cube and summing up the obtained products, we find that the total number of incidences of the vertices to the elements of higher dimensionality in the n - cube is determined by the expression

$$f_0(C_n^1 + C_n^2 + ... + C_n^n) = f_0(2^n - 1).$$

This expression coincides with the expression of the total number of incidences of vertices to elements of higher dimension in the n - simplex. The numerical difference between these expressions consists in a different value of the vertices in the n - cube and in the n - simplex, i.e. in the meaning of $f_0(n)$.

The number of edges in n - cube is equal $f_1(n) = 2^{n-1} C_n^1$ (see (12)). The number of two - dimension elements in n - cube incident to any of the edges of a cube, by its definition, is equal to $k_{d_1^i d_2} = C_{n-1}^1, i = 1 \div f_1$.

The number of three - dimension elements incident to any of the edges of a cube is equal $k_{d_1^i d_3} = C^2_{n-1}, i = 1 \div f_1$. We can also record the number of elements of any dimension incident to any of the edges of the cube up to the cube itself $k_{d_1^i d_n} = C^{n-1}_{n-1}, i = 1 \div f_1$.

Multiplying each of these incidence coefficients by the number of edges in the *n* - cube and summing up the obtained products, we find that the total number of incidences of the edges to the elements of higher dimensionality in the *n* - cube is determined by the expression

$$f_1(C^1_{n-1} + C^2_{n-1} + ... + C^{n-1}_{n-1}) = f_1(2^{n-1} - 1).$$

This expression coincides with the expression of the total number of incidences of edges to elements of higher dimension in the *n* - simplex. The numerical difference between these expressions consists in a different value of the edges in the *n* - cube and in the *n* - simplex, i.e. in the meaning of $f_1(n)$.

Continuing to write expressions for the total numbers of incidents of elements of different dimensions to elements of a higher dimension in a *n* - cube, summing up these numbers, we obtain a general expression for the number of incidences of elements of a lower dimension in relation to elements of a higher dimension in *n* - cube

$$f_0(n)(2^n - 1) + f_1(n)(2^{n-1} - 1) + f_2(n)(2^{n-2} - 1) + ... + f_{n-1}(n), f_d(n) = C^{d+1}_{n+1}, d = 0 \div (n-1). \quad (21)$$

It is easy to make sure that the calculations of the incidence flow in the *n* - cube according to expression (21) at values *n*=2÷4 coincide with the values of these quantities obtained earlier by analyzing the images of *n* - cube at these values *n*.

Let $k_{d_h d_j^i}$ be the number of elements of dimension *h* in a *n* - cube belonging to some one element of dimension *j* (*h*<*j*), that is d_j^i. Obviously, this number is equal to $f_h(j)$, and $i=1 \div f_j(n)$. The product $f_h(j) f_j(n)$ corresponds to the number of elements of dimension *h* belonging to all elements of dimension *j*. For a *n* - cube, this product is equal to

$$f_h(j) f_j(n) = 2^{j-h} C^h_j 2^{n-j} C^j_n = 2^{n-h} \frac{n!}{(n-j)!(j-h)!h!}.$$

Let us compare this number with the number of elements of dimension *j*, which have elements of dimension *h* in the *n* - cube

$$k_{d_h^i d_j} f_h(n) = C_{n-h}^{j-h} 2^{n-h} C_n^h = 2^{n-h} \frac{n!}{(n-h)!h!} \frac{(n-h)!}{(n-j)!(j-h)!}.$$

Obviously, these numbers are equal to each other. This proves that the number of elements of dimension *h* in a *n* - cube belonging to all dimensions of *j* (*j*>*h*) in a *n* - cube is equal to the number of elements of dimension *j*, which have elements of dimension *h* as elements.

Thus, the total number of incidents of elements of a smaller dimension with respect to elements of a higher dimension is equal to the total number of incidents of elements of a higher dimension with respect to elements of a smaller dimension. This the total number define expression (21).

It is significant that equality of incidence (information) flows exists despite the difference in the incidence coefficients of elements of a lower dimension in relation to elements of a higher dimension and the incidence coefficients of elements of a higher dimension in relation to elements of a smaller dimension. Therefore, the incidence conservation law found in the previous section for the *n* - simplex is also valid for the *n* - cube. Moreover, the analytical form of this law is preserved, expressed through functions $f_i(n)$ that determine the values of the numbers of the elements of dimension *i* in the polytope of dimension *n*.

INCIDENT CONSERVATION LAW FOR N - CROSS - POLYTOPE

In *n* - cross - polytope (Zhizhin, 2013, 2014, 2019 a) the number of elements with dimension *d* is

$$f_d(n) = 2^{d+1} C_n^{n-1-d}, d = 0 \div n. \tag{22}$$

For example, if $d = 0$, so

$$f_0(n) = 2C_n^{n-1} = 2n. \tag{23}$$

This is the number of vertices in n - cross - polytope. If $d = 1$, so

$$f_1(n) = 2^2 C_n^{n-2} = 2\frac{n!}{(n-2)!} = 2(n-1)n. \tag{24}$$

This is the number of edges in n - cross - polytope. If $d = 2$, so

$$f_2(n) = 2^3 C_n^{n-3} = \frac{4}{3}(n-2)(n-1)n. \tag{25}$$

This is the number of two - dimension faces in n - cross - polytope. If $d = 3$, so

$$f_3(n) = 2^4 C_n^{n-4} = \frac{2}{3}(n-3)(n-2)(n-1)n. \tag{26}$$

This is the number of three - dimension faces in n - cross - polytope.

The smallest dimension of the cross - polytope is 4. From (23) - (26) it follows that in this polytope there are 8 vertices, 24 edges, 32 two - dimensional faces, 16 three - dimensional faces (Figure 4).

Figure 4. The 4-cross-polytope

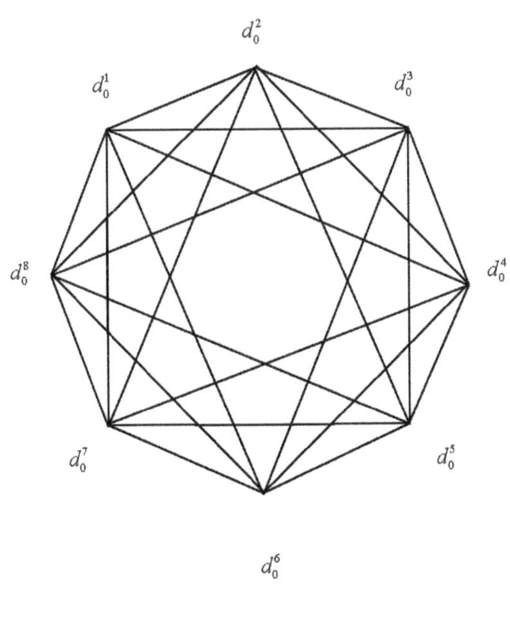

The factors of incidents (from smaller dimension to larger) are

$$k_{d_0^i d_1} = 6, i = 1 \div 8; k_{d_0^i d_2} = 12, i = 1 \div 8; k_{d_0^i d_3} = 8, i = 1 \div 8; k_{d_0^i d_4} = 1, i = 1 \div 8;$$
$$k_{d_1^i d_2} = 4, i = 1 \div 24; k_{d_1^i d_3} = 4, i = 1 \div 24; k_{d_1^i d_4} = 1, i = 1 \div 24; k_{d_2^i d_3} = 2,$$
$$i = 1 \div 32; k_{d_2^i d_4} = 1, i = 1 \div 32; k_{d_3^i d_4} = 1, i = 1 \div 16.$$

Sum up the incidence coefficients for all vertices, edges, two - dimension faces and three -dimension faces of the 4 - cross - polytope

$$\sum_{i=1}^{8} k_{d_0^i d_1} = 6 \cdot 8 = 48, \sum_{i=1}^{8} k_{d_0^i d_2} = 12 \cdot 8 = 96, \sum_{i=1}^{8} k_{d_0^i d_3} = 8 \cdot 8 = 64, \sum_{i=1}^{8} k_{d_0^i d_4} = 8,$$
$$\sum_{i=1}^{24} k_{d_1^i d_2} = 4 \cdot 24 = 96, \sum_{i=1}^{24} k_{d_1^i d_3} = 4 \cdot 24 = 96, \sum_{i=1}^{24} k_{d_1^i d_4} = 24,$$
$$\sum_{i=1}^{32} k_{d_2^i d_3} = 2 \cdot 32 = 64, \sum_{i=1}^{32} k_{d_2^i d_4} = 32, \sum_{i=1}^{16} k_{d_3^i d_4} = 16.$$
$$\sum_{i=1}^{8} k_{d_0^i d_1} + \sum_{i=1}^{8} k_{d_0^i d_2} + \sum_{i=1}^{8} k_{d_0^i d_3} + \sum_{i=1}^{8} k_{d_0^i d_4} + \sum_{i=1}^{24} k_{d_1^i d_2} + \sum_{i=1}^{24} k_{d_1^i d_3}$$
$$+ \sum_{i=1}^{24} k_{d_1^i d_4} + \sum_{i=1}^{32} k_{d_2^i d_3} + \sum_{i=1}^{16} k_{d_3^i d_4} + \sum_{i=1}^{32} k_{d_2^i d_4} = 544.$$

The factors of incidents (from larger dimension to smaller, Figure 4) are

$$k_{d_0 d_1^i} = 2, i = 1 \div 24; k_{d_0 d_2^i} = 3, i = 1 \div 32; k_{d_0 d_3^i} = 4, i = 1 \div 16; k_{d_0 d_4} = 8; k_{d_1 d_2^i} = 3,$$
$$i = 1 \div 32; k_{d_1 d_3^i} = 6, i = 1 \div 16; k_{d_1 d_4} = 24; k_{d_2 d_3^i} = 4, \mathrm{i} = 1 \div 8; k_{d_2 d_4} = 32; k_{d_3 d_4} = 16.$$

Sum up the incidence coefficients for all elements of the 4 - cube with dimension larger of zero

$$\sum_{i=1}^{24} k_{d_0 d_1^i} + \sum_{i=1}^{32} k_{d_0 d_2^i} + \sum_{i=1}^{32} k_{d_1 d_2^i} + \sum_{i=1}^{16} k_{d_0 d_3^i} + \sum_{i=1}^{16} k_{d_1 d_3^i} + \sum_{i=1}^{16} k_{d_2 d_3^i} + k_{d_0 d_4} + k_{d_1 d_4} + k_{d_2 d_4} + k_{d_3 d_4} = 544. \tag{27}$$

Comparing (26) and (27) you can see that the sum of incidents in a 4 - cross - polytope from elements with a lower dimension to elements with a higher dimension is equal to the sum of incidents from elements with a higher dimension to elements with a lower dimension. Thus, the sum of incidents retains its value when changes the direction of the relationship between the elements.

Theorem 3. In any cross - polytope of dimension n, the sum of all incidents of elements of a lower dimension with respect to all elements of a higher dimension is equal to the sum of all incidents of elements of a higher dimension with respect to all elements of a lower dimension and equals the sum of the series

$$f_0(n)3^{n-1} + f_1(n)3^{n-2} + f_2(n)3^{n-3} + ... + f_{n-1}(n)3^0, f_d(n) = C_n^{n-1-d}2^{1+d}, d = 0 \div (n-1).$$

Proof. According to equation (23), each cross - polytope of dimension n has $2n = f_0(n)$ vertices. The peculiarity of a cross - polytope is that each of its vertices has an opposite vertex, with which it is not connected by an edge. Moreover, there is one edge between this vertex and all other vertices. We subtract from the total number of vertices two vertices (the selected vertex and its opposite) 2n - 2. This is the possible number of edges emanating from the selected vertex. Thus, the incidence coefficient of the edges of any vertex is

$$k_{d_0^i d_1} = 2(n-1) = 2\,C_{n-1}^1, i = 1 \div f_0(n).$$

A cross - polytope of any dimension can be depicted as a projection on a two - dimensional plane (Zhizhin, 2019a). In this image, all its vertices are located on a circle, with the selected vertex and its opposite vertex located symmetrically relative to the center of the circle. A mentally drawn line through these two vertices halves the circle (Figure 4). Therefore, the number of variants the location of the edges from the selected vertex to the other vertices in one of the halves of the circle is n - 1, i.e. the number of combinations of n - 1 vertices one by one. Since there are two halves of a circle, then for the total number of vertex selection options for edge formation, this number of combinations should be multiplied by 2. This is the meaning of the expression for the incidence rate of any vertex in the n - cross - polytope with respect to the edge. To further prove the theorem 3 and clarify the nature of the formation of the coefficients of incidence in the n - cross - polytope, we will use this technique. We arrange the vertices of the 4 - cross - polytope in two lines, so that the vertices unlinked by an edge form vertical pairs: each vertex in the top line is not connected by an edge to the vertex in the bottom line located strictly under this vertex in the top line

$$\begin{matrix} d_0^1 & d_0^2 & d_0^3 & d_0^4 \\ d_0^5 & d_0^6 & d_0^7 & d_0^8 \end{matrix}. \tag{28}$$

We choose some vertex in the top line (28), for example d_0^1 . It cannot be connected by an edge with a vertex d_0^5 in the bottom line, but it can be connected by an edge with other vertices of the bottom line. The number of such options is 3, i.e. with $n = 4$ is $3 = C_{n-1}^1 = C_3^1$. However, a vertex d_0^1 may be connected by an edge with any of the remaining three vertices in the top line (28). Therefore, the incidence coefficient of a vertex with respect to an edge in 4 - cross - polytope

$$k_{d_0^i d_1} = 2\,\mathrm{C}_{n-1}^1 = 6, i = 1 \div 8.$$

When considering the belonging of a vertex d_0^1 to two - dimensional faces, it is necessary to determine the number of options for the participation of two vertices in the top line (28) (without a vertex d_0^1) and in the bottom line (28) (without a vertex d_0^5). This will be the number $2C_3^1$. In addition, as the vertices of the triangle, there may be vertices located in the upper and lower lines (28) in a cross - section way. There are 6 such variants in (28), i.e. more $2C_3^1$. Therefore, the incidence coefficient of a vertex with respect to a two - dimensional element in 4 - cross - polytope is

$$k_{d_0^i d_2} = 2^2\,\mathrm{C}_{n-1}^2 = 12, i = 1 \div 8.$$

Multiplying this number by the number of vertices in 4 - cross - polytope (8) you can get the total number of incidents of vertices to two - dimensional elements in 4 - cross - polytope equal to 96. This number coincides with the corresponding number, defined earlier in Figure 4. Combination d_0^1 with three the different vertices from (28) can get the incidence coefficient of vertex to the three - dimension body (tetrahedron) for condition absent in combination opposite vertices. Can to show that this combination is 8,

$$k_{d_0^1 d_3} = 2^3 C_3^3 = 8, i = 1 \div 8.$$

Multiplying this number by the number of vertices in 4 - cross - polytope (8) you can get the total number of incidents of vertices to three - dimensional elements in 4 - cross - polytope equal to 64. This number coincides with the corresponding number, defined earlier in Figure 4. Combination two vertices, for example d_0^1 and d_0^2 (the edge $d_0^1\ d_0^2$) with two the different vertices from (28) you can get the incidence coefficient of edge to three - dimension body (tetrahedron) for condition absent in combination opposite vertices. Can to show that this combination is 4,

$$k_{d_1^1 d_3} = 2^2 C_2^2 = 4, i = 1 \div 8.$$

Multiplying this number by the number of edges in 4 - cross - polytope (8) you can get the total number of incidents of edges to three - dimensional elements in 4 - cross - polytope equal to 96. This number coincides with the corresponding number, defined earlier in Figure 4. Combination two vertices (the edge $d_0^1\ d_0^2$) with one the vertex from (28) can get the incidence coefficient of edge to triangle for condition absent in combination opposite vertices. Can to show that this combination is 4,

$$k_{d_1^1 d_2} = 2C_2^1 = 4, i = 1 \div 8.$$

Multiplying this number by the number of edges in 4 - cross - polytope (8) you can get the total number of incidents of edges to triangle in 4 - cross - polytope equal to 24. This number coincides with the corresponding number, defined earlier in Figure 4.

In general case can write of $2n$ vertices of the n - cross - polytope in two lines

$$\begin{matrix} d_0^1 & d_0^2 & d_0^3 & \dots & d_0^n \\ d_0^5 & d_0^6 & d_0^7 & \dots & d_0^{2n} \end{matrix}, \qquad (29)$$

where vertex in the top line is not connected by an edge to the vertex in the bottom line located strictly under this vertex in the top line. The incidence coefficient of a vertex with respect to an edge in n - cross - polytope is

$$k_{d_0^i d_1} = 2\,\mathrm{C}_{n-1}^1, i = 1 \div f_0(n).$$

The incidence coefficient of a vertex with respect to a two - dimensional element in *n* - cross - polytope is

$$k_{d_0^i d_2} = 2^2\,\mathrm{C}_{n-1}^2, i = 1 \div f_0(n).$$

The incidence coefficient of a vertex with respect to a three - dimensional edge in *n* - cross - polytope is

$$k_{d_0^i d_3} = 2^3\,\mathrm{C}_{n-1}^3, i = 1 \div f_0(n).$$

Go on can to say the incidence coefficient of a vertex with respect to a (*n* – 1) - dimensional element in *n* - cross - polytope is

$$k_{d_0^i d_{n-1}} = 2^{n-1}\,\mathrm{C}_{n-1}^{n-1}, i = 1 \div f_0(n).$$

Obviously, that the incidence coefficient of a vertex with respect to *n* - cross - polytope is

$$k_{d_0^i d_n} = 1, i = 1 \div f_0(n).$$

Multiplying of the incidence coefficients of a vertex with respect to different dimension elements in *n* - cross - polytope and sum the product you can get the common express for the number of incident vertices to elements of different dimension in *n* - cross - polytope

$$f_0(2^0 + 2^1 C_{n-1}^{\,1} + 2^2 C_{n-1}^2 + ... + 2^{n-1} C_{n-1}^{n-1}) = f_0(n)\sum_{i=0}^{n-1} 2^i C_{n-1}^i = f_0(n) 3^{n-1}.$$

In this way you can get the common express for the number of incident edges to elements of different dimension more one in *n* - cross - polytope

$$f_1(2^0 + 2^1 C^{\,1}_{n-2} + 2^2 C^2_{n-2} + ... + 2^{n-2} C^{n-2}_{n-2}) = f_1(n) \sum_{i=0}^{n-2} 2^i C^i_{n-2} = f_1(n) 3^{n-2},$$

and go on.

In result, one can get the common express for the sum of all incidents of elements of a lower dimension with respect to all elements of a higher dimension in n - cross - polytope

$$f_0(n) 3^{n-1} + f_1(n) 3^{n-2} + ... + f_{n-1} 3^0 = \sum_{i=0}^{n-1} f_i(n) 3^{n-1-i}, f_i(n) = C_n^{n-1-i} 2^{1+i}. \tag{30}$$

Let $k_{d_h d_j^i}$ be the number of elements of dimension h in a n - cross - polytope belonging to some one element of dimension j ($h<j$), that is d_j^i . Obviously, this number is equal to $f_h(j)_s$ for simplex, and $i=1 \div f_j(n)_{cr}$ for n - cross - polytope. So, elements of cross - polytope are simplexes the product $f_h(j)_s f_j(n)_{cr}$ corresponds to the number of elements of dimension h belonging to all elements of dimension j for a simplex. This product is equal to $f_d(n) = C_{n+1}^{d+1}$

$$C_{j+1}^{h+1} 2^{1+j} C_n^{n-1-j} = 2^{1+j} \frac{(j+1)!}{(h+1)!(j-h)!} \frac{n!}{(n-1-j)!(1+j)!}.$$

Let us compare this number with the number of elements of dimension j, which have elements of dimension h in the n - cross - polytope

$$k_{d_h^i d_j} f_h(n)_{cr} = C_n^{n-1-h} 2^{1+h} C_{n-h-1}^{j-h} 2^{j-h} = 2^{1+j} \frac{n!}{(n-1-h)!(1+h)!} \frac{(n-h-1)!}{(n-j-1)!(j-h)!}.$$

Obviously, these numbers are equal to each other. This proves that the number of elements of dimension h in a n - cross - polytope belonging to all dimensions of j ($j>h$) in a n - cross - polytope is equal to the number of elements of dimension j, which have elements of dimension h as elements.

Thus, the total number of incidents of elements of a smaller dimension with respect to elements of a higher dimension is equal to the total number of incidents of elements of a higher dimension with respect to elements of a smaller dimension. This the total number define expression (30).

INCIDENT CONSERVATION LAW FOR CLUSTERS CHEMICAL COMPOUND

Incident Conservation Law Regular and Irregular Polytopes

The polytopes n - simplex, n - cube and n - cross - polytopes considered in early was topological regular. Each vertex in any of it was incident same number of edges. However, the length of edges not put equal. Therefore, the polytopes are not metrical regular. Obviously, incident conservation law is for regular and topological regular polytopes. Now consider incident conservation law for irregular and same - regular polytopes.

Incident Conservation Law for Quadrangular Three - Dimension Pyramid

In this pyramid (Figure 5) the number of vertices are f_0=5, the number of edges are f_1=8, the number of two - dimension faces are f_2=5.

Figure 5. The quadrangular three-dimension pyramid

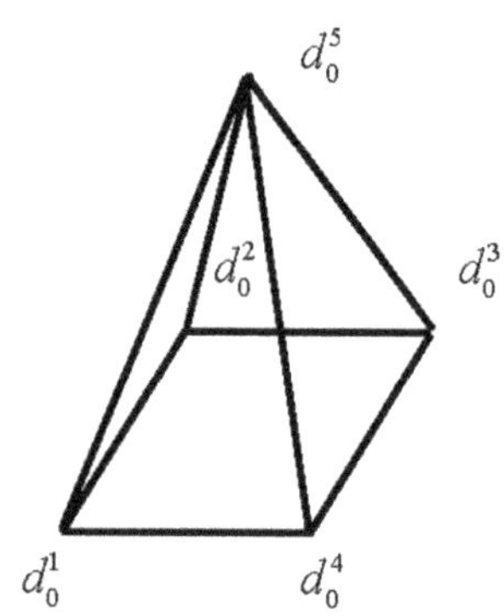

The incident coefficients at vertices of bases to edges are equal 3 $(k_{d_0^i d_1} = 3, i = 1 \div 4)$, the incident coefficient at top of the pyramid to edges is equal 4 $(k_{d_0^5 d_1} = 4)$. The incident coefficients at vertices of bases to two - dimension faces are equal 3 $(k_{d_0^i d_2} = 3, i = 1 \div 4)$, the incident coefficient at top of the pyramid to two - dimension faces is equal 4 $(k_{d_0^5 d_2} = 4)$. The incident coefficients at edges of pyramid to two - dimension faces are equal 2 $(k_{d_1^i d_2} = 2, i = 1 \div 8)$. The incident coefficients of all elements of pyramid to pyramid are 1

$$(k_{d_0^i d_3} - 1, i = 1 \div 5; k_{d_1^i d_3} = 1, i = 1 \div 8; k_{d_2^i d_3} = 1, i = 1 \div 5).$$

Multiplying this incident coefficients by the number elements of correspond dimension you can get

$$\sum_{i=1}^{5} k_{d_0^i d_1} = 16, \sum_{i=1}^{5} k_{d_0^i d_2} = 16, \sum_{i=1}^{5} k_{d_0^i d_3} = 5, \sum_{i=1}^{8} k_{d_1^i d_2} = 16, \sum_{i=1}^{5} k_{d_2^i d_3} = 5, \sum_{i=1}^{8} k_{d_1^i d_3} = 8.$$

Sum up the incidence coefficients for all elements of the pyramid (from smaller dimension to larger)

$$\sum_{i=1}^{5} k_{d_0^i d_1} + \sum_{i=1}^{5} k_{d_0^i d_2} + \sum_{i=1}^{5} k_{d_0^i d_3} + \sum_{i=1}^{8} k_{d_1^i d_2} + \sum_{i=1}^{5} k_{d_2^i d_3} + \sum_{i=1}^{8} k_{d_1^i d_3} = 64. \tag{31}$$

The factors of incidents (from larger dimension to smaller) are

$$k_{d_0 d_1^i} = 2, i = 1 \div 8; k_{d_0 d_2^i} = 3, i = 1 \div 4$$

(for triangle faces of the pyramid), $k_{d_0 d_2} = 4$ (for bases of the pyramid), $k_{d_0 d_3} = 5$, $k_{d_1 d_2^i} = 3, i = 1 \div 4$ (for triangle faces of the pyramid), $k_{d_1 d_2} = 4$ (for bases of the pyramid), $k_{d_1 d_3} = 8$, $k_{d_2 d_3} = 5$. Sum up the incidence coefficients for all elements of the pyramid (from larger dimension to smaller)

$$\sum_{i=1}^{8} k_{d_0 d_1^i} + \sum_{i=1}^{4} k_{d_0 d_2^i} + \sum_{i=1}^{4} k_{d_1 d_2^i} + k_{d_0 d_2} + k_{d_0 d_3} + k_{d_1 d_2} + k_{d_1 d_3} + k_{d_2 d_3} = 64. \tag{32}$$

Comparing (31) and (32) you can see that the sum of incidents in a pyramid from elements with a lower dimension to elements with a higher dimension is equal to the sum of incidents from elements with a higher dimension to elements with a lower dimension. Thus, the sum of incidents retains its value when changes the direction of the relationship between the elements.

Incident Conservation Law for the Mackay Cluster

Consider the Mackay cluster (Figure 1, Chapter 2). In it, the values of the numbers of elements of different dimensions are determined by the equalities

f_0=24, f_1=72, f_2=70, f_3=22, f_4=1.

We calculate the values of the incidence coefficients of the elements of lower dimension with respect to the elements of higher dimension. The factors of incidents (from smaller dimension to larger) are

$k_{d_0^i d_1} = 6, i = 1 \div 24; k_{d_0^i d_2} = 10, i = 1 \div 24; k_{d_0^i d_3} = 6, i = 1 \div 24; k_{d_0^i d_4} = 1, i = 1 \div 24; k_{d_1^i d_2} = 3, i = 1 \div 60$ (edges on icosahedron);

$k_{d_1^i d_2} = 5, i = 1 \div 12$ (edges in connected of icosahedrons);

$k_{d_1^i d_3} = 3, i = 1 \div 60$ (edges on icosahedron);

$k_{d_1^i d_2} = 5, i = 1 \div 12$ (edges in connected of icosahedrons);

$k_{d_1^i d_4} = 1, i = 1 \div 72; k_{d_2^i d_3} = 2, i = 1 \div 70; k_{d_2^i d_4} = 1, i = 1 \div 70; k_{d_3^i d_4} = 1, i = 1 \div 22.$

Sum up the incidence coefficients for all vertices, edges, two dimension faces and three dimension faces of the Mackay cluster

$\sum_{i=1}^{24} k_{d_0^i d_1} = 6 \cdot 24 = 144, \sum_{i=1}^{24} k_{d_0^i d_2} = 10 \cdot 24 = 240, \sum_{i=1}^{24} k_{d_0^i d_3} = 6 \cdot 24 = 144, \sum_{i=1}^{24} k_{d_0^i d_4} = 24, \sum_{i=1}^{60} k_{d_1^i d_2} = 3 \cdot 60 = 180,$ (edges on icosahedron);

$$\sum_{i=1}^{5} k_{d_1^i d_2} = 5 \cdot 12 = 60 \text{ (edges in connected of icosahedrons);}$$

$$\sum_{i=1}^{60} k_{d_1^i d_3} = 3 \cdot 60 = 180 \text{ (edges on icosahedron);}$$

$$\sum_{i=1}^{12} k_{d_1^i d_3} = 5 \cdot 12 = 60 \text{ (edges in connected of icosahedrons);}$$

$$\begin{gathered}\sum_{i=1}^{72} k_{d_1^i d_4} = 72, \sum_{i=1}^{70} k_{d_2^i d_3} = 2 \cdot 70 = 140, \sum_{i=1}^{70} k_{d_2^i d_4} = 70, \sum_{i=1}^{22} k_{d_3^i d_4} = 22 \\ \sum_{i=1}^{24} k_{d_0^i d_1} + \sum_{i=1}^{24} k_{d_0^i d_2} + \sum_{i=1}^{24} k_{d_0^i d_3} + \sum_{i=1}^{24} k_{d_0^i d_4} + \sum_{i=1}^{72} k_{d_1^i d_2} + \sum_{i=1}^{72} k_{d_1^i d_3} + \sum_{i=1}^{72} k_{d_1^i d_4} \\ + \sum_{i=1}^{70} k_{d_2^i d_3} + \sum_{i=1}^{70} k_{d_2^i d_4} + \sum_{i=1}^{22} k_{d_3^i d_4} = 1336.\end{gathered} \tag{33}$$

The factors of incidents (from larger dimension to smaller) are

$k_{d_0 d_1^i} = 2, i = 1 \div 72; k_{d_0 d_2^i} = 3, i = 1 \div 40$ (at triangle faces);

$k_{d_0 d_2^i} = 4, i = 1 \div 30$ (at quadrangular faces);

$k_{d_0 d_3^i} = 6, i = 1 \div 20$ (at prisms);

$k_{d_0 d_3^i} = 12, i = 1 \div 2$ (at icosahedrons);

$k_{d_0 d_4} = 24; k_{d_1 d_2^i} = 3, i = 1 \div 40$ (at triangle);

$k_{d_1 d_2^i} = 4, i = 1 \div 30$ (at quadrangular faces);

$k_{d_1 d_3^i} = 3, i = 1 \div 60$ (edges on icosahedron);

$k_{d_1 d_2^i} = 5, i = 1 \div 12$ (edges in connected of icosahedrons);

$k_{d_1 d_4} = 1, i = 1 \div 72; k_{d_2 d_3^i} = 2, i = 1 \div 70; k_{d_2 d_4} = 1, i = 1 \div 70; k_{d_3 d_4} = 1, i = 1 \div 22.$

Sum up the incidence coefficients for all vertices, edges, two - dimension faces and three - dimension faces of the Mackay cluster

$$\sum_{i=1}^{72} k_{d_0 d_1^i} = 2 \cdot 72 = 144, \sum_{i=1}^{70} k_{d_0 d_2^i} = 3 \cdot 40 + 4 \cdot 30 = 240, \sum_{i=1}^{22} k_{d_0 d_3^i} = 6 \cdot 20 + 2 \cdot 12 = 144,$$
$$\sum_{i=1}^{24} k_{d_0 d_4} = 24, \sum_{i=1}^{70} k_{d_1 d_2^i} = 3 \cdot 40 + 4 \cdot 30 = 240, \sum_{i=1}^{22} k_{d_1 d_3^i} = 9 \cdot 20 + 2 \cdot 20 = 240, \sum_{i=1}^{72} k_{d_1 d_4} = 72,$$
$$\sum_{i=1}^{22} k_{d_2 d_3^i} = 5 \cdot 20 + 2 \cdot 20 = 140, \sum_{i=1}^{70} k_{d_2 d_4} = 70, \sum_{i=1}^{22} k_{d_3 d_4} = 22,$$
$$\sum_{i=1}^{72} k_{d_0 d_1^i} + \sum_{i=1}^{70} k_{d_0 d_2^i} + \sum_{i=1}^{22} k_{d_0 d_3^i} + \sum_{i=1}^{24} k_{d_0 d_4} + \sum_{i=1}^{70} k_{d_1 d_2^i} + \sum_{i=1}^{22} k_{d_1 d_3^i} + \sum_{i=1}^{72} k_{d_1 d_4}$$
$$+ \sum_{i=1}^{22} k_{d_2 d_3^i} + \sum_{i=1}^{70} k_{d_2 d_4} + \sum_{i=1}^{22} k_{d_3 d_4} = 1336. \quad (34)$$

Comparing (33) and (34) you can see that the sum of incidents in the Mackay cluster from elements with a lower dimension to elements with a higher dimension is equal to the sum of incidents from elements with a higher dimension to elements with a lower dimension. Thus, the sum of incidents retains its value when changes the direction of the relationship between the elements.

The Law of Conservation of Incidence in Clusters with a Central Metal Atom

Let us verify the implementation of the law of conservation of incidence in clusters with a central metal atom. As an example, take an octahedron with a center. A metal atom is located in the center; ligands are located at the vertices of the octahedron. In this polytope (Figure 6) the values of the numbers of elements of different dimensions are determined by the equalities

Figure 6. The octahedron with a centrum

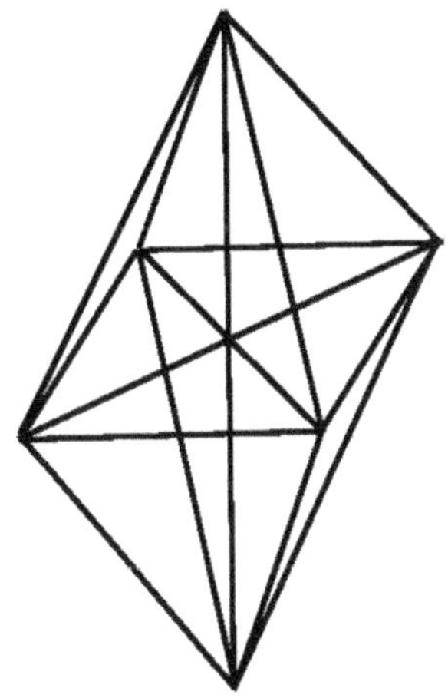

$f_0=7, f_1=18, f_2=16, f_3=9, f_4=1.$

We calculate the values of the incidence coefficients of the elements of lower dimension with respect to the elements of higher dimension. The factors of incidents (from smaller dimension to larger) are

$k_{d_0^i d_1} = 5, i = 1 \div 6$ (for vertices of octohedron);

$k_{d_0 d_1} = 6$ (for atom at the center);

$k_{d_0^i d_2} = 8, i = 1 \div 6$ (for vertices of octohedron);

$k_{d_0 d_2} = 12$ (for atom at the center);

$k_{d_0^i d_3} = 5, i = 1 \div 6$ (for vertices of octohedron);

$k_{d_0 d_3} = 8$ (for atom at the center);

$k_{d_0^i d_4} = 1, i = 1 \div 7;$

$k_{d_1^i d_2} = 4, i = 1 \div 6$ (edges from the center);

$k_{d_1^i d_2} = 3$ (for edges of octahedron);

$k_{d_1^i d_3} = 4, i = 1 \div 6$ (edges from the center);

$k_{d_1^i d_3} = 3$ (for edges of octahedron);

$k_{d_1^i d_4} = 1, i = 1 \div 18; k_{d_2^i d_3} = 2, i = 1 \div 16; k_{d_2^i d_4} = 1, i = 1 \div 16; k_{d_3^i d_4} = 1, i = 1 \div 9.$

Sum up the incidence coefficients for all vertices, edges, two - dimension faces and three - dimension faces

$$\sum_{i=1}^{7} k_{d_0^i d_1} + \sum_{i=1}^{7} k_{d_0^i d_2} + \sum_{i=1}^{7} k_{d_0^i d_3} + \sum_{i=1}^{7} k_{d_0^i d_4} + \sum_{i=1}^{18} k_{d_1^i d_2} + \sum_{i=1}^{18} k_{d_1^i d_3} + \sum_{i=1}^{18} k_{d_1^i d_4} + \sum_{i=1}^{16} k_{d_2^i d_3} + \sum_{i=1}^{16} k_{d_2^i d_4} + \sum_{i=1}^{9} k_{d_3^i d_4} = 324. \quad (35)$$

The factors of incidents (from larger dimension to smaller) are

$k_{d_0 d_1^i} = 2, i = 1 \div 18; k_{d_0 d_2^i} = 3, i = 1 \div 16; k_{d_0^i d_3} = 4, i = 1 \div 9$ (at tetrahedrons)

$k_{d_0 d_3^i} = 6, i = 1$ (at octahedron)

$k_{d_0 d_4} = 7; k_{d_1 d_2^i} = 3, i = 1 \div 16;$

$k_{d_1 d_3^i} = 6, i = 1 \div 8; k_{d_1 d_3^i} = 12, i = 1$ (for octahedron)

$k_{d_1 d_4} = 18; k_{d_2 d_3^i} = 4, i = 1 \div 8; k_{d_2 d_3} = 8$ (for octahedron)

$k_{d_2d_4} = 16; k_{d_3d_4} = 9.$

Sum up the incidence coefficients for all vertices, edges, two - dimension faces and three -dimension faces

$$\sum_{i=1}^{18} k_{d_0d_1^i} + \sum_{i=1}^{16} k_{d_0d_2^i} + \sum_{i=1}^{10} k_{d_0d_3^i} + k_{d_0d_4} + \sum_{i=1}^{16} k_{d_1d_2^i} + \sum_{i=1}^{9} k_{d_1d_3^i} + k_{d_1d_4} + \sum_{i=1}^{9} k_{d_2d_3^i} + k_{d_2d_4} + k_{d_0d_4} = 324. \tag{36}$$

Comparing (35) and (36) you can see that the sum of incidents in the octahedron with center of elements with a lower dimension to elements with a higher dimension is equal to the sum of incidents from elements with a higher dimension to elements with a lower dimension. Thus, the sum of incidents retains its value when changes the direction of the relationship between the elements.

CONCLUSION

In the modern scientific world, information exchange is of great importance. This is characteristic for various fields of knowledge. In mathematics, there is the concept of incidence, i.e. belonging of some mathematical element to another mathematical element. If mathematical elements are models of some material bodies, then incidence can be interpreted as the transfer of information from one material body to another. In the framework of this monograph, the polytopes of higher dimension are the mathematical elements that model material bodies. With their help, the structure of clusters of chemical compounds is described, i.e. bonded to each other by chemical bonds of atoms, molecules, functional groups. The geometric elements in this case are the faces of different dimensions of polytopes. Of interest is the question of transferring information from elements of lower dimension to elements of higher dimension and vice versa from elements of higher dimension to elements of lower dimension. This chapter first establishes the existence of integral equality in relation to the issue of the transmission of information by elements of lower and higher dimensions that describe natural objects (Zhizhin, 2019c). This integral equality is called the law of conservation of incidents. At the beginning, using simple geometric examples, a direct calculation of incidents from images of geometric shapes shows that the sum of incidents of elements of lower dimension with respect to elements of higher dimension is equal to the sum of incidents of elements of higher dimension with respect to elements of lower dimension. Then, the fulfillment of the law of conservation of incidents for the *n* - simplex of the *n* - cube and the *n* - cross - polytope is proved in general terms. It is shown that the law of conservation of incidents is valid for both regular bodies and irregular bodies, which can be clusters of chemical compounds. It is significant that the flow of incidents from elements of lower dimension to elements of higher dimension and vice versa increases sharply with increasing dimension of the polytope. It is significant that the flow of incidents from elements of lower dimension to elements of higher dimension and vice versa increases sharply with increasing dimension of the polytope. Interestingly, the flow of incidents (information) is of greatest importance to the *n* - cross - polytope. If we compare the values of these flows calculated from the obtained expressions, then with the same dimension value (13) for 13 - simplex this value is $4.75 \bullet 10^6$, for 13 - cube this value is a thousand times greater ($1.2 \bullet 10^9$), and in the 13 - cross - polytope is 1.5 times more than the 13 - cube and is $1.8 \bullet 10^9$. In the monograph (Zhizhin, 2019b), that the binding of helices in a DNA molecule with the help of nitrogen bases occurs in a cross - polytope of dimension 13. This corresponds to a significant flow of information between the elements of the DNA molecule. The latter can serve as a mathematical basis for the recently discovered epigenetic principle of the transmission of hereditary information without changing the sequence of genes in DNA and RNA molecules (Zhizhin, 2019d).

REFERENCES

Lindquist, S. (2016). Luminidependens (LD) is an arabidopsis protein with prion behavior. *Proceedings of the National Academy of Sciences of the United States of America*, *113*(21), 6065–6070. doi:10.1073/pnas.1604478113 PMID:27114519

Mancuso, S. (2017). *PLANT REVOLUTION. Le piante hanno già inventato il nostro futuro*. Firenze, Milano: GUIUTI EDITORE.

Mancuso, S., & Viola, A. (2013). *Sensibilità e intelligenza del mondo vegetale*. Firenze, Milano: GUIUTI EDITORE.

Sanbonmatsu, K. Y. (2016). COOLAIR antisense RNAs from evolutionarily conserved elaborate secondary structures. *Cell Reports*, *16*(12), 3087–3096. doi:10.1016/j.celrep.2016.08.045 PMID:27653675

Zhizhin, G. V. (2013). Relations for the numbers of faces of various dimensions in the tower of n dimensional convex regular polytopes. In *Proceedings of the 9th All - Russian School "Mathematical Studies in the Natural Sciences"*. Apatity: Geological Institute, Russian Academy of Sciences.

Zhizhin, G. V. (2014). *World - 4D*. St. Petersburg: Polytechnic Service.

Zhizhin, G. V. (2018). *Chemical Compound Structures and the Higher Dimensional of Molecules: Emerging Research and Opportunities*. Hershey, PA: IGI Global. doi:10.4018/978-1-5225-4108-0

Zhizhin, G. V. (2019a). *The Geometry of Higher - Dimensional Polytopes*. Hershey, PA: IGI Global. doi:10.4018/978-1-5225-6968-8

Zhizhin, G. V. (2019b). *Attractors and Higher Dimensions in Population and Molecular Biology: Emerging Research and Opportunities*. Hershey, PA: IGI Global. doi:10.4018/978-1-5225-9651-6

Zhizhin, G. V. (2019c). The Law of Conservation of Incidents in the Space of Nanoworld. *International Journal of Chemoinformatics and Chemical Engineering*, *8*(1), 25–46. doi:10.4018/IJCCE.2019010103

Zhizhin, G. V. (2019d). The Polytope of Hereditary Information the Structure, Location, Signification. *Biochemistry and Modern Applications*, *2*(1), 56–62. doi:10.33805/2638-7735.124

KEY TERMS AND DEFINITIONS

Dimension of the Space: The member of independent parameters needed to describe the change in position of an object in space.

Incidence Coefficients of Elements of Higher Dimension With Respect to Elements of Lower Dimension: The number of elements of a given lower dimension that are included in a particular element of a higher dimension.

Incidence Coefficients of Elements of Lower Dimension With Respect to Elements of Higher Dimension: The number of elements of a certain higher dimension to which the given element of a lower dimension belongs.

N-Cross-Polytope: The convex polytope of dimension *n* in which opposite related of centrum edges does not have connection of edge.

N-Cube: The convex polytope of dimension *n* in which each vertex incident to *n* edges.

N-Simplex: The convex polytope of dimension *n* in which each vertex is joined by edges with all remain vertices of polytope.

Nanocluster: A nanometric set of connected atoms, stable either in isolation state or in building unit of condensed matter.

Polytope: Polyhedron in the space of higher dimension.

Chapter 8
Mathematical Design of Nanomaterials From Clusters of Higher Dimension

ABSTRACT

Using concrete examples of clusters of chemical compounds of various types (intermetallic clusters, metal chains with ligands, polyhedral metal clusters with ligands), it is shown how nanomaterials are formed from individual clusters by multiplying their geometric structure by other geometric elements of different dimensions. The considered examples correspond to nanomaterials with a structure of limited complexity. However, the mathematical apparatus developed on the basis of the geometry of high-dimensional polytopes allows, in principle, to describe and study and design nanomaterials of this type of any complexity and any dimension. In particular, nanomaterials with the simultaneous use of elements with different metric characteristics can be attributed to such nanomaterials.

INTRODUCTION

In the monograph Zhizhin G.V. (Zhizhin, 2018) it was established that the fundamental area in the filling of the n - dimensional space is the polytopic prismahedron, which is a prism or a complex of prisms with bases in the form of polytopes of dimension n. It is polytopes provide the filling of the n - dimensional space with a face into the face without gaps. Thus, the polytopic

DOI: 10.4018/978-1-7998-3784-8.ch008

prismahedron is the stereohedron, the existence of which was suggested by B.N. Delone (Delone, 1961; Delone & Sandakova, 1961). However, Delone never succeeded in constructing any particular stereohedron in a space with a dimension greater than three. Considering, that clusters are of higher dimension, the question of creating nanomaterials consisting of clusters leads to the need to look for solutions to this problem in a space of higher dimension. Here we digress from practical methods of creating nanomaterials associated with special methods of mechanical and physical impact on clusters. The problem is considered mathematically about how clusters, as objects of higher dimension, must interact with each other to form a nanomaterial. The mathematical apparatus for contacting polytopes of higher dimension for the formation of continuous arrays consisting of these polytopes was developed in the monograph "Geometry of High - Dimensional Polytopes" (Zhizhin, 2019a). It examined the correct and topologically correct polytopes of the highest dimension. It is shown that polytopes filling the space form polytopic prismahedrons (Zhizhin, 2015). The main operation for creating polytopic prismahedrons is the production of polytopes. By its design, a polytopic prismahedron is the product of a polytope to another polytope, or in a particular case to a one - dimensional segment. Pontryagin mentioned the structures resulting from the product of a polyhedron by a one - dimensional segment the cylinder ones (Pontryagin, 1976). We can say that the product of a polytope by a segment is a prism (Robertson, 1984) with a base in form of a polytope. To distinguish it from a usual three - dimensional prism, can to call it polytopic prism. Ziegler noted that the product of polytopes is not a simplex even if the factors are simplexes, so the polytopes are of considerable interest (Ziegler, 1995). In the chapter, the process of forming polytopic prismahedrons is used for clusters different types discussed in previous chapters.

POLYTOPIC PRISMAHEDRON FROM MACKAY CLUSTERS

Chapter 2 examined clusters of intermetallic compounds. It is proved that the Mackay cluster, consisting of two icosahedrons with a common center (Figure 1 in Chapter 2) has dimension 4. In determining polytopic prismahedrons from clusters, we will use the following theorem (Zhizhin, 2019a), preserving the accepted notation,

Theorem 1. (Zhizhin, 2015)

If there are convex polytopes of dimensions n and m, respectively denoted P^n and Q^m (or simply P and Q), then their product $P^n \times Q^m$, (or simply $\times$, when it is clear what polytopes are multiplied) has a face $F_{\times}^{k}$ with numbers

$$f_{\times}^{k} - \sum_{i=0}^{j} f_{P}^{k-i} f_{Q}^{i}, \qquad (1)$$

$j=k$, $j=k$ if $0 \leq k < m$; $j=m$, if $m \leq k \leq n+m$; $n \geq m$.

Here the symbol f indicates the number of faces, the superscript of f and F indicates the dimension of faces, the lower index indicates belonging of a face to the respective polytope.

Consider the polytopic prismahedron formed by the product of the Mackay cluster in a one -dimensional segment. The Mackay cluster (P) with two icosahedrons (Chapter 2) has the following numbers of elements of different dimensions

$$f_{P}^{0} = 24, f_{P}^{1} = 72, f_{P}^{2} = 70, f_{P}^{3} = 22, f_{P}^{4} = 1.$$

The one - dimensional segment Q^1 has $f_{Q}^{0} = 2, f_{Q}^{1} = 1$. The product of polytopes $P=P^4$ and Q^1 according to (1) it is determined by the equalities:

the number of faces with dimension zero is

$$f_{x}^{0} = f_{P}^{0} \cdot f_{Q}^{0} = 24 \cdot 2 = 48\,;$$

the number of faces with dimension one is

$$f_{x}^{1} = f_{P}^{1} \cdot f_{Q}^{0} + f_{P}^{0} \cdot f_{Q}^{1} = 72 \cdot 2 + 24 \cdot 1 = 144 + 24 = 168;$$

the number of faces with dimension two is

$$f_{x}^{2} = f_{P}^{2} \cdot f_{Q}^{0} + f_{P}^{1} \cdot f_{Q}^{1} = 70 \cdot 2 + 72 \cdot 1 = 212; \qquad (2)$$

the number of faces with dimension three is

$$f_x^3 = f_P^3 \cdot f_Q^0 + f_P^2 f_Q^1 = 22 \cdot 2 + 70 \cdot 1 = 112;$$

the number of faces with dimension four is

$$f_x^4 = f_P^4 \cdot f_Q^0 + f_P^3 \cdot f_Q^1 = 1 \cdot 2 + 22 \cdot 1 = 24;$$

the number of faces with dimension five is

$$f_x^5 = f_P^4 \cdot f_Q^1 = 1 \cdot 1 = 1.$$

Thus, the product of a Mackay cluster with two icosahedrons with a common center on a one -dimensional segment is a polytopic prismahedron of dimension 5. The form of this prismahedron is shown in Figure 1. The totality of values determines the structure of this product.

Figure 1. The product of a Mackay cluster with two icosahedrons with a common center

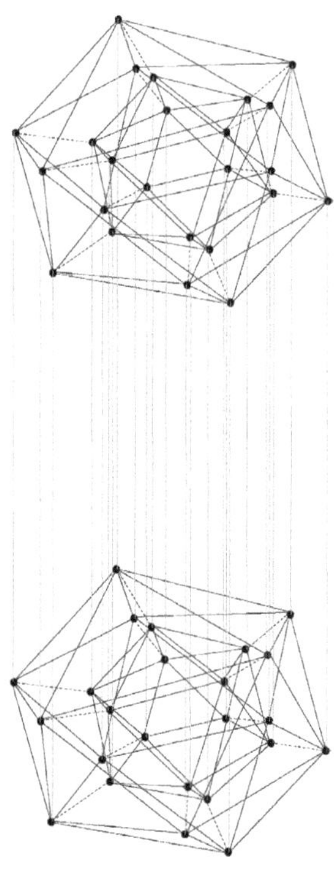

Consider the polytopic prismahedron formed by the product of the Mackay cluster of two icosahedrons with a common center on a triangle. In this case, the first factor in the product is still P^4, and the second factor is Q^2 with faces $f_Q^0 = 3, f_Q^1 = 3, f_Q^2 = 1.$

The product P^4 and Q^2 according to (1) has the faces:

the number of faces with dimension zero is

$$f_x^0 = f_P^0 \cdot f_Q^0 = 24 \cdot 3 = 72;$$

the number of faces with dimension one is

$$f_x^1 = f_P^1 \cdot f_Q^0 + f_P^0 \cdot f_Q^1 = 72 \cdot 3 + 24 \cdot 3 = 216 + 72 = 288;$$

the number of faces with dimension two is

$$f_x^2 = f_P^2 \cdot f_Q^0 + f_P^1 \cdot f_Q^1 + f_P^0 \cdot f_Q^2 = 70 \cdot 3 + 72 \cdot 3 + 24 \cdot 1 = 450; \tag{3}$$

the number of faces with dimension three is

$$f_x^3 = f_P^3 \cdot f_Q^0 + f_P^2 f_Q^1 + f_P^1 \cdot f_Q^2 = 22 \cdot 3 + 70 \cdot 3 + 72 \cdot 1 = 348;$$

the number of faces with dimension four is

$$f_x^4 = f_P^4 \cdot f_Q^0 + f_P^3 \cdot f_Q^1 + f_P^2 \cdot f_Q^2 = 1 \cdot 3 + 22 \cdot 3 + 70 \cdot 1 = 139;$$

the number of faces with dimension five is

$$f_x^5 = f_P^4 \cdot f_Q^1 + f_P^3 f_Q^2 = 1 \cdot 3 + 22 \cdot 1 = 25;$$

the number of faces with dimension six is

$$f_x^6 = f_P^4 \cdot f_Q^2 = 1 \cdot 1 = 1.$$

Thus, the product of a Mackay cluster with two icosahedrons with a common center on a triangle is a polytopic prismahedron of dimension 6 with faces defined by equalities (3).

You can also get a polytopic prismahedron formed by the product of the Mackay cluster with two icosahedrons with a common center on the tetrahedron. In this case, the first factor is still P^4, and the second factor is Q^3 with faces

$$f_Q^0 = 4, f_Q^1 = 6, f_Q^2 = 4, f_Q^3 = 1.$$

The product of polytopes P^4 and Q^3 according to (1) it is determined by the equalities:

the number of faces with dimension zero is

$$f_x^0 = f_P^0 \cdot f_Q^0 = 24 \cdot 4 = 64;$$

the number of faces with dimension one is

$$f_x^1 = f_P^1 \cdot f_Q^0 + f_P^0 \cdot f_Q^1 = 72 \cdot 4 + 24 \cdot 6 = 432;$$

the number of faces with dimension two is

$$f_x^2 = f_P^2 \cdot f_Q^0 + f_P^1 \cdot f_Q^1 + f_P^0 \cdot f_Q^2 = 70 \cdot 4 + 72 \cdot 6 + 24 \cdot 4 = 808; \tag{4}$$

the number of faces with dimension three is

$$f_x^3 = f_P^3 \cdot f_Q^0 + f_P^2 f_Q^1 + f_P^1 \cdot f_Q^2 + f_P^0 \cdot f_Q^3 = 22 \cdot 4 + 70 \cdot 6 + 72 \cdot 4 + 24 \cdot 1 = 820;$$

the number of faces with dimension four is

$$f_x^4 = f_P^4 \cdot f_Q^0 + f_P^3 \cdot f_Q^1 + f_P^2 \cdot f_Q^2 + f_P^1 \cdot f_Q^3 = 1 \cdot 4 + 22 \cdot 6 + 70 \cdot 4 + 72 \cdot 1 = 488;$$

the number of faces with dimension five is

$$f_x^5 = f_P^4 \cdot f_Q^1 + f_P^3 \cdot f_Q^2 + f_P^2 \cdot f_Q^3 = 1 \cdot 6 + 22 \cdot 4 + 70 \cdot 1 = 164;$$

the number of faces with dimension six is

$$f_x^6 = f_P^4 \cdot f_Q^2 + f_P^3 \cdot f_Q^3 = 1 \cdot 4 + 22 \cdot 1 = 26;$$

the number of faces with dimension seven is

$$f_x^7 = f_P^4 \cdot f_Q^3 = 1 \cdot 1 = 1.$$

Thus, the product of a Mackay cluster with two icosahedrons with a common center on a tetrahedron is a polytopic prismahedron of dimension 7 with faces defined by equalities (4).

Obviously, in this way it is possible to obtain nanomaterial with any finite number of clusters, if the Mackay cluster is replaced by convex polytopes with a given number of vertices.

POLYTOPIC PRISMAHEDRONS OF CLUSTERS CONSISTING OF N SHELLS IN THE FORM OF ICOSAHEDRONS WITH A COMMON CENTER

In Chapter 2, it was shown that in clusters of *n* icosahedrons with a common center, the numbers of faces of different dimensions are determined by the equalities

$$\begin{gathered} f_0 = 12n, f_1 = 30n + 12(n-1), f_2 = 20n + 30(n-1), \\ f_3 = 20(n-1) + n, f_4 = C_n^2, f_5 = C_n^3, \ldots, f_{n+1} = C_n^{n-1}. \end{gathered} \tag{5}$$

The dimension of a cluster of *n* shells in the form of icosahedrons is defined in Chapter 2; it is 2+*n*. Consider the product of this cluster on a one - dimensional segment. So, we are considering the product $P^{2+n} \times Q^1$. The factors of this product have, according to (1), faces

$$\begin{gathered} f_P^0 = 12n, f_P^1 = 30n + 12(n-1), f_P^2 = 20n + 30(n-1), \\ f_P^3 = 20(n-1) + n, f_P^4 = C_n^2, f_P^5 = C_n^3, \ldots, f_P^{n+1} = C_n^{n-1}, f_Q^0 = 2, f_Q^1 = 1. \end{gathered}$$

In the product $P^{2+n} \times Q^1$ according to (1) is determined by the equalities:

the number of faces with dimension zero is

$$f_x^0 = f_P^0 \cdot f_Q^0 = 24n;$$

the number of faces with dimension one is

$$f_x^1 = f_P^1 \cdot f_Q^0 + f_P^0 \cdot f_Q^1 = 60n + 24(n-1) + 12n = 72n + 24(n-1);$$

the number of faces with dimension two is

$$f_x^2 = f_P^2 \cdot f_Q^0 + f_P^1 \cdot f_Q^1 = 40n + 60(n-1) + 30n + 12(n-1) = 70n + 72(n-1); \tag{6}$$

the number of faces with dimension three is

$$f_x^3 = f_P^3 \cdot f_Q^0 + f_P^2 f_Q^1 = 40(n-1) + 2n + 20n + 30(n-1) = 22n + 70(n-1);$$

the number of faces with dimension four is

$$f_x^4 = f_P^4 \cdot f_Q^0 + f_P^3 \cdot f_Q^1 = 2C_n^2 + 20(n-1) + n;$$

the number of faces with dimension five is

$$f_x^5 = f_P^5 \cdot f_Q^0 + f_P^4 \cdot f_Q^1 = 2C_4^3 + C_n^2;$$

the number of faces with dimension six is

$$f_x^6 = f_P^6 \cdot f_Q^0 + f_P^5 \cdot f_Q^1 = 2C_n^4 + C_n^3;$$

the number of faces with dimension $2 + n$ is

$$f_x^{2+n} = f_P^{2+n} \cdot f_Q^0 + f_P^{1+n} \cdot f_Q^1 = 1 \cdot 2 + C_n^{n-1} \cdot 1 = 2 + C_n^{n-1} = 2 + n;$$

the number of faces with dimension 3 + *n* is

$$f_x^{3+n} = f_P^{2+n} \cdot f_Q^1 = 1 \cdot 1 = 1.$$

Thus, the product $P^{2+n} \times Q^1$ is a polytopic prismahedron of dimension 3 + *n* with faces defined by equalities (6).

Consider a polytopic prismahedron formed of the product of cluster from *n* icosahedron shells on the triangle. In this case the second factor is Q^2 with faces $f_Q^0 = 3, f_Q^1 = 3, f_Q^2 = 1$. The product $P^{2+n} \times Q^2$ according (1) has a faces:

the number of faces with dimension zero is

$$f_x^0 = f_P^0 \cdot f_Q^0 = 36n;$$

the number of faces with dimension one is

$$f_x^1 = f_P^1 \cdot f_Q^0 + f_P^0 \cdot f_Q^1 = [30n + 12(n-1)]3 + 36n = 126n + 36(n-1);$$

the number of faces with dimension two is

$$f_x^2 = f_P^2 \cdot f_Q^0 + f_P^1 \cdot f_Q^1 + f_P^0 \cdot f_Q^2$$
$$= [20n + 30(n-1)]3 + [30n + 12(n-1)]3 + 12n = 162n + 126(n-1);$$

the number of faces with dimension three is

$$f_x^3 = f_P^3 \cdot f_Q^0 + f_P^2 f_Q^1 + f_P^1 \cdot f_Q^2$$
$$= [20(n-1) + n]3 + [20n + 30(n-1)]3 + 30n + (n-1)12 = 93n + 162(n-1);$$

the number of faces with dimension four is

$$f_x^4 = f_P^4 \cdot f_Q^0 + f_P^3 \cdot f_Q^1 + f_P^2 \cdot f_Q^2$$
$$= 3C_n^2 + [20(n-1) + n]3 + 20n + 30(n-1) = 3C_n^2 + 23n + 90(n-1);$$

the number of faces with dimension five is

$$f_x^5 = f_P^5 \cdot f_Q^0 + f_P^4 \cdot f_Q^1 = 2C_4^3 + C_n^2; \tag{7}$$

the number of faces with dimension six is

$$f_x^6 = f_P^6 \cdot f_Q^0 + f_P^5 \cdot f_Q^1 = 2C_n^4 + C_n^3;$$

the number of faces with dimension 2 + n is

$$f_x^{2+n} = f_P^{2+n} \cdot f_Q^0 + f_P^{1+n} \cdot f_Q^1 + f_P^n \cdot f_Q^2 = 1 \cdot 3 + C_n^{n-3} \cdot 3 + C_n^{n-2};$$

the number of faces with dimension 3 + n is

$$f_x^{3+n} = f_P^{2+n} \cdot f_Q^1 + f_P^{1+n} f_Q^2 = 1 \cdot 3 + C_n^{n-1};$$

the number of faces with dimension 4 + n is

$$f_x^{4+n} = f_P^{2+n} \cdot f_Q^2 = 1 \cdot 1 = 1.$$

Thus, the product $P^{2+n} \times Q^2$ is a polytopic prismahedron of dimension 4 + n with faces defined by equalities (7).

Obviously, in this way it is possible to obtain nanomaterial with any finite number of clusters from n - icosahedrons, if multiply the n - icosahedron by a convex polytope with a given number of vertices.

NANOMATERIALS FROM CLUSTERS OF γ - BRASS

In Chapter 2 was shown that minimum dimension of the cluster γ - brass is 4. This cluster has 8 vertices and is cross - polytope. Next dimension in a number of the clusters of γ - brass is 7. This cluster has 14 vertices and tape cross - polytope too. From cluster of γ - brass with dimension 4 can obtain a polytopic prismahedron with dimension 5 if multiply it on a one - dimension segment. The form of this prismahedron is shown in Figure 2.

Figure 2. The product of the cluster of γ - brass with dimension 4 on a one - dimension segment

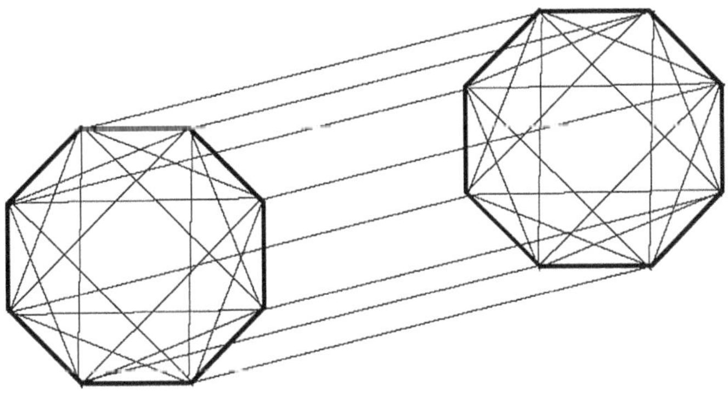

The 4 - cross - polytope has faces (Zhizhin, 2019 a)

f_0=8, f_1=24, f_2=32, f_4=1.

In this case the product $P^4 \times Q^1$ in according with (1) has faces:

the number of faces with dimension zero is

$$f_x^0 = f_P^0 \cdot f_Q^0 = 8 \cdot 2 = 16;$$

the number of faces with dimension one is

$$f_x^1 = f_P^1 \cdot f_Q^0 + f_P^0 \cdot f_Q^1 = 24 \cdot 2 + 8 \cdot 1 = 56;$$

the number of faces with dimension two is

$$f_x^2 = f_P^2 \cdot f_Q^0 + f_P^1 \cdot f_Q^1 = 32 \cdot 2 + 24 \cdot 1 = 88; \quad (8)$$

the number of faces with dimension three is

$$f_x^3 = f_P^3 \cdot f_Q^0 + f_P^2 f_Q^1 = 16 \cdot 2 + 32 \cdot 1 = 64;$$

the number of faces with dimension four is

$$f_x^4 = f_P^4 \cdot f_Q^0 + f_P^3 \cdot f_Q^1 = 1 \cdot 2 + 16 \cdot 1 = 18;$$

the number of faces with dimension five is

$$f_x^5 = f_P^4 \cdot f_Q^1 = 1 \cdot 1 = 1.$$

Thus, the product of the 4 - cross - polytope by one - dimension segment is polytopic prismahedron with dimension 5 and faces dimension of equality (8). Can to multiply the 4 - cross - polytope on a triangle (polytope Q^2). So, the product P^4 and Q^2 according to (1) has the faces:

the number of faces with dimension zero is

$$f_x^0 = f_P^0 \cdot f_Q^0 = 8 \cdot 3 = 24;$$

the number of faces with dimension one is

$$f_x^1 = f_P^1 \cdot f_Q^0 + f_P^0 \cdot f_Q^1 = 24 \cdot 3 + 8 \cdot 3 = 96;$$

the number of faces with dimension two is

$$f_x^2 = f_P^2 \cdot f_Q^0 + f_P^1 \cdot f_Q^1 + f_P^0 \cdot f_Q^2 = 32 \cdot 3 + 24 \cdot 3 + 8 \cdot 1 = 176; \quad (9)$$

the number of faces with dimension three is

$$f_x^3 = f_P^3 \cdot f_Q^0 + f_P^2 f_Q^1 + f_P^1 \cdot f_Q^2 = 16 \cdot 3 + 32 \cdot 3 + 24 \cdot 1 = 168;$$

the number of faces with dimension four is

$$f_x^4 = f_P^4 \cdot f_Q^0 + f_P^3 \cdot f_Q^1 + f_P^2 \cdot f_Q^2 = 1 \cdot 3 + 16 \cdot 3 + 32 \cdot 1 = 83;$$

the number of faces with dimension five is

$$f_x^5 = f_P^4 \cdot f_Q^1 + f_P^3 f_Q^2 = 1 \cdot 3 + 16 \cdot 1 = 19;$$

the number of faces with dimension six is

$$f_x^6 = f_P^4 \cdot f_Q^2 = 1 \cdot 1 = 1.$$

Thus, the product of the 4 - cross - polytope by a triangle is polytopic prismahedron with dimension 6 and faces dimension of equality (9). Next on a dimension cluster of γ - brass is 7 - cross - polytope. It has the faces (Zhizhin, 2019a)

$$f_0=14, f_1=84, f_2=280, f_3=560, f_4=672, f_5=448, f_6=128, f_7=1.$$

The product of 7 - cross - polytope by a one - dimension segment $P^7 \times Q^1$ has faces:

the number of faces with dimension zero is

$$f_x^0 = f_P^0 \cdot f_Q^0 = 14 \cdot 2 = 28;$$

the number of faces with dimension one is

$$f_x^1 = f_P^1 \cdot f_Q^0 + f_P^0 \cdot f_Q^1 = 84 \cdot 2 + 14 \cdot 1 = 182;$$

the number of faces with dimension two is

$$f_x^2 = f_P^2 \cdot f_Q^0 + f_P^1 \cdot f_Q^1 = 280 \cdot 2 + 84 \cdot 1 = 644; \tag{10}$$

the number of faces with dimension three is

$$f_x^3 = f_P^3 \cdot f_Q^0 + f_P^2 f_Q^1 = 560 \cdot 2 + 280 \cdot 1 = 1400;$$

the number of faces with dimension four is

$$f_x^4 = f_P^4 \cdot f_Q^0 + f_P^3 \cdot f_Q^1 = 672 \cdot 2 + 560 \cdot 1 = 1904;$$

the number of faces with dimension five is

$$f_x^5 = f_P^5 \cdot f_Q^0 + f_P^4 f_Q^1 = 448 \cdot 2 + 672 \cdot 1 = 1568;$$

the number of faces with dimension six is

$$f_x^6 = f_P^6 \cdot f_Q^0 + f_P^5 \cdot f_Q^1 = 128 \cdot 2 + 448 \cdot 1 = 704;$$

the number of faces with dimension seven is

$$f_x^7 = f_P^7 \cdot f_Q^0 + f_P^6 \cdot f_Q^1 = 1 \cdot 2 + 128 \cdot 1 = 130;$$

the number of faces with dimension eight is

$$f_x^8 = f_P^7 \cdot f_Q^1 = 1 \cdot 1 = 1.$$

Thus, the product of the cluster γ - brass in as 7 - cross - polytope by one - dimension segment is polytopic prismahedron with dimension 8 and faces defined of equality (10). Can to multiply the 7 - cross - polytope on a triangle (polytope Q^2). So, the product P^7 and Q^2 according to (1) has the faces:

the number of faces with dimension zero is

$$f_x^0 = f_P^0 \cdot f_Q^0 = 14 \cdot 3 = 42;$$

the number of faces with dimension one is

$$f_x^1 = f_P^1 \cdot f_Q^0 + f_P^0 \cdot f_Q^1 = 84 \cdot 3 + 14 \cdot 3 = 294;$$

the number of faces with dimension two is

$$f_x^2 = f_P^2 \cdot f_Q^0 + f_P^1 \cdot f_Q^1 + f_P^0 \cdot f_Q^2 = 280 \cdot 3 + 84 \cdot 3 + 14 \cdot 1 = 1106; \tag{11}$$

the number of faces with dimension three is

$$f_x^3 = f_P^3 \cdot f_Q^0 + f_P^2 f_Q^1 + f_P^1 \cdot f_Q^2 + f_P^0 \cdot f_Q^3 = 560 \cdot 3 + 280 \cdot 3 + 84 \cdot 1 = 2604;$$

the number of faces with dimension four is

$$f_x^4 = f_P^4 \cdot f_Q^0 + f_P^3 \cdot f_Q^1 + f_P^2 \cdot f_Q^2 + f_P^1 \cdot f_Q^3 = 672 \cdot 3 + 560 \cdot 3 + 280 \cdot 1 = 3976;$$

the number of faces with dimension five is

$$f_x^5 = f_P^5 \cdot f_Q^0 + f_P^4 \cdot f_Q^1 + f_P^3 \cdot f_Q^2 = 448 \cdot 3 + 672 \cdot 3 + 560 \cdot 1 = 3920;$$

the number of faces with dimension six is

$$f_x^6 = f_P^6 \cdot f_Q^0 + f_P^5 \cdot f_Q^1 + f_P^4 \cdot f_Q^2 = 128 \cdot 3 + 448 \cdot 3 + 672 \cdot 1 = 2400;$$

the number of faces with dimension seven is

$$f_x^7 = f_P^7 \cdot f_Q^0 + f_P^6 \cdot f_Q^1 + f_P^5 \cdot f_Q^2 = 1 \cdot 3 + 128 \cdot 3 + 448 \cdot 1 = 835;$$

the number of faces with dimension eight is

$$f_x^8 = f_P^7 \cdot f_Q^1 + f_P^6 \cdot f_Q^2 = 1 \cdot 3 + 128 \cdot 1 = 131;$$

the number of faces with dimension nine is

$$f_x^9 = f_P^7 \cdot f_Q^2 = 1 \cdot 1 = 1.$$

Thus, the product of the cluster of γ - brass as 7 - cross - polytope by a triangle is polytopic prismahedron with dimension 9 and faces defined of equality (11).

Obviously, the multiplication of clusters of γ - brass in the form of cross - polytopes of even greater dimension (with an increase in the number of shells) by one - dimensional segments and triangles will also lead to the formation of polytopic prismahedron with faces determined by relations (1). An increase in the number of vertices in the polytope, which plays the role of the second factor in the product, will lead to an increase in the number of clusters of γ - brass in the resulting nanomaterial.

In the Chapter 2 shows that known clusters can approximately describe n - icosahedral clusters containing an atom in a common center. Such clusters can also be used to obtain nanomaterials, multiplying them by convex polytopes of different dimensions. As shown in Chapter 2, such clusters with an arbitrary number of shells n have faces

$$f_0 = 12n + 1, f_1 = 42n, f_2 = 50n, f_3 = 21n, f_4 = C^2_{n+1}, f_5 = C^3_{n+1}, \ldots, f_{n+2} = C^n_{n+1}.$$

The dimension of this clusters equal $d = n + 3$. Multiply this cluster P^{n+3} for arbitrary n by a one - dimensional segment Q^1. Then, according to (1), the result of the work is a polytope with faces:

the number of faces with dimension zero is

$$f_x^0 = f_P^0 \cdot f_Q^0 = (12n + 1) \cdot 2;$$

the number of faces with dimension one is

$$f_x^1 = f_P^1 \cdot f_Q^0 + f_P^0 \cdot f_Q^1 = 42n \cdot 2 + (12n + 1) \cdot 1 = 85n + 1;$$

the number of faces with dimension two is

$$f_x^2 = f_P^2 \cdot f_Q^0 + f_P^1 \cdot f_Q^1 = 50n \cdot 2 + 42n \cdot 1 = 142n; \quad (12)$$

the number of faces with dimension three is

$$f_x^3 = f_P^3 \cdot f_Q^0 + f_P^2 f_Q^1 = 21n \cdot 2 + 50n \cdot 1 = 92n;$$

the number of faces with dimension four is

$$f_x^4 = f_P^4 \cdot f_Q^0 + f_P^3 \cdot f_Q^1 = 2C^2_{n+1} + 21n;$$

the number of faces with dimension five is

$$f_x^5 = f_P^5 \cdot f_Q^0 + f_P^4 \cdot f_Q^1 = 2C_{n+1}^3 + C_{n+1}^2;$$

the number of faces with dimension $2 + n$ is

$$f_x^{2+n} = f_P^{2+n} \cdot f_Q^0 + f_P^{1+n} \cdot f_Q^1 = C_{n+1}^n \cdot 2 + C_{n+1}^{n-1};$$

the number of faces with dimension $3 + n$ is

$$f_x^{3+n} = f_P^{3+n} \cdot f_Q^0 + f_P^{2+n} \cdot f_Q^1 = 2 + C_{n+1}^n;$$

the number of faces with dimension $4 + n$ is

$$f_x^{4+n} = f_P^{3+n} \cdot f_Q^1 = 1 \cdot 1 = 1.$$

Thus, the product of cluster n - icosahedron with an atom at center by one - dimension segment is a polytopic prismahedron of dimension $n + 4$ with faces defined equality (12).

Multiply a cluster of n - icosahedron with an atom at center by triangle. Then according (1) a result of the product will be a polytope with faces:

the number of faces with dimension zero is

$$f_x^0 = f_P^0 \cdot f_Q^0 = (12n + 1) \cdot 3;$$

the number of faces with dimension one is

$$f_x^1 = f_P^1 \cdot f_Q^0 + f_P^0 \cdot f_Q^1 = 42n \cdot 3 + (12n + 1) \cdot 3 = 162n + 3;$$

the number of faces with dimension two is

$$f_x^2 = f_P^2 \cdot f_Q^0 + f_P^1 \cdot f_Q^1 + f_P^0 \cdot f_Q^2 = 50n \cdot 3 + 42n \cdot 3 + 12n + 1 = 288n + 1; \tag{13}$$

the number of faces with dimension three is

$$f_x^3 = f_P^3 \cdot f_Q^0 + f_P^2 f_Q^1 + f_P^1 \cdot f_Q^2 = 21n \cdot 3 + 50n \cdot 3 + 42n \cdot 1 = 255n;$$

the number of faces with dimension four is

$$f_x^4 = f_P^4 \cdot f_Q^0 + f_P^3 \cdot f_Q^1 + f_P^2 f_Q^2 = C_{n+1}^2 \cdot 3 + 21n \cdot 3 + 50n = 3C_{n+1}^2 + 113n;$$

the number of faces with dimension five is

$$f_x^5 = f_P^5 \cdot f_Q^0 + f_P^4 \cdot f_Q^1 + f_P^3 \cdot f_Q^2 = C_{n+1}^3 \cdot 3 + C_{n+1}^2 \cdot 3 + 21n;$$

the number of faces with dimension $3 + n$ is

$$f_x^{3+n} = f_P^{3+n} \cdot f_Q^0 + f_P^{2+n} \cdot f_Q^1 + f_P^{1+n} \cdot f_Q^2 = 3 + C_{n+1}^n \cdot 3 + C_{n+1}^{n-1};$$

the number of faces with dimension $4 + n$ is

$$f_x^{4+n} = f_P^{4+n} \cdot f_Q^0 + f_P^{3+n} \cdot f_Q^1 + f_P^{2+n} \cdot f_Q^2 = 1 \cdot 3 + 1 \cdot 3 + C_{n+1}^n \cdot 1 = 6 + C_{n+1}^n;$$

the number of faces with dimension $5 + n$ is

$$f_x^{5+n} = f_P^{3+n} \cdot f_Q^2 = 1 \cdot 1 = 1.$$

Thus, the product of cluster n - icosahedron with an atom at center by triangle is a polytopic prismahedron of dimension $n + 5$ with faces defined equality (13).

Obviously, it is possible to obtain nanomaterials with any finite number of clusters from n - icosahedrons if multiply the n - icosahedron with a central atom by a convex polytope with any given number of vertices.

NANOMATERIALS FROM WURTZITE

In Chapter 3 was shown that minimum dimension of the cluster wurtzite is 4 (Zhizhin, 2019b). This cluster has faces

Figure 3. The product of the cluster wurtzite with dimension 4 on a one - dimension segment

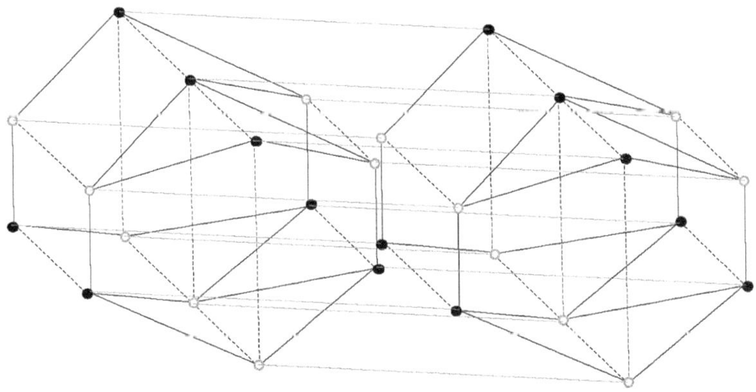

$f_0=14, f_1=29, f_2=21, f_3=6, f_4=1.$

From of the cluster wurtzite with dimension 4 can obtain a polytopic prismahedron with dimension 5 if multiply it on a one - dimension segment. The form of this prismahedron is shown in Figure 3.

In this case the product $P^4 \times Q^1$ in according with (1) has faces:

the number of faces with dimension zero is

$$f_x^0 = f_P^0 \cdot f_Q^0 = 14 \cdot 2 = 28;$$

the number of faces with dimension one is

$$f_x^1 = f_P^1 \cdot f_Q^0 + f_P^0 \cdot f_Q^1 = 29 \cdot 2 + 14 \cdot 1 = 72;$$

the number of faces with dimension two is

$$f_x^2 = f_P^2 \cdot f_Q^0 + f_P^1 \cdot f_Q^1 = 21 \cdot 2 + 29 \cdot 1 = 71; \quad (14)$$

the number of faces with dimension three is

$$f_x^3 = f_P^3 \cdot f_Q^0 + f_P^2 f_Q^1 = 6 \cdot 2 + 21 \cdot 1 = 33;$$

the number of faces with dimension four is

$$f_x^4 = f_P^4 \cdot f_Q^0 + f_P^3 \cdot f_Q^1 = 1 \cdot 2 + 6 \cdot 1 = 8;$$

the number of faces with dimension five is

$$f_x^5 = f_P^4 \cdot f_Q^1 = 1 \cdot 1 = 1.$$

Thus, the product of the wurtzite by one - dimension segment is polytopic prismahedron with dimension 5 and faces dimension of equality (14). Can to multiply the wurtzite on a triangle (polytope Q^2). So, the product P^4 and Q^2 according to (1) has the faces:

the number of faces with dimension zero is

$$f_x^0 = f_P^0 \cdot f_Q^0 = 14 \cdot 3 = 42;$$

the number of faces with dimension one is

$$f_x^1 = f_P^1 \cdot f_Q^0 + f_P^0 \cdot f_Q^1 = 29 \cdot 3 + 14 \cdot 3 = 129;$$

the number of faces with dimension two is

$$f_x^2 = f_P^2 \cdot f_Q^0 + f_P^1 \cdot f_Q^1 + f_P^0 \cdot f_Q^2 = 21 \cdot 3 + 29 \cdot 3 + 14 \cdot 1 = 164; \tag{15}$$

the number of faces with dimension three is

$$f_x^3 = f_P^3 \cdot f_Q^0 + f_P^2 f_Q^1 + f_P^1 \cdot f_Q^2 = 6 \cdot 3 + 21 \cdot 3 + 29 \cdot 1 = 110;$$

the number of faces with dimension four is

$$f_x^4 = f_P^4 \cdot f_Q^0 + f_P^3 \cdot f_Q^1 + f_P^2 \cdot f_Q^2 = 1 \cdot 3 + 6 \cdot 3 + 21 \cdot 1 = 42;$$

the number of faces with dimension five is

$f_x^5 = f_P^4 \cdot f_Q^1 + f_P^3 f_Q^2 = 1 \cdot 3 + 6 \cdot 1 = 9;$

the number of faces with dimension six is

$f_x^6 = f_P^4 \cdot f_Q^2 = 1 \cdot 1 = 1.$

Thus, the product of the wurtzite by triangle is polytopic prismahedron with dimension 6 and faces dimension of equality (15).

NANOMATERIALS FROM FLUORITE

In Chapter 3 was shown that minimum dimension of the cluster fluorite is 4. This cluster has faces

$f_0=14, f_1=36, f_2=30, f_3=8, f_4=1.$

From of the cluster fluorite with dimension 4 can obtain a polytopic prismahedron with dimension 5 if multiply it on a one - dimension segment. The form of this prismahedron is shown in Figure 4.

Figure 4. The product of the cluster fluorite with dimension 4 on a one - dimension segment

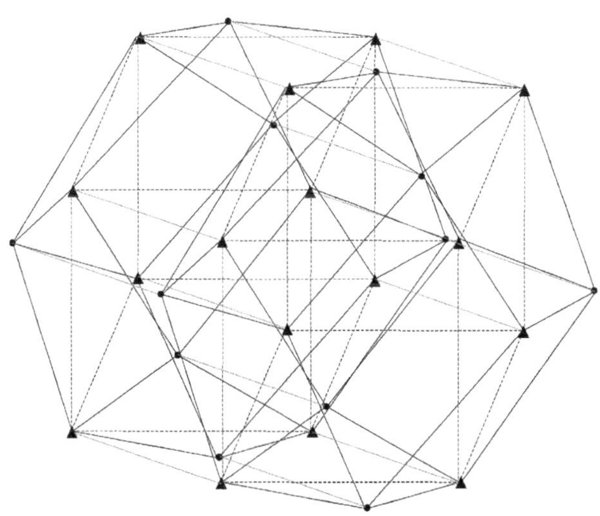

In this case the product $P^4 \times Q^1$ in according with (1) has faces:

the number of faces with dimension zero is

$$f_x^0 = f_P^0 \cdot f_Q^0 = 14 \cdot 2 = 28;$$

the number of faces with dimension one is

$$f_x^1 = f_P^1 \cdot f_Q^0 + f_P^0 \cdot f_Q^1 = 36 \cdot 2 + 14 \cdot 1 = 86;$$

the number of faces with dimension two is

$$f_x^2 = f_P^2 \cdot f_Q^0 + f_P^1 \cdot f_Q^1 = 30 \cdot 2 + 36 \cdot 1 = 96; \qquad (16)$$

the number of faces with dimension three is

$$f_x^3 = f_P^3 \cdot f_Q^0 + f_P^2 f_Q^1 = 8 \cdot 2 + 30 \cdot 1 = 46;$$

the number of faces with dimension four is

$$f_x^4 = f_P^4 \cdot f_Q^0 + f_P^3 \cdot f_Q^1 = 1 \cdot 2 + 8 \cdot 1 = 10;$$

the number of faces with dimension five is

$$f_x^5 = f_P^4 \cdot f_Q^1 = 1 \cdot 1 = 1.$$

Thus, the product of the fluorite by one - dimension segment is polytopic prismahedron with dimension 5 and faces dimension of equality (16). Can to multiply the fluorite on a triangle (polytope Q^2). So, the product P^4 and Q^2 according to (1) has the faces:

the number of faces with dimension zero is

$$f_x^0 = f_P^0 \cdot f_Q^0 = 14 \cdot 3 = 42;$$

the number of faces with dimension one is

$$f_x^1 = f_P^1 \cdot f_Q^0 + f_P^0 \cdot f_Q^1 = 36 \cdot 3 + 14 \cdot 3 = 150;$$

the number of faces with dimension two is

$$f_x^2 = f_P^2 \cdot f_Q^0 + f_P^1 \cdot f_Q^1 + f_P^0 \cdot f_Q^2 = 30 \cdot 3 + 36 \cdot 3 + 14 \cdot 1 = 212; \tag{17}$$

the number of faces with dimension three is

$$f_x^3 = f_P^3 \cdot f_Q^0 + f_P^2 f_Q^1 + f_P^1 \cdot f_Q^2 = 8 \cdot 3 + 30 \cdot 3 + 36 \cdot 1 = 150;$$

the number of faces with dimension four is

$$f_x^4 = f_P^4 \cdot f_Q^0 + f_P^3 \cdot f_Q^1 + f_P^2 \cdot f_Q^2 = 1 \cdot 3 + 8 \cdot 3 + 30 \cdot 1 = 57;$$

the number of faces with dimension five is

$$f_x^5 = f_P^4 \cdot f_Q^1 + f_P^3 f_Q^2 = 1 \cdot 3 + 8 \cdot 1 = 11;$$

the number of faces with dimension six is

$$f_x^6 = f_P^4 \cdot f_Q^2 = 1 \cdot 1 = 1.$$

Thus, the product of the fluorite by triangle is polytopic prismahedron with dimension 6 and faces dimension of equality (17).

NANOMATERIALS FROM CLUSTERS B_6Cl_6

In Chapter 3 was shown that dimension of the cluster B_6Cl_6 is 4. This cluster has faces

f_0=12, f_1=30, f_2=28, f_3=10, f_4=1.

From of the cluster B_6Cl_6 with dimension 4 can obtain a polytopic prismahedron with dimension 5 if multiply it on a one - dimension segment. The form of this prismahedron is shown in Figure 5.

Figure 5. The product of the cluster B_6Cl_6 with dimension 4 on a one - dimension segment

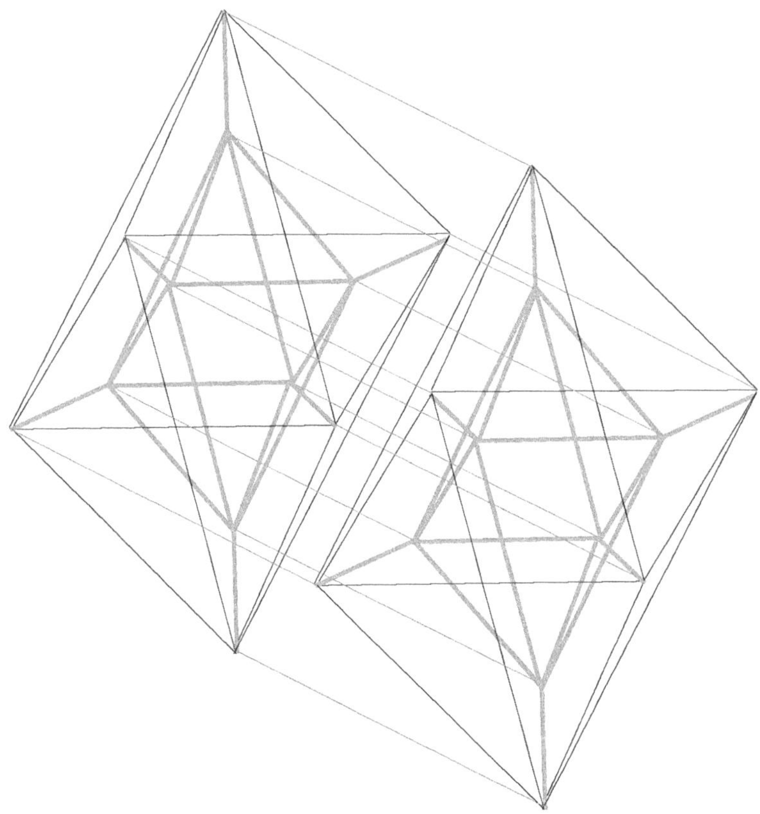

In this case the product $P^4 \times Q^1$ in according with (1) has faces:

the number of faces with dimension zero is

$$f_x^0 = f_P^0 \cdot f_Q^0 = 12 \cdot 2 = 24;$$

the number of faces with dimension one is

$$f_x^1 = f_P^1 \cdot f_Q^0 + f_P^0 \cdot f_Q^1 = 30 \cdot 2 + 12 \cdot 1 = 72;$$

the number of faces with dimension two is

$$f_x^2 = f_P^2 \cdot f_Q^0 + f_P^1 \cdot f_Q^1 = 28 \cdot 2 + 30 \cdot 1 = 86; \tag{18}$$

the number of faces with dimension three is

$$f_x^3 = f_P^3 \cdot f_Q^0 + f_P^2 f_Q^1 = 10 \cdot 2 + 28 \cdot 1 = 48;$$

the number of faces with dimension four is

$$f_x^4 = f_P^4 \cdot f_Q^0 + f_P^3 \cdot f_Q^1 = 1 \cdot 2 + 10 \cdot 1 = 12;$$

the number of faces with dimension five is

$$f_x^5 = f_P^4 \cdot f_Q^1 = 1 \cdot 1 = 1.$$

Thus, the product of the cluster B_6Cl_6 by one - dimension segment is polytopic prismahedron with dimension 5 and faces dimension of equality (18). Can to multiply the fluorite on a triangle (polytope Q^2). So, the product P^4 and Q^2 according to (1) has the faces:

the number of faces with dimension zero is

$$f_x^0 = f_P^0 \cdot f_Q^0 = 12 \cdot 3 = 36;$$

the number of faces with dimension one is

$$f_x^1 = f_P^1 \cdot f_Q^0 + f_P^0 \cdot f_Q^1 = 30 \cdot 3 + 12 \cdot 3 = 126;$$

the number of faces with dimension two is

$$f_x^2 = f_P^2 \cdot f_Q^0 + f_P^1 \cdot f_Q^1 + f_P^0 \cdot f_Q^2 = 28 \cdot 3 + 30 \cdot 3 + 12 \cdot 1 = 186; \tag{19}$$

the number of faces with dimension three is

$$f_x^3 = f_P^3 \cdot f_Q^0 + f_P^2 f_Q^1 + f_P^1 \cdot f_Q^2 = 10 \cdot 3 + 28 \cdot 3 + 30 \cdot 1 = 144;$$

the number of faces with dimension four is

$$f_x^4 = f_P^4 \cdot f_Q^0 + f_P^3 \cdot f_Q^1 + f_P^2 \cdot f_Q^2 = 1 \cdot 3 + 10 \cdot 3 + 28 \cdot 1 = 61;$$

the number of faces with dimension five is

$$f_x^5 = f_P^4 \cdot f_Q^1 + f_P^3 f_Q^2 = 1 \cdot 3 + 10 \cdot 1 = 13;$$

the number of faces with dimension six is

$$f_x^6 = f_P^4 \cdot f_Q^2 = 1 \cdot 1 = 1.$$

Thus, the product of the cluster B_6Cl_6 by triangle is polytopic prismahedron with dimension 6 and faces dimension of equality (19).

NANOMATERIALS FROM LINEAR HOMO - ELEMENT METAL CHAINS WITH LIGANDS

In Chapter 4 was shown that dimension of the cluster linear homo - element metal chains with ligands (Figure 6 in Chapter 4) is 4. This cluster has faces

f_0=10, f_1=21, f_2=16, f_3=5, f_4=1.

From this cluster with dimension 4 can obtain a polytopic prismahedron with dimension 5 if multiply it on a one - dimension segment.

In this case the product $P^4 \times Q^1$ in according with (1) has faces:

the number of faces with dimension zero is

$$f_x^0 = f_P^0 \cdot f_Q^0 = 10 \cdot 2 = 20;$$

the number of faces with dimension one is

$$f_x^1 = f_P^1 \cdot f_Q^0 + f_P^0 \cdot f_Q^1 = 21 \cdot 2 + 10 \cdot 1 = 52;$$

the number of faces with dimension two is

$$f_x^2 = f_P^2 \cdot f_Q^0 + f_P^1 \cdot f_Q^1 = 16 \cdot 2 + 21 \cdot 1 = 53; \tag{20}$$

the number of faces with dimension three is

$$f_x^3 = f_P^3 \cdot f_Q^0 + f_P^2 f_Q^1 = 5 \cdot 2 + 16 \cdot 1 = 26;$$

the number of faces with dimension four is

$$f_x^4 = f_P^4 \cdot f_Q^0 + f_P^3 \cdot f_Q^1 = 1 \cdot 2 + 5 \cdot 1 = 7;$$

the number of faces with dimension five is

$$f_x^5 = f_P^4 \cdot f_Q^1 = 1 \cdot 1 = 1.$$

Thus, the product of the cluster linear homo - element metal chains with ligands by one - dimension segment is polytopic prismahedron with dimension 5 and faces dimension of equality (20). Can to multiply the cluster linear homo - element metal chains with ligands on a triangle (polytope Q^2). So, the product P^4 and Q^2 according to (1) has the faces:

the number of faces with dimension zero is

$$f_x^0 = f_P^0 \cdot f_Q^0 = 10 \cdot 3 = 30;$$

the number of faces with dimension one is

$$f_x^1 = f_P^1 \cdot f_Q^0 + f_P^0 \cdot f_Q^1 = 21 \cdot 3 + 10 \cdot 3 = 93;$$

the number of faces with dimension two is

$$f_x^2 = f_P^2 \cdot f_Q^0 + f_P^1 \cdot f_Q^1 + f_P^0 \cdot f_Q^2 = 16 \cdot 3 + 21 \cdot 3 + 10 \cdot 1 = 121; \tag{21}$$

the number of faces with dimension three is

$$f_x^3 = f_P^3 \cdot f_Q^0 + f_P^2 f_Q^1 + f_P^1 \cdot f_Q^2 = 5 \cdot 3 + 16 \cdot 3 + 21 \cdot 1 = 84;$$

the number of faces with dimension four is

$$f_x^4 = f_P^4 \cdot f_Q^0 + f_P^3 \cdot f_Q^1 + f_P^2 \cdot f_Q^2 = 1 \cdot 3 + 5 \cdot 3 + 16 \cdot 1 = 34;$$

the number of faces with dimension five is

$$f_x^5 = f_P^4 \cdot f_Q^1 + f_P^3 f_Q^2 = 1 \cdot 3 + 5 \cdot 1 = 8;$$

the number of faces with dimension six is

$$f_x^6 = f_P^4 \cdot f_Q^2 = 1 \cdot 1 = 1.$$

Thus, the product of the cluster linear homo - element metal chains with ligands by triangle is polytopic prismahedron with dimension 6 and faces dimension of equality (21).

NANOMATERIALS FROM METALLIC CLUSTERS WITH LIGANDS AND POLYHEDRAL CORE

In Chapter 6 was shown that dimension of the metallic cluster with ligands and polyhedral core $Ir_4(CO)_{12}$ (Figure 2 in Chapter 6) is 4. This cluster has faces

f_0=16, f_1=42, f_2=42, f_3=16, f_4=1.

From this cluster with dimension 4 can obtain a polytopic prismahedron with dimension 5 if multiply it on a one - dimension segment.

In this case the product $P^4 \times Q^1$ in according with (1) has faces:

the number of faces with dimension zero is

$$f_x^0 = f_P^0 \cdot f_Q^0 = 16 \cdot 2 = 32;$$

the number of faces with dimension one is

$$f_x^1 = f_P^1 \cdot f_Q^0 + f_P^0 \cdot f_Q^1 = 42 \cdot 2 + 16 \cdot 1 = 100;$$

the number of faces with dimension two is

$$f_x^2 = f_P^2 \cdot f_Q^0 + f_P^1 \cdot f_Q^1 = 42 \cdot 2 + 42 \cdot 1 = 126; \tag{22}$$

the number of faces with dimension three is

$$f_x^3 = f_P^3 \cdot f_Q^0 + f_P^2 f_Q^1 = 16 \cdot 2 + 42 \cdot 1 = 74;$$

the number of faces with dimension four is

$$f_x^4 = f_P^4 \cdot f_Q^0 + f_P^3 \cdot f_Q^1 = 1 \cdot 2 + 16 \cdot 1 = 18;$$

the number of faces with dimension five is

$$f_x^5 = f_P^4 \cdot f_Q^1 = 1 \cdot 1 = 1.$$

Thus, the product of the metallic cluster with ligands and polyhedral core $Ir_4(CO)_{12}$ by one - dimension segment is polytopic prismahedron with dimension 5 and faces dimension of equality (22). Can to multiply the cluster linear homo - element metal chains with ligands on a triangle (polytope Q^2). So the product P^4 and Q^2 according to (1) has the faces:

the number of faces with dimension zero is

$$f_x^0 = f_P^0 \cdot f_Q^0 = 16 \cdot 3 = 48;$$

the number of faces with dimension one is

$$f_x^1 = f_P^1 \cdot f_Q^0 + f_P^0 \cdot f_Q^1 = 42 \cdot 3 + 16 \cdot 3 = 174;$$

the number of faces with dimension two is

$$f_x^2 = f_P^2 \cdot f_Q^0 + f_P^1 \cdot f_Q^1 + f_P^0 \cdot f_Q^2 = 42 \cdot 3 + 42 \cdot 3 + 16 \cdot 1 = 268; \tag{23}$$

the number of faces with dimension three is

$$f_x^3 = f_P^3 \cdot f_Q^0 + f_P^2 f_Q^1 + f_P^1 \cdot f_Q^2 = 16 \cdot 3 + 42 \cdot 3 + 42 \cdot 1 = 216;$$

the number of faces with dimension four is

$$f_x^4 = f_P^4 \cdot f_Q^0 + f_P^3 \cdot f_Q^1 + f_P^2 \cdot f_Q^2 = 1 \cdot 3 + 16 \cdot 3 + 42 \cdot 1 = 93;$$

the number of faces with dimension five is

$$f_x^5 = f_P^4 \cdot f_Q^1 + f_P^3 f_Q^2 = 1 \cdot 3 + 16 \cdot 1 = 19;$$

the number of faces with dimension six is

$$f_x^6 = f_P^4 \cdot f_Q^2 = 1 \cdot 1 = 1.$$

Thus, the product of the metallic cluster with ligands and polyhedral core $Ir_4(CO)_{12}$ by triangle is polytopic prismahedron with dimension 6 and faces defined of the equality (23).

CONCLUSION

Using concrete examples of clusters of chemical compounds of various types (intermetallic clusters, metal chains with ligands, polyhedral metal clusters with ligands), it is shown how nanomaterials are formed from individual clusters by multiplying their geometric structure by other geometric elements of different dimensions. Moreover, each cluster is geometrically a convex polytope of higher dimension. Formulas are obtained that make it possible for each cluster to determine the number of elements of different dimensions in a nanomaterial of finite size. This determines the structure of the nanomaterial. Each geometric element can be considered as a product of geometric elements of certain dimensions. Therefore, from the work of two polytopes, revealing each polytope as the product of certain geometric elements, we can proceed to the work of several polytopes (more than two). By consecutively revealing each factor by Theorem 1, it is possible to obtain formulas for calculating the numbers of elements of different dimensions, which are sums of the product of the numbers of elements of different dimensions with the number of factors more than two. This is of importance when building the structure of nanomaterials, which can be quite complex and include elements with different metric characteristics. For example, the lengths of the one - dimensional segments by which the polytopes are multiplied at each step may differ from the lengths of the one - dimensional segments in the previous step. Moreover, Theorem 1 can be used in all cases of constructing nanomaterials of any complexity.

It should be emphasized that three methods can be specified for filling the space of higher dimension with polytopes of higher dimension with the formation of a continuously distributed body. The first way is hierarchical filling of space (Zhizhin, 2010, 2012, 2014). It consists in filling the space with an expanding body. At each step of expansion, the body remains similar to itself. Such a large - scale increase in bodies is characteristic of scaling processes (Kadanoff, 1966; Wilson, 1971a, b). The second method is translational filling of space with a body of higher dimension (Zhizhin, 2015). In these two methods, the body dimension at each step of its change remains unchanged. The second method is devoted to the second method of filling the space associated with the formation of nanomaterials. In the same chapter, the third way was considered to fill the space with a cluster with the formation at each step of its increase, in essence, a new cluster of greater and greater dimension. This cluster, in general, is not similar to the original cluster. It is the result of multiplying it at each step by various geometric elements.

REFERENCES

Delone, B. (1961). Proof of the main theorem of the theory of stereohedrons. *Reports of the Academy of Sciences of the USSR*, *138*(6), 1270–1272.

Delone, B., & Sandakova, N. (1961). Theory of stereohedrons. *Proceedings of the Mathematical Institute. V.A. Steklov*, *64*, 28 - 51.

Kadanoff, L. P. (1966). Scaling Laws for Ising Models Near τ_c*. *Physics*, *2*(6), 263–272. doi:10.1103/PhysicsPhysiqueFizika.2.263

Pontryagin, L. S. (1976). *The foundations of combinatorial topology*. Moscow: Science.

Robertson, S. A. (1984). *Polytopes and Symmetry*. Cambridge: Cambridge University Press.

Wilson, R. G. (1971b). Renornalization Group and Critical Phenomena. II. Phase - space Cell Analysis of Critical Behavior. *Phys. Rev. B.*, *4*(9), 3184–3205. doi:10.1103/PhysRevB.4.3184

Wilson, R. G.Renormalization Group and the Kadanoff Scaling Picture. (1971a). Renornalization Group and Critical Phenomena. I. Renormalization Group and the Kadanoff Scaling Picture. *Phys. Rev. B.*, *4*(9), 3174–3183. doi:10.1103/PhysRevB.4.3174

Zhizhin, G. V. (2010). *Geometrical bases of the dissipative structures*. St. Petersburg: Polytechnika.

Zhizhin, G. V. (2012, October). *Hierarchical filling of spaces with polytopes.* Paper presented at "St. Petersburg Scientific Forum: Science and Human Progress". 7th St. - Petersburg meeting of Nobel Prize laureates, St. Petersburg, Russia.

Zhizhin, G. V. (2014). *World - 4D*. St. Petersburg: Polytechnic Service.

Zhizhin, G. V. (2015, November). *Polytopic prismahedrons - fundamental regions of the n -dimension nanostructures.* Paper presented at The International conference "Nanoscience in Chemistry, Physics, Biology and Mathematics", Cluj-Napoca, Romania.

Zhizhin, G. V. (2018). *Chemical Compounds Structures and the Higher Dimension of Molecules: Emerging Research and Opportunities*. Hershey, PA: IGI Global. doi:10.4018/978-1-5225-4108-0

Zhizhin, G. V. (2019a). *The Geometry of Higher - Dimensional Polytopes*. Hershey, PA: IGI Global. doi:10.4018/978-1-5225-6968-8

Zhizhin, G. V. (2019b). *Attractors and Higher Dimensions in Population and Molecular Biology: Emerging Research and Opportunities*. Hershey, PA: IGI Global. doi:10.4018/978-1-5225-9651-6

Ziegler, G. M. (1995). *Lectures on Polytopes*. Springer. doi:10.1007/978-1-4613-8431-1

KEY TERMS AND DEFINITIONS

Dimension of the Space: The member of independent parameters needed to describe the change in position of an object in space.

Nanocluster: A nanometric set of connected atoms, stable either in isolation state or in building unit of condensed matter.

Polytope: Polyhedron in the space of higher dimension.

Polytopic Prismahedron: Is the product of a polytope to another polytope, or in a particular case to a one-dimensional segment.

Chapter 9
Nanostructures as Tillings of Higher Dimension Spaces

ABSTRACT

It is proved that clusters in the form of the polytopic prismahedrons have the necessary properties for partitioning the n-dimensional spaces of a face into a face, that is, they satisfy the conditions for solving the eighteenth Hilbert problem of the construction of n-dimensional spaces from congruent figures. Moreover, they create extended nanomaterials, in principle, of any size. General principles and an analytical method for constructing n-dimensional spaces with the help of polytopic prismahedrons are developed. On the example of specific types of the polytopic prismahedrons (tetrahedral prism, triangular prismahedron), the possibility of such constructions is analytically proved. It was found that neighboring polytopic prismahedrons in these constructions can have common geometric elements of any dimension less than n or do not have common elements.

INTRODUCTION

In Chapter 1, when studying intermetallic nanostructures, it was found that the space of intermetallic nanostructures has the highest dimension and the fundamental region of these nanostructures is the golden hyper-rombohedron whose dimension is greater than 3 (Zhizhin, 2014c; Zhizhin, 2015; Zhizhin & Diudea, 2016; Zhizhin, Khalaj, & Diudea, 2016). The golden hyper-rombohedron of intermetallic compounds forms a partition of higher -

DOI: 10.4018/978-1-7998-3784-8.ch009

dimensional space. It is a type of nanocluster. Other types of intermetallic clusters were considered in Chapter 2. Various types of metal clusters with ligands were discussed in Chapters 3 - 6. All these clusters have the highest dimension. However, not all types of clusters of chemical compounds can serve as a fundamental area of partitions of spaces of higher dimension, i.e. fill the space of the highest dimension by broadcasting, adjoining each other along whole faces. Only polytopes of the highest dimension, which are polytopic prismahedrons, possess such properties (Zhizhin, 2019). In particular, the golden hyper-rombohedron is a polytopic prismahedron. In the previous chapter, the process of formation of a polytopic prismahedron from any cluster of chemical compounds was considered. To do this, you need to multiply the original cluster as a convex closed polytope by a geometric element with a dimension other than zero. Continuing to multiply the cluster by various geometric elements, you can get clusters of larger size and so fill the space. However, at each step of increasing the size of the cluster, its dimension also increases. This is not a partition of space. At each step, multiplication changes the appearance of the cluster. This chapter discusses the partitioning of higher - dimensional spaces by polytopic prismahedrons, which can be obtained from clusters of chemical compounds, with the formation of nanomaterials.

The problem of completing space by polyhedrons is one of the fundamental problems of mathematics, which has long attracted the attention of scientists. In 1900, D. Hilbert formulated 23 mathematical problems that require solution (Hilbert, 1901). One of these problems (eighteenth) was devoted to this question. It was formulated as follows: "Construction of space from congruent polyhedrons". This problem is especially complicated in the case of n -dimensional spaces (Delone, 1969), and up to the end it has not been solved to this day under these conditions. In 1961, Delone proved that if we require that polytopes in n - dimensional space be adjacent along entire (n - 1) faces, then for any n there are only finitely many topologically different partitions of space into polytopes. These partitions are called normal. Delone and Sandakova (1961) obtained a finite algorithm, in principle (according to the authors), which allows one to find all such partitions for a given n. In this case, the polytopes themselves (stereohedrons by the Delone terminus) of these partitions for a fixed n can be only of a finite number of topologically different types. However, unfortunately, the authors did not obtain concrete examples of normal partitions of higher - dimensional space into polytopes and the specific types of corresponding polytopes. The question of the existence of fundamental domains in n - dimensional space is quite nontrivial. The main definition of the fundamental domain it was formulated by Delone (1969). The fundamental domain of a group of motions is a set of points of space such that, firstly, all its points are not equivalent to each other with respect to the group of motions, and secondly, any point of space is equivalent to some point of this region relative to the group of motions. What kind of these fundamental areas for a space with a dimension greater than three remains is uncertain to this day. Here we must also keep in mind the existence of polytopes, whose congruential copies can fill a gapless space, but each of them is not a fundamental area (Reingard, 1928).

The partitions of higher dimensional space first it was given by Zhizhin (2018). He showed that polytopes of partitions of higher - dimensional space are the polytopic prismahedrons. Moreover, the normality condition in these partitions is not obligatory, i.e. in the n - dimensional space there are polytopes adjacent along entire (n - 1) faces, but there are polytopes adjacent along the faces of another dimension.

One begin our consideration of the problem of partitioning a space of higher dimension from the reduction of concrete examples of these partitions.

EXAMPLES OF PARTITIONS OF HIGHER DIMENSIONAL SPACE

Partition of Space by the Polytopic Prism

One of the simplest types of the polytopic prismahedrons is the tetrahedral prism, i.e. a product tetrahedron by the segment. Tetrahedral coordination is widespread in clusters of chemical compounds. Its structural formula has form (Zhizhin, 2019)

$$P_4^3(4F_3^2) \times P_2^1 = P_8^4[4P_6^3(3F_4^2, 2F_3^2), 2F_4^3(4F_3^2)]. \tag{1}$$

Here, the subscript in the polytope P and his face F indicates the number of vertices, and the superscript indicates the dimension of the corresponding polytope or faces. The right side of (1) describing the structural formula of the product, the facet indicated by the symbol of the polytope to specify which polytopes of dimension n - 1 is composed work polytopes. Thus, P_4^3

- the tetrahedron, P_2^1 - the segment, P_6^3 - the triangular prism, P_3^2 - the triangle, P_4^2 - the quadrilateral, P_8^4 - the tetrahedral prism. The dimension of the tetrahedral prism is equal to 4, it has 8 vertices, 16 edges, 14 faces two - dimensional, 6 three - dimensional faces (2 tetrahedrons, 4 triangular prisms). Image tetrahedral prism is shown in Figure 1 in Chapter 4.

Figure 1. The tetrahedral prism

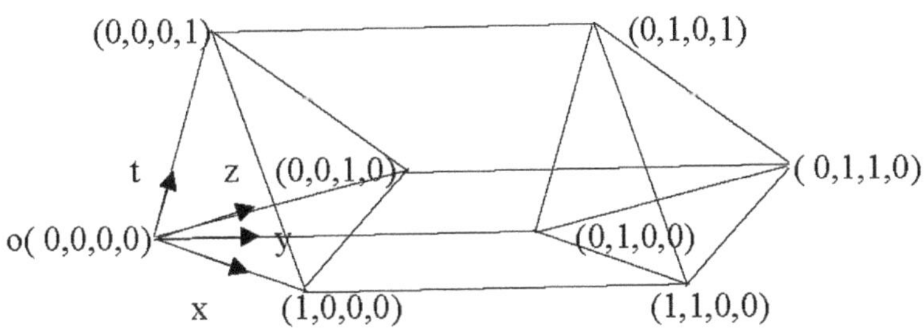

Can to introduce one of the vertices of the tetrahedral prism origin of the four - dimensional space (*x, y, z, t*). Orient the coordinates, such as indicated in Figure 1.

Assume that the length of each edge is equal to 1. Then, each node tetrahedral prism can be associated with a set of integers (Figure 1). Translating tetrahedral prism along the coordinates *x, y, z, t,* we obtain the lattice vertices. Let A_0 tetrahedral prism with the values of vertex coordinates in Figure 1. Then

$$A_0 = [(0,0,0,0),(1,0,0,0),(0,1,0,0),(0,0,1,0),(0,0,0,1),(1,1,0,0),(0,1,1,0),(0,1,0,1)], \quad (2)$$

$$A_1 = A_0(x+1) = [(1,0,0,0),(2,0,0,0),(1,1,0,0),(1,0,1,0),(1,0,0,1),(2,1,0,0),(1,1,1,0),(1,1,0,1)],$$

$$A_2 = A_0(z+1) = [(0,0,1,0),(1,0,1,0),(0,1,1,0),(0,0,2,0),(0,0,1,1),(1,1,1,0),(0,1,2,0),(0,1,1,1)],$$

$$A_3 = A_0(x+1,z+1) = A_1(z+1) =$$
$$[(1,0,1,0),(2,0,1,0),(1,1,1,0),(1,0,2,0),(1,0,1,1,),(2,1,1,0),(1,1,2,0),(1,1,1,1)],$$

$$A_4 = A_0(y+1,z+1) = A_2(y+1) =$$
$$[(0,1,1,0),(1,1,1,0),(0,2,1,0),(0,1,2,0),(0,1,1,1),(1,2,1,0),(0,2,2,0),(0,2,1,1)],$$

$$A_5 = A_0(x+1,y+1,z+1) = A_3(y+1) =$$
$$[(1,1,1,0),(2,1,1,0),(1,2,1,0),(1,1,2,0),(1,1,1,1),(2,2,1,0),(1,2,2,0),(1,2,1,1)],$$

$$A_6 = A_0(y+1) = [(0,1,0,0),(1,1,0,0),(0,2,0,0),(0,1,1,0),(0,1,0,1),(1,2,0,0),(0,2,1,0),(0,2,0,1)],$$

$$A_7 = A_0(y+1,x+1) = A_1(y+1) =$$
$$[(1,1,0,0),(2,1,0,0),(1,2,0,0),(1,1,1,0),(1,1,0,1),(2,2,0,0),(1,2,1,0),(1,2,0,1)],$$

$$A_8 = A_0(t+1) =$$
$$[(0,0,0,1),(1,0,0,1),(0,1,0,1),(0,0,1,1),(0,0,0,2),(1,1,0,1),(0,1,1,1),(0,1,0,2)],$$

$$A_9 = A_0(y+1,t+1) = A_6(t+1) =$$
$$[(0,1,0,1),(1,1,0,1),(0,2,0,1),(0,1,1,1),(0,1,0,2),(1,2,0,1),(0,2,1,1),(0,2,0,2)],$$

$$A_{10} = A_0(x+1,y+1,t+1) = A_7(t+1) =$$
$$[(1,1,0,1),(2,1,0,1),(1,2,0,1),(1,1,1,1),(1,1,0,2),(2,2,0,1),(1,2,1,1),(1,2,0,2)],$$

$$A_{11} = A_0(x+1,t+1) = A_1(t+1) =$$
$$[(1,0,0,1),(2,0,0,1),(1,1,0,1),(1,0,1,1),(1,0,0,2),(2,1,0,1),(1,1,1,1),(1,1,0,2)],$$

$$A_{12} = A_0(y+1,z+1) = A_2(t+1) =$$
$$[(0,0,1,1),(1,0,1,1),(0,1,1,1),(0,0,2,1),(0,0,1,2),(1,1,1,1),(0,1,2,1),(0,1,1,2)],$$

$$A_{13} = A_0(z+1,y+1,t+1) = A_4(t+1) =$$
$$[(0,1,1,1),(1,1,1,1),(0,2,1,1),(0,1,2,1),(0,2,2,1),(0,1,1,2),(1,2,1,1),(0,2,1,2)],$$

$$A_{14} = A_0(x+1,y+1,z+1,t+1) = A_5(t+1) =$$
$$[(1,1,1,1),(2,1,1,1),(1,2,1,1),(1,1,2,1),(1,1,1,2),(2,2,1,1),(1,2,2,1),(1,2,1,2)],$$

$A_{15} = A_0(x+1, t+1, z+1) = A_3(t+1) =$
$[(1,0,1,1),(2,0,1,1),(1,1,1,1),(1,0,2,1),(1,0,1,2),(2,1,1,1),(1,1,2,1),(1,1,1,2)]$.

Representing the tetrahedral prisms $A_0 \div A_{15}$ dots in three - dimensional space,

Figure 2. The hypercube from 16 tetrahedral prisms

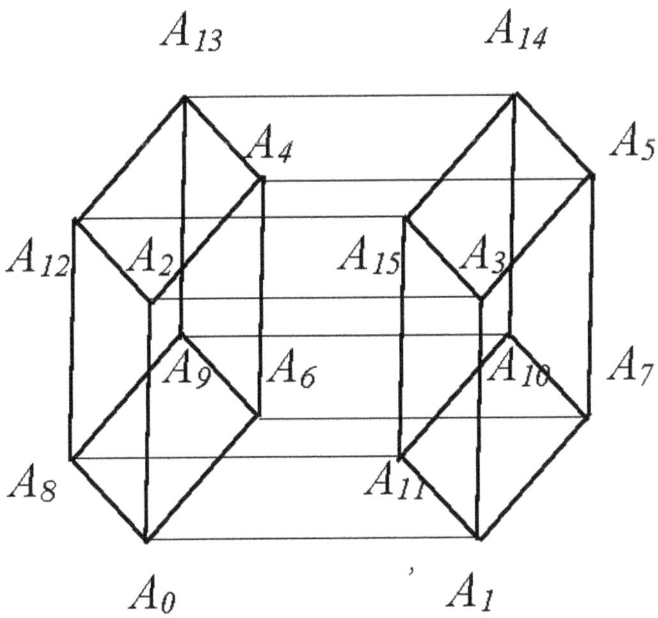

can to get the hypercube. Moreover, the edges of the hypercube correspond to possible changes in the values of one of the coordinates of the vertices of the tetrahedral prism unit (Figure 2).

In addition, each edge of the hypercube in Figure 2 can be considered as an element of the overall two tetrahedral prisms, connected by an edge. Using the coordinate expression tetrahedral prisms (2) can be analytically determined. In Table 1 geometry elements are common to each pair of the tetrahedral prisms connected by an edge in Figure 2 are listed.

Diagonals flat faces in the hypercube correspond to a simultaneous change in the unit values of the two coordinates of the vertices of the tetrahedral prisms (Table 2). Diagonals 8 cubes in the hypercube in Figure 2 correspond to a simultaneous change in the unit of some three coordinates. Common elements of the tetrahedral prisms with such change vertex coordinates is either a vertex or the empty set. When you change the same unit coordinates all four common elements in the tetrahedral prisms not.

Between A_i type tetrahedral prisms arranged tetrahedral prism type B_i (see

Figure 3. The partition space by the tetrahedral prisms

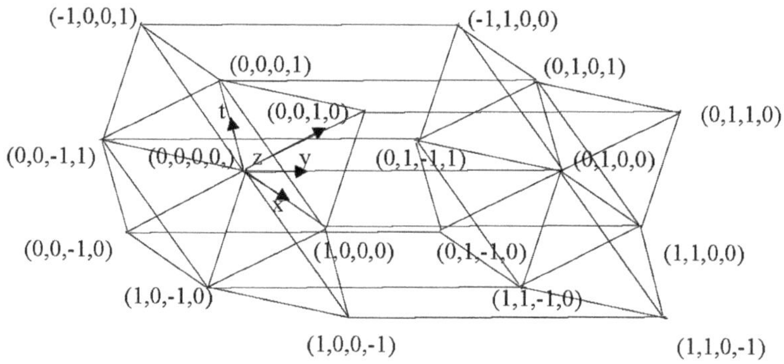

Figure 3) having a closest to them tetrahedral prisms A_i general quadrangular two - dimensional face.

The tetrahedral prisms A_i and B_i have common edges, which are parties to the general quadrilateral faces. The tetrahedral prisms A_i and B_i are connected symmetry transformation - turning the 180^0 around a common edge of the tetrahedron. Let B_0 tetrahedral prism type B_i, adjacent tetrahedral prism A_0, contains the x - axis, will translate the prism B_0 on edge length for all the coordinates x, y, z, t four - dimensional space. Then we obtain a lattice of the tetrahedral prisms B_i.

$$B_0 = [(0,0,0,0),(0,1,0,0),(1,1,0,0),(1,0,0,0),(1,0,-1,0),(1,0,0,-1),(1,1,0,-1),(1,1,-1,0)], \tag{3}$$

$$B_1 = B_0(x+1)$$
$$= [(1,0,0,0),(1,1,0,0),(2,1,0,0),(2,0,0,0),(2,0,-1,0),(2,0,0,-1),(2,1,0,-1),(2,1,-1,0)],$$

$$B_2 = B_0(z+1)$$
$$= [(0,0,1,0),(0,1,1,0),(1,1,1,0),(1,0,1,0),(1,0,0,0),(1,0,1,-1),(1,1,1,-1),(1,1,0,0)],$$

$$B_3 = B_0(x+1, z+1) = B_1(z+1) =$$
$$[(1,0,1,0),(1,1,1,0),(2,1,1,0),(2,0,1,0),(2,0,0,0,),(2,0,1,-1),(2,1,1,-1),(2,1,0,0)],$$

$$B_4 = B_0(y+1, z+1) = B_2(y+1) =$$
$$[(0,1,1,0),(0,2,1,0),(1,2,1,0),(1,1,1,0),(1,1,0,0),(1,1,1,-1),(1,2,1,-1),(1,2,0,0)],$$

$$B_5 = B_0(\mathrm{x}+1, y+1, z+1) = B_3(y+1) =$$
$$[(1,1,1,0),(1,2,1,0),(2,2.1.0),(2,1,1,0),(2,1,0,0),(2,1,1,\quad 1),(2,2,1,\quad 1),(2,2,0,0)],$$

$$B_6 = B_0(y+1)$$
$$= [(0,1,0,0),(0,2,0,0),(1,2,0,0),(1,1,0,0),(1,1,-1,0),(1,1,0,-1),(1,2,0,-1),(1,2,-1,0)],$$

$$B_7 = B_0(y+1, \mathrm{x}+1) = B_1(y+1) =$$
$$[(1,1,0,0),(1,2,0,0),(2,2,0,0),(2,1,0,0),(2,1,-1,0),(2,1,0,-1),(2,2,0,-1),(2,2,-1,0)],$$

$$B_8 = B_0(\mathrm{t}+1) =$$
$$[(0,0,0,1),(0,1,0,1),(1,1,0,1),(1,0,0,1),(1,0,-1,1),(1,0,0,1),(1,1,0,0),(1,1,-1,1)],$$

$$B_9 = B_0(y+1, \mathrm{t}+1) = B_6(\mathrm{t}+1) =$$
$$[(0,1,0,1),(0,2,0,1),(1,2,0,1),(1,1,0,1),(1,1,-1,1),(1,1,0,0),(1,2,0,0),(1,2,-1,1)],$$

$$B_{10} = B_0(\mathrm{x}+1, y+1, \mathrm{t}+1) = B_7(\mathrm{t}+1) =$$
$$[(1,1,0,1),(1,2,0,1),(2,2,0,1),(2,1,0,1),(2,1,-1,1),(2,1,0,0),(2,2,0,0),(2,2,-1,1)],$$

$$B_{11} = B_0(\mathrm{x}+1, \mathrm{t}+1) = B_1(\mathrm{t}+1) =$$
$$[(1,0,0,1),(1,1,0,1),(2,1,0,1),(2,0,0,1),(2,0,-1,1),(2,0,0,0),(2,1,0,0),(2,1,-1,1)],$$

$$B_{12} = B_0(y+1, z+1) = B_2(t+1) =$$
$$[(0,0,1,1),(0,1,1,1),(1,1,1,1),(1,0,1,1),(1,0,0,1),(1,0,1,0),(1,1,1,0),(1,1,0,1)],$$

$$B_{13} = B_0(z+1, y+1, t+1) = B_4(t+1) =$$
$$[(0,1,1,1),(0,2,1,1),(1,2,1,1),(1,1,1,1),(1,1,0,1),(1,1,1,0),(1,2,1,0),(1,2,0,1)],$$

$$B_{14} = B_0(x+1, y+1, z+1, t+1) = B_5(t+1) =$$
$$[(1,1,1,1),(1,2,1,1),(2,2,1,1),(2,1,1,1),(2,1,0,1),(2,1,1,0),(2,2,1,0),(2,2,0,1)],$$

$$B_{15} = B_0(x+1, t+1, z+1) = B_3(t+1) =$$
$$[(1,0,1,1),(1,1,1,1),(2,1,1,1),(2,0,1,1),(2,0,0,1),(2,0,1,0),(2,1,1,0),(2,1,0,1)].$$

It is presented tetrahedral prisms $B_0 \div B_{15}$ points in the 4 - dimensional space can to get up hypercube in Figure 2 with change notation A_i on B_i. From the construction it is follows that tetrahedral prism B_i can have common elements such as tetrahedrons, edges and vertices. This is easily seen by analytical expressions (3). In addition, it follows from the construction (Figure 3) that the tetrahedral prism A_i and B_i can have common elements only tetrahedrons, edges and vertices. Triangular prisms, which are present in tetrahedral prisms of both types cannot be common for these tetrahedral prisms.

They only deal with each other on a flat quadrilateral. This also easily seen from the formulas (2) and (3). This situation is significantly different from the conditions of normality, to accept in general theory of stereohedrons. The provided construction proves that the tetrahedral prism is a fundamental area in the 4 - dimensional space, as it completely fills the space in the translation of the prism in all the coordinates of space and turning it into 180^0.

Since all the vertices of the tetrahedral prisms are equal and filling the space with their translation is uniform, then the partition created by them is the right one space.

As follows from the division of space by the tetrahedral prisms (Figure 3), the geometric shape of the tetrahedral prism *B* (Figure 4) differs from the geometric shape of the tetrahedral prism *A*.

These differences are related to the different angle of the one - dimensional segment, as a factor of the tetrahedral prism, to the tetrahedron in these tetrahedral prisms. Moreover, the angles between tetrahedrons and one - dimensional segment are coordinated so that tetrahedral prisms *A* and *B* can adjoin to each other along a common two - dimensional face having the

Table 1. Common elements of the tetrahedral prisms

Edge of the Hypercube	Common Element of the Tetrahedral Prisms	Edge of the Hypercube	Common Element of the Tetrahedral Prisms
$A_0 A_1$	edge (1,0,0,0)(1,1,0,0)	$A_5 A_{14}$	edge (1,1,1,1)(1,2,1,1)
$A_0 A_2$	edge (0,1,1,0)(0,0,1,0)	$A_6 A_7$	vertex (1,1,1,0)
$A_0 A_6$	tetrahedron (0,1,0,0)(1,1,0,0)(0,1,0,1)(0,1,0,1)	$A_6 A_9$	edge (0,1,0,1)(0,2,0,1)
$A_0 A_8$	edge (0,1,0,1)(0,0,0.1)	A_7A_{10}	vertex (1,1,0,1)
A_1A_3	edge (1,0,1,0)(1,1,1,0)	$A_8 A_{11}$	edge (1,0,0,1)(1,1,0,1)
A_7A_1	tetrahedron (1,1,0,0,)(2,1,0,0)(1,1,1,0)(1,1,0,1)	A_8A_9	tetrahedron (0,1,0,1)(1,1,0,1)(0,1,1,1) (0,1,0,2)
A_1A_{11}	edge (1,0,0,1)(1,1,0,1)	A_8A_{12}	edge (0,0,1,1)(0,1,1,1)
A_2A_3	edge (1,0,1,0)(1,1,0,0)	A_9A_{10}	edge (1,1,0,1)(1,2,0,1)
A_2A_4	tetrahedron (0,1,1,0)(1,1,1,0)(0,1,1,1)(0,1,2,0)	A_9A_{13}	edge (1,2,1,1)(0,1,1,1)
A_2A_{12}	edge (0,1,1,0)(0,0,1,1)	$A_{11}A_{10}$	edge (1,2,0,1)(1,1,0,1)
A_3A_5	tetrahedron (1,1,1,0)(2,1,1,0)(1,1,2,0)(0,1,1,1)	$A_{14}A_{10}$	edge (1,1,1,1)(1,2,1,1)
A_3A_{15}	edge (1,0,1,1)(1,1,1,1)	$A_{11}A_{15}$	edge (1,0,1,1)(1,1,1,1)
A_4A_5	edge (1,1,1,0)(1,2,1,0)	$A_{13}A_{12}$	edge (0,1,1,1)(1,1,1,1)
A_4A_6	edge (0,1,1,0)(0,2,1,0)	$A_{12}A_{15}$	edge (1,0,1,1)(1,1,1,1)
A_4A_{13}	vertex (0,1,1,1)	$A_{13}A_{14}$	tetrahedron (1,1,1,1)(1,1,2,1)(1,2,1,1) (1,2,2,1)
A_7A_5	edge (1,1,1,0)(1,2,1,0)	$A_{14}A_{15}$	tetrahedron (1,1,1,1)(2,1,1,1)(1,1,2,1) (1,1,1,1)

Table 2. Common elements of the tetrahedral prisms

Diagonal Flat Face of the Hypercube	Common Element of the Tetrahedral Prisms	Diagonal Flat Face of the Hypercube	Common Element of the Tetrahedral Prism
$A_8 A_{13}$, $A_8 A_{13}$	vertex (0,1,1,1)	$A_8 A_{10}$, $A_9 A_{11}$	vertex (1,1,1,1)
$A_6 A_2$, $A_0 A_4$	vertex (0,1,1,0)	$A_{14} A_7$	∅
$A_8 A_2$	edge (0,1,1,1)(0,0,1,1)	$A_5 A_{10}$	edge (1,1,1,1)(1,2,1,1)
$A_0 A_{12}$	∅	$A_{13}A_{14}$, $A_5 A_{15}$	vertex (1,1,1,1)
$A_4 A_9$	vertex (0,1,1,1)(0,0,1,1)	A_3A_{11}	vertex (1,0,1,1)(1,1,1,1)
A_6A_{13}	∅	$A_1 A_{15}$	∅
A_5A_{13}	edge (1,1,1,1)(1,2,1,1)	A_7A_{11}, $A_1 A_{10}$	vertex (1,1,0,1)
$A_4 A_{14}$	∅	$A_5 A_1$, $A_7 A_3$	vertex (1,1,1,0)
A_5A_{10}	edge (1,2,1,1)(1,1,1,1)	A_3A_{12}	edge (1,0,1,1)(1,1,1,1)
A_7A_{14}	∅	A_2A_{15}	∅
A_7A_9	edge (1,1,0,1)(1,2,0,1)	$A_{11}A_0$	∅
$A_6 A_{10}$	∅	A_1A_8	edge (1,0,0,1)(1,1,0,1)
A_9A_{14}	∅	A_4A_3, $A_5 A_2$	edge (1,1,1,0)
$A_{13}A_{10}$	edge (1,2,1,1)(1,1,1,1)	$A_1 A_6$, $A_0 A_7$	vertex (1,1,0,0)
A_5A_6	∅	$A_6 A_8$, $A_0 A_9$	vertex (0,1,0,1)
A_4A_7	edge (1,1,1,0)(1,2,1,0)	$A_{11}A_{14}$, $A_{10} A_{15}$	vertex (1,1,1,1)
$A_{11}A_{12}$	edge (1,1,1,1)(1,0,1,1)	A_8A_{15}	∅
$A_{14}A_{12}$	vertex (1,1,1,1)	$A_{13}A_{15}$	edge (1,1,2,1)(1,1,1,1)

Figure 4. The tetrahedral prism B

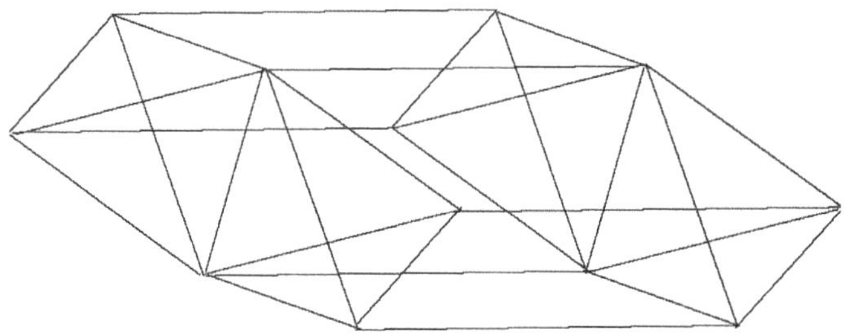

form of a parallelogram. The polytope depicted in the figure consists of four tetrahedral prisms: two tetrahedral prisms of type *A* and two tetrahedral prisms of type *B*. The tetrahedral prisms *A* and *B* have common two - dimensional quadrangular faces, and each of the pairs of tetrahedral prisms *A* and *B* have the common one - dimensional segment. This polytope, as is easy to see, is the fundamental domain of the partition of a four - dimensional space by tetrahedral prisms, since the translation of this region covers all four - dimensional space. Moreover, translation in the direction of a one - dimensional segment, which is a factor in the product of tetrahedrons per segment, leads to overlapping of regions with four tetrahedrons, and translation in directions perpendicular to this segment leads to overlapping of regions along two - dimensional faces. Moreover, each point of this region does not have an equivalent in the region with respect to the enumerated group of motions. Thus, for the first time the concrete fundamental region of four - dimensional space constructed has a complex form in the case of tetrahedral prisms and is a complex of tetrahedral prisms. Each individual tetrahedral prism is not a fundamental domain.

The fundamental area of the partition of a four - dimensional space by tetrahedral prisms can be conditionally represented as a set of tetrahedral prisms $\begin{matrix} BA \\ AB \end{matrix}$ in which the diagonal elements are joined to each other along the edge, and the elements along the vertical and horizontally join each other along two - dimensional quadrangular surfaces connected along this edge. The fundamental region with adjacent fundamental regions is bordered by two - dimensional quadrangular surfaces and by the totality of four tetrahedrons.

Partition of Space of the Triangular Prismahedron

The triangular prismahedron is the product of a triangle by a triangle. The dimension of this polytope equal 4, it has 9 vertices and 6 triangle prisms (Zhizhin, 2019).

One direct the coordinates of the four - dimensional space along the edges of the triangular prismahedron (see the Figure 5).

Figure 5. The triangular prismahedron

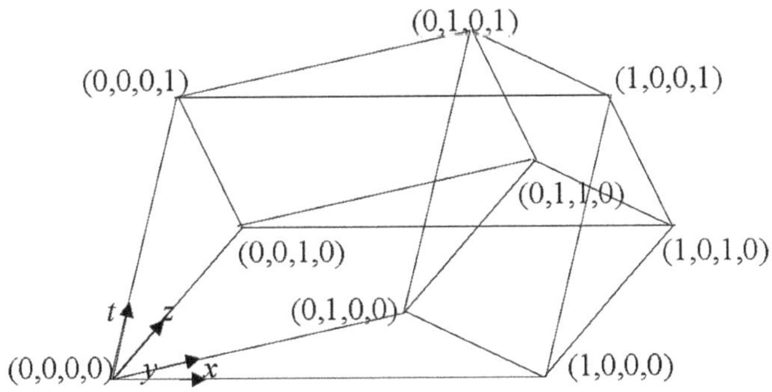

Assume that the length of each edge is equal to 1. Then, each node triangular prismahedron can be associated with a set of integers (Figure 5). Translating the triangular prismahedron along the coordinates *x, y, z, t*, can to obtain the lattice vertices. Let E_0 triangular prismahedron with the values of vertex coordinates in Figure 5. Then

$$E_0 = [(0,0,0,0),(1,0,0,0),(0,1,0,0),(0,0,1,0),(0,0,0,1),(1,0,1,0),(0,1,1,0),(1,0,0,1),(0,1,0,1], \quad (4)$$

$$E_1 = E_0(x+1)$$
$$= [(1,0,0,0),(2,0,0,0),(1,1,0,0),(1,0,1,0),(1,0,0,1),(2,0,1,0),(1,1,1,0),(2,0,0,1),(1,1,0,1)],$$

$$E_2 = E_0(z+1)$$
$$= [(0,0,1,0),(1,0,1,0),(0,1,1,0),(0,0,2,0),(0,0,1,1),(1,0,2,0),(0,1,2,0),(1,0,1,1),(0,1,1,1)],$$

$$E_3 = E_0(x+1, z+1) = E_1(z+1) =$$
$$[(1,0,1,0),(2,0,1,0),(1,1,1,0),(1,0,2,0),(1,0,1,1,),(2,0,2,0),(1,1,2,0),(2,0,1,1),(1,1,1,1)],$$

$$E_4 = E_0(y+1, z+1) = E_2(y+1) =$$
$$[(0,1,1,0),(1,1,1,0),(0,2,1,0),(0,1,2,0),(0,1,1,1),(1,1,2,0),(0,2,2,0),(1,1,1,1),(0,2,1,1)],$$

$$E_5 = E_0(x+1, y+1, z+1) = E_3(y+1) =$$
$$[(1,1,1,0),(2,1,1,0),(1,2,1,0),(1,1,2,0),(1,1,1,1),(2,1,2,0),(1,2,2,0),(1,2,1,1),(1,2,1,1)],$$

$$E_6 = E_0(y+1)$$
$$= [(0,1,0,0),(1,1,0,0),(0,2,0,0),(0,1,1,0),(0,1,0,1),(1,1,1,0),(0,2,1,0),(1,1,0,1),(0,2,0,1)],$$

$$E_7 = E_0(y+1, \mathrm{x}+1) = E_1(y+1) =$$
$$[(1,1,0,0),(2,1,0,0),(1,2,0,0),(1,1,1,0),(1,1,0,1),(2,1,1,0),(1,2,1,0),(2,1,0,1),(1,2,0,1)],$$

$$E_8 = E_0(t+1) =$$
$$[(0,0,0,1),(1,0,0,1),(0,1,0,1),(0,0,1,1),(0,0,0,2),(1,0,1,1),(0,1,1,1),(1,0,0,2),(0,1,0,2)],$$

$$E_9 = E_0(y+1, t+1) = E_6(t+1) =$$
$$[(0,1,0,1),(1,1,0,1),(0,2,0,1),(0,1,1,1),(0,1,0,2),(1,1,1,1),(0,2,1,1),(1,1,0,2),(0,2,0,2)],$$

$$E_{10} = E_0(x+1, y+1, t+1) = E_7(t+1) =$$
$$[(1,1,0,1),(2,1,0,1),(1,2,0,1),(1,1,1,1),(1,1,0,2),(2,1,1,1),(1,2,1,1),(2,1,0,2),(1,2,0,2)],$$

$$E_{11} = E_0(x+1, t+1) = E_1(t+1) =$$
$$[(1,0,0,1),(2,0,0,1),(1,1,0,1),(1,0,1,1),(1,0,0,2),(2,0,1,1),(1,1,1,1),(2,0,0,2),(1,1,0,2)],$$

$$E_{12} = E_0(y+1, z+1) = E_2(t+1) =$$
$$[(0,0,1,1),(1,0,1,1),(0,1,1,1),(0,0,2,1),(0,0,1,2),(1,0,2,1),(0,1,2,1),(1,0,1,2),(0,1,1,2)],$$

$$E_{13} = E_0(z+1, y+1, t+1) = E_4(t+1) =$$
$$[(0,1,1,1),(1,1,1,1),(0,2,1,1),(0,1,2,1),(0,1,1,2),(0,1,1,2),(1,1,2,1),(1,1,1,2),(0,2,1,2)],$$

$$E_{14} = E_0(x+1, y+1, z+1, t+1) = E_5(t+1) =$$
$$[(1,1,1,1),(2,1,1,1),(1,2,1,1),(1,1,2,1),(1,1,1,2),(2,1,2,1),(1,2,2,1),(2,1,1,2),(1,2,1,2)],$$

$$E_{15} = E_0(x+1, t+1, z+1) = E_3(t+1) =$$
$$[(1,0,1,1),(2,0,1,1),(1,1,1,1),(1,0,2,1),(1,0,1,2),(2,0,2,1),(1,1,2,1),(2,0,1,2),(1,1,1,2)].$$

Representing the triangular prismahedrons $E_0 \div E_{15}$ dots in three-dimensional space, can to get the hypercube. Moreover, the edges of the hypercube correspond to possible changes in the values of one of the coordinates of the vertices of the triangular prismahedron unit (Figure 6).

Figure 6. The hypercube from 16 triangular prismahedrons

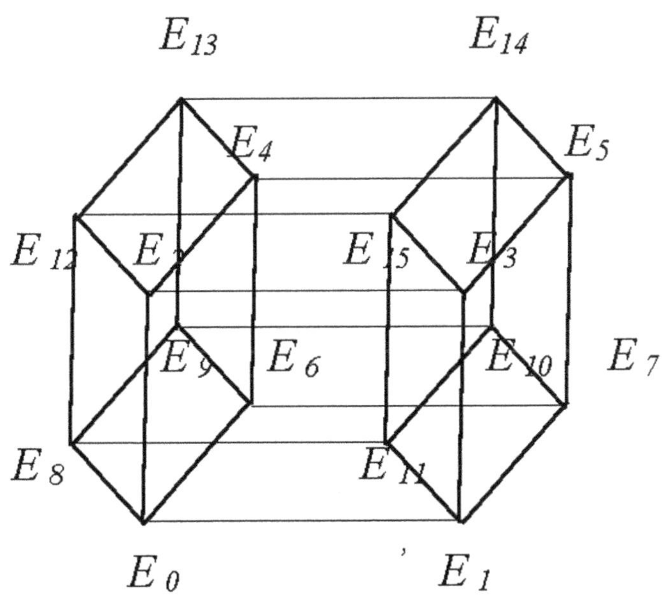

In addition, each edge of the hypercube in Figure 6 can be considered as an element of the overall two triangular prismahedrons, connected by an edge. Using the coordinate expression triangular prismahedrons (4) can be analytically determined. In Table 3 geometry elements are common to each pair of triangular prismahedrons connected by an edge in Figure 6 are listed. From Table 3 it follows that triangular prismahedrons of type *E* when translating along the coordinates of a four-dimensional space form a network, adjoining each other along entire triangles. The direct construction of this network shows (see Figure 7) that between these prismhedrons formed closed areas of three more types, which are also triangular prismahedrons *E*, but differing in the different arrangement of the triangles, as the factors of the product.

Table 3. Common elements of the triangular prismahedrons $E_0 \div E_{15}$

Edge of the Hypercube	Common Element of the Triangular Prismahedrons	Edge of the Hypercube	Common Element of the Triangular Prismahedrons
$E_0 E_1$	triangle (1,0,0,0)(1,0,1,0)(1,0,0,0)	$E_5 E_{14}$	triangle (1,1,1,1)(1,2,1,1)(2,1,1,1)
$E_0 E_2$	triangle (0,1,1,0)(0,0,1,0)(1,0,1,0)	$E_6 E_7$	triangle (1,1,1,0)(1,1,0,0)(1,1,0,1)
$E_0 E_6$	triangle (0,1,0,0)(0,1,1,0)(0,1,0,1)	$E_6 E_9$	triangle (0,1,0,1)(0,2,0,1)(1,1,0,1)
$E_0 E_8$	triangle (0,1,0,1)(0,0,0,1)(1,0,0,1)	E_7E_{10}	triangle (1,1,0,1)(2,1,0,1)(1,2,0,1)
E_1E_3	triangle (1,0,1,0)(1,1,1,0)(2,0,1,0)	$E_8 E_{11}$	triangle (1,0,0,1)(1,0,1,1)(1,0,0,2)
E_7E_1	triangle (1,1,0,0,)(1,1,1,0)(1,1,0,1)	E_8E_9	triangle (0,1,0,1)(0,1,1,1)(1,0,0,2)
E_1E_{11}	triangle (1,0,0,1)(1,1,0,1)(2,0,0,1)	E_8E_{12}	triangle (0,0,1,1)(1,0,1,1)(1,0,0,2)
E_2E_3	triangle (1,0,1,0)(1,0,2,0)(1,0,1,1)	E_9E_{10}	triangle (1,1,0,1)(1,1,0,2)(1,1,1,1)
E_2E_4	triangle (0,1,1,0)(0,1,1,1)(0,1,2,0)	E_9E_{13}	triangle (0,2,1,1)(0,1,1,1)(1,1,1,1)
E_2E_{12}	triangle (0,1,1,1)(0,0,1,1)(1,0,1,1)	$E_{11}E_{10}$	triangle (1,1,0,1)(1,1,0,2)(1,1,1,1)
E_3E_5	triangle (1,1,1,0)(1,1,2,0)(0,1,1,1)	$E_{14}E_{10}$	triangle (1,1,1,1)(1,2,1,1)(2,1,1,1)
E_3E_{15}	triangle (1,0,1,1)(1,1,1,1)(2,0,1,1)	$E_{11}E_{15}$	triangle (1,0,1,1)(1,1,1,1)(2,0,1,1)
E_4E_5	triangle (1,1,1,0)(1,1,2,0)(1,1,1,1)	$E_{13}E_{12}$	triangle (0,1,1,1)(0,1,2,1)(0,1,1,2)
E_4E_6	triangle (0,1,1,0)(0,2,1,0)(1,1,1,0)	$E_{12}E_{15}$	triangle (1,0,1,1)(1,0,2,1)(1,1,1,2)
E_4E_{13}	triangle (0,1,1,1)(1,1,1,1)(0,2,1,1)	$E_{13}E_{14}$	triangle (1,1,1,1)(1,1,2,1)(1,1,1,2)
E_7E_5	triangle (1,1,1,0)(1,2,1,0)(2,1,1,0)	$E_{14}E_{15}$	triangle (1,1,1,1)(1,1,2,1)(1,1,1,2)

The types of these triangular prismahedrons (denoted by *L, C, D*) are presented separately in the Figure 8, Figure 9, Figure 10 and are indicated by a different color.

The triangular prismahedrons *L, C, D*, as well as triangular prismahedrons *E* form networks in which they adjoin each other along entire triangles. This follows from indicated the triangular prismahedrons *L, C, D* in coordinates of vertices of network and tables 4, 5, 6 of common elements in each of network.

The indicated of the triangular prismahedrons type *L*

Figure 7. The partition space by the triangular prismahedrons E

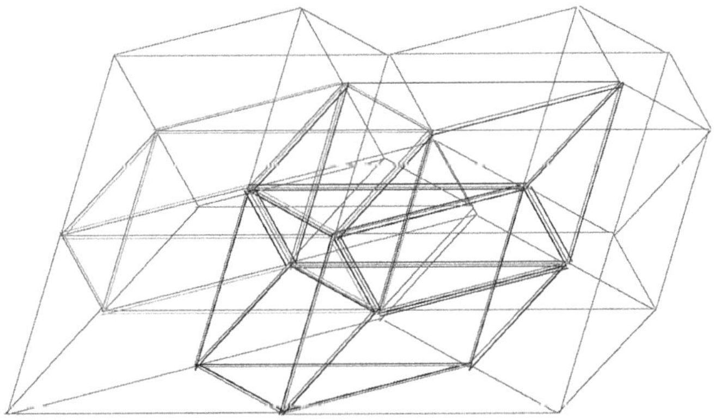

Figure 8. The triangular prismahedron L

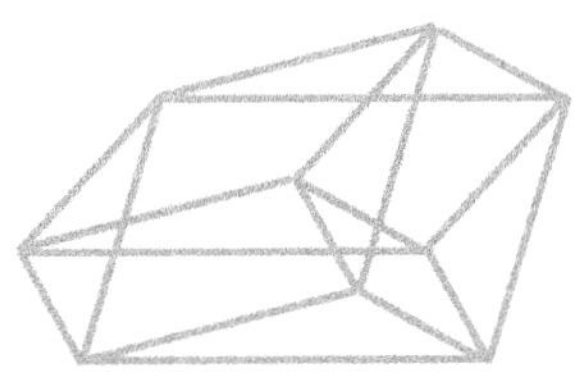

Figure 9. The triangular prismahedron C

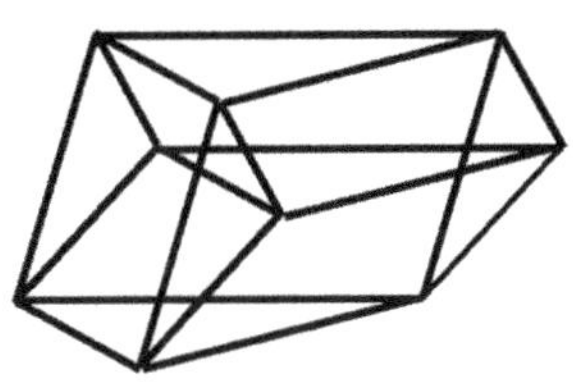

Figure 10. The triangular prismahedron D

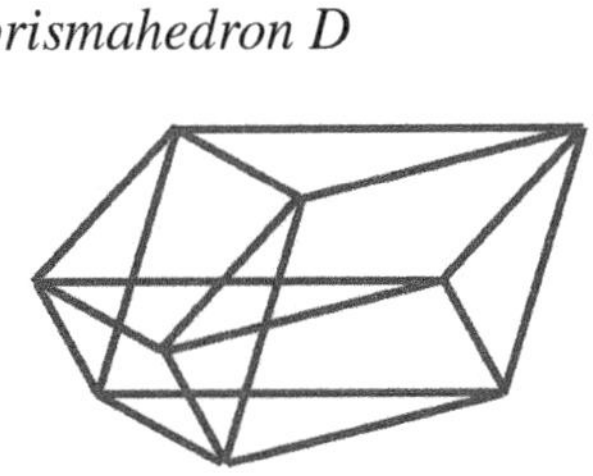

$$L_0 = [(0,0,1,0),(0,0,0,1),(0,0,1,1),(1,0,1,0),(0,1,1,0),(0,1,0,1),(1,0,0,1),(0,1,1,1),(1,0,1,1),$$

$$L_1 = L_0(x+1)$$
$$= [(1,0,1,0),(1,0,0,1),(1,0,1,1),(2,0,1,0),(1,1,1,0),(1,1,0,1),(2,0,0,1),(1,1,1,1),(2,0,1,1)],$$

$$L_2 = L_0(z+1)$$
$$= [(1,0,2,0),(1,0,1,1),(1,0,2,0),(2,0,2,0),(1,1,2,0),(1,1,1,1),(2,0,1,1),(1,1,2,1),(2,0,2,1)],$$

$$L_3 = L_0(x+1,z+1) = L_1(z+1) =$$
$$[(1,0,2,0),(1,0,1,1),(1,0,2,1),(2,0,2,0),(1,1,2,0,),(1,1,1,1),(2,0,1,1),(1,1,2,1),(2,0,2,1)],$$

$$L_4 = L_0(y+1,z+1) = L_2(y+1) =$$
$$[(1,1,2,0),(1,1,1,1),(1,1,2,1),(2,1,2,0),(1,2,2,0),(1,2,1,1),(2,1,1,1),(1,2,2,1),(2,1,2,1)],$$

$$L_5 = L_0(x+1,y+1,z+1) = L_3(y+1) =$$
$$[(1,1,2,0),(1,1,1,1),(1,1,2,1),(2,1,2,0),(1,2,2,0),(1,2,1,1),(2,1,1,1),(1,2,2,1),(2,1,2,1)],$$

$$L_6 = L_0(y+1)$$
$$= [(0,1,1,0),(0,1,0,1),(0,1,1,1),(1,1,1,0),(0,2,1,0),(0,2,0,1),(1,1,0,1),(0,2,1,1),(1,1,1,1)],$$

$$L_7 = L_0(y+1,x+1) = L_1(y+1) =$$
$$[(1,1,1,0),(1,1,0,1),(1,1,1,1),(2,1,1,0),(1,2,1,0),(1,2,0,1),(2,1,0,1),(1,2,1,1),(2,1,1,1)],$$

$$L_8 = L_0(t+1) =$$
$$[(0,0,1,1),(0,0,0,2),(0,0,1,2),(1,0,1,2),(0,1,1,1),(0,1,0,2),(1,0,0,2),(0,1,1,2),(1,0,1,2)],$$

$$L_9 = L_0(y+1,t+1) = L_6(t+1) =$$
$$[(0,1,1,1),(0,1,0,2),(0,1,1,2),(1,1,1,1),(0,2,1,1),(0,2,0,2),(1,1,0,2),(0,2,1,2),(1,1,1,2)],$$

$$L_{10} = L_0(x+1, y+1, t+1) = L_7(t+1) =$$
$$[(1,1,1,1),(1,1,0,2),(1,1,1,2),(2,1,1,1),(1,2,1,1),(1,2,0,2),(2,1,0,2),(1,2,1,2),(2,1,1,2)],$$

$$L_{11} = L_0(x+1, t+1) = L_1(t+1) =$$
$$[(1,0,1,1),(1,0,0,2),(1,0,1,2),(2,0,1,1),(1,1,1,1),(1,1,0,2),(2,0,0,2),(1,1,1,2),(2,0,1,2)],$$

$$L_{12} = L_0(y+1, z+1) = L_2(t+1) =$$
$$[(1,0,2,1),(1,0,1,2),(1,0,2,2),(2,0,2,1),(1,1,2,1),(1,1,1,2),(2,0,1,2),(1,1,2,2),(2,0,2,2)],$$

$$L_{13} = L_0(z+1, y+1, t+1) = L_4(t+1) =$$
$$[(1,1,2,1),(1,1,1,2),(1,1,2,2),(2,1,2,1),(1,2,2,1),(1,2,1,2),(2,1,1,2),(1,2,2,2),(2,1,2,2)],$$

$$L_{14} = L_0(x+1, y+1, z+1, t+1) = L_5(t+1) =$$
$$[(1,1,2,1),(1,1,1,2),(1,1,2,2),(2,1,2,1),(1,2,2,1),(1,2,1,2),(2,1,1,2),(1,2,2,2),(2,1,2,2)],$$

$$L_{15} = L_0(x+1, t+1, z+1) = L_3(t+1) =$$
$$[(1,0,2,1),(1,0,1,2),(1,0,2,2),(2,0,2,1),(1,1,2,1),(1,1,1,2),(2,0,1,2),(1,1,2,2),(2,0,2,2)].$$

The indicated of the triangular prismahedrons type C

$$C_0 = [(0,1,0,0),(1,0,0,0),(0,1,0,1),(0,1,1,0),(1,0,1,0),(1,1,0,0),(1,0,0,1),(1,1,1,0),(1,1,0,1)],$$

$$C_1 = C_0(x+1)$$
$$= [(1,1,0,0),(2,0,0,0),(1,1,0,1),(1,1,1,0),(2,0,1,0),(2,1,0,0),(2,0,0,1),(2,1,1,0),(2,1,0,1)],$$

$$C_2 = C_0(z+1)$$
$$= [(0,1,1,0),(1,0,1,0),(0,1,1,1),(0,1,2,0),(1,0,2,0),(1,1,1,0),(1,0,1,1),(1,1,2,0),(1,1,1,1)],$$

$$C_3 = C_0(x+1, z+1) = C_1(z+1)$$
$$= [(1,1,1,0),(2,0,1,0),(1,1,1,1),(1,1,2,0),(2,0,2,0,),(2,1,1,0),(2,0,1,1),(2,1,2,0),(2,1,1,1)],$$

$$C_4 = C_0(y+1, z+1) = C_2(y+1)$$
$$= [(0,2,1,0),(1,1,1,0),(0,2,1,1),(0,2,2,0),(1,1,2,0),(1,2,1,0),(1,1,1,1),(1,2,2,0),(1,2,1,1)],$$

$$C_5 = C_0(x+1, y+1, z+1) = C_3(y+1) =$$
$$[(1,2,1,0),(2,1,1,0),(1,2,1,1),(1,2,2,0),(2,1,2,0),(2,2,1,0),(2,1,1,1),(2,2,2,0),(2,2,1,1)],$$

$$C_6 = C_0(y+1)$$
$$= [(2,2,0,0),(1,1,0,0),(0,2,0,1),(0,2,1,0),(1,1,1,0),(1,2,0,0),(1,1,0,1),(1,2,1,0),(1,2,0,1)],$$

$$C_7 = C_0(y+1, x+1) = C_1(y+1) =$$
$$[(1,2,0,0),(2,1,0,0),(1,2,0,1),(1,2,1,0),(2,1,1,0),(2,2,0,0),(2,1,0,1),(2,2,1,0),(2,2,0,1)],$$

$$C_8 = C_0(t+1) =$$
$$[(0,1,0,1),(1,0,0,1),(0,1,0,2),(0,1,1,1),(1,0,1,1),(1,1,0,1),(1,0,0,2),(1,1,1,1),(1,1,0,2)],$$

$$C_9 = C_0(y+1, t+1) = C_6(t+1) =$$
$$[(0,2,0,1),(1,1,0,1),(0,2,0,2),(0,2,1,1),(1,1,1,1),(1,2,0,1),(1,1,0,2),(1,2,1,1),(1,2,0,2)],$$

$$C_{10} = C_0(x+1, y+1, t+1) = C_7(t+1) =$$
$$[(1,2,0,1),(2,1,0,1),(1,2,0,2),(1,2,1,1),(2,1,1,1),(2,2,0,1),(2,1,0,2),(2,2,1,1),(2,2,0,2)],$$

$$C_{11} = C_0(x+1, t+1) = C_1(t+1) =$$
$$[(1,1,0,1),(2,0,0,1),(1,1,0,2),(1,1,1,1),(2,0,1,1),(2,1,0,1),(2,0,0,2),(2,1,1,1),(2,1,0,2)],$$

$$C_{12} = C_0(y+1, z+1) = C_2(t+1) =$$
$$[(0,1,1,1),(1,0,1,1),(0,1,1,2),(0,1,2,1),(1,0,2,1),(1,1,1,1),(1,0,1,2),(1,1,2,1),(1,1,1,2)],$$

$$C_{13} = C_0(z+1, y+1, t+1) = C_4(t+1) =$$
$$[(0,2,1,1),(1,1,1,1),(0,2,1,2),(0,2,2,1),(1,1,2,1),(1,2,1,1),(1,1,1,2),(1,2,2,1),(1,2,1,2)],$$

$$C_{14} = C_0(x+1, y+1, z+1, t+1) = C_5(t+1) =$$
$$[(1,2,1,1),(2,1,1,1),(1,2,1,2),(1,2,2,1),(2,1,2,1),(2,2,1,1),(2,1,1,2),(2,2,2,1),(2,2,1,2)],$$

$$C_{15} = C_0(x+1, t+1, z+1) = C_3(t+1) =$$
$$[(1,1,1,1),(2,0,1,1),(1,1,1,2),(1,1,2,1),(2,0,2,1),(2,1,1,1),(2,0,1,2),(2,1,2,1),(2,1,1,2)].$$

The indicated of the triangular prismahedrons type D

$$D_0 = [(0,1,1,0),(1,0,1,0),(1,0,0,1),(0,1,0,1),(0,1,1,1),(1,0,1,1),(1,1,0,1),(1,1,1,0),(1,1,1,1)],$$

$$D_1 = D_0(x+1)$$
$$= [(1,1,1,0),(2,0,1,0),(2,0,0,1),(1,1,0,1),(1,1,1,1),(2,0,1,1),(2,1,0,1),(2,1,1,0),(2,1,1,1)],$$

$$D_2 = D_0(z+1)$$
$$= [(0,1,2,0),(1,0,2,0),(1,0,1,1),(0,1,1,1),(0,1,2,1),(1,1,1,1),(1,1,2,0),(1,1,2,1),(1,0,2,1)],$$

$$D_3 = D_0(x+1, z+1) = D_1(z+1) =$$
$$[(1,1,2,0),(2,0,2,0),(2,0,1,1),(1,1,1,1),(1,1,2,1,),(2,0,2,1),(2,1,1,1),(2,1,2,0),(2,1,2,1)],$$

$$D_4 = D_0(y+1, z+1) = D_2(y+1) =$$
$$[(0,2,2,0),(1,1,2,0),(1,1,1,1),(0,2,1,1),(0,2,2,1),(1,1,2,1),(1,2,1,1),(1,2,2,0),(1,2,2,1)],$$

$$D_5 = D_0(x+1, y+1, z+1) = D_3(y+1) =$$
$$[(1,2,2,0),(2,1,2,0),(2,1,1,1),(1,2,1,1),(1,2,2,1),(2,1,2,1),(2,2,1,1),(2,2,2,0),(2,2,2,1)],$$

$$D_6 = D_0(y+1)$$
$$= [(0,2,1,0),(1,1,1,0),(1,1,0,1),(0,2,0,1),(0,2,1,1),(1,1,1,1),(1,2,0,1),(1,2,1,0),(1,2,1,1)],$$

$$D_7 = D_0(y+1, x+1) = D_1(y+1) =$$
$$[(1,2,1,0),(2,1,1,0),(2,1,0,1),(1,2,0,1),(1,2,1,1),(2,1,1,1),(2,2,0,1),(2,2,1,0),(2,2,1,1)],$$

$$D_8 = D_0(t+1) =$$
$$[(0,1,1,1),(1,0,1,1),(1,0,0,2),(0,1,0,2),(0,1,1,2),(1,0,1,2),(1,1,0,2),(1,1,1,1),(1,1,1,2)],$$

$$D_9 = D_0(y+1, t+1) = D_6(t+1) =$$
$$[(0,2,1,1),(1,1,1,1),(1,1,0,2),(0,2,0,2),(0,2,1,2),(1,1,1,2),(1,2,0,2),(1,2,1,1),(1,2,1,2)],$$

$$D_{10} = D_0(x+1, y+1, t+1) = D_7(t+1) =$$
$$[(1,2,1,1),(2,1,1,1),(2,1,0,2),(1,2,0,2),(1,2,1,2),(2,1,1,2),(2,2,1,1),(2,2,1,2),(2,1,1,2)],$$

$$D_{11} = D_0(x+1, t+1) = D_1(t+1) =$$
$$[(1,1,1,1),(2,0,1,1),(2,0,0,2),(1,1,0,2),(1,1,1,2),(2,0,1,2),(2,1,0,2),(2,1,1,1),(2,1,1,2)],$$

$$D_{12} = D_0(y+1, z+1) = D_2(t+1) =$$
$$[(1,1,1,1),(2,0,1,1),(2,0,0,2),(1,1,0,2),(1,1,1,2),(2,0,1,2),(2,1,0,2),(2,1,1,1),(2,1,1,2)],$$

$$D_{13} = D_0(z+1, y+1, t+1) = D_4(t+1) =$$
$$[(0,2,2,1),(1,1,2,1),(1,1,1,2),(0,2,1,2),(0,2,2,2),(1,1,2,2),(1,2,1,2),(1,2,2,1),(1,2,2,2)],$$

$$D_{14} = D_0(x+1, y+1, z+1, t+1) = D_5(t+1) =$$
$$[(1,2,2,1),(2,1,2,1),(2,1,1,2),(1,2,1,2),(1,2,2,2),(2,1,2,2),(2,2,1,2),(2,2,2,1),(2,2,2,2)],$$

$$D_{15} = D_0(x+1, t+1, z+1) = D_3(t+1) =$$
$$[(1,1,2,1),(2,0,2,1),(2,0,1,2),(1,1,1,2),(1,1,2,2),(2,0,2,2),(2,1,1,2),(2,1,2,1),(2,1,2,2)].$$

Defining the intersections of the triangular prismahedrons in each of the sets of prismahedrons *E, L, C* or *D* from their coordinate expressions it is easy to determine that the common part of any pair of triangular prismahedrons belonging to a given set of prismahedrons (*E, L, C* or *D*) is some triangular face of the chosen pair of prismahedrons. Representing the triangular prismahedrons

$L_0 \div L_{15}$, $C_0 \div C_{15}$, $D_0 \div D_{15}$

dots in three - dimensional space, for each type triangular prismahedrons can to get the hypercube. Moreover, the edges of the hypercube correspond to possible changes in the values of one of the coordinates of the vertices of the triangular prismahedrons of given type unit (see Figure 6 for the triangular prismahedrons of type *E*).

Using the coordinate expressions of the triangular prismahedrons, can establish that triangular prismahedrons belonging to different types can have common geometric elements of different dimensions that enter into triangular prismahedrons. Their intersections can be empty (that is, they do not have common elements), for example, $E_2 \cap D_1 = \emptyset$. Their intersection can be single edge, for example,

$E_3 \cap L_0$=(1,0,1,0) (1,0,1,1),

$E_3 \cap C_1$=(1,0,1,0) (1,1,1,0), $E_2 \cap D_0$=(0,1,1,0) (1,0,1,1).

Their intersection can be square face, for example,

$E_7 \cap D_1$ = (1,1,1,0)(1,1,0,1)(2,1,1,0)(2,1,0,1),

$E_3 \cap D_2$= (1,0,2,0)(1,0,1,1)(1,1,2,0)(1,1,1,1).

Their intersection can be triangular prism also, for example,

$$E_3 \cap L_1 = (1,0,1,0)(2,0,1,0)(1,1,1,0)(1,0,1,1)(2,0,1,1)(1,1,1,1),$$

$$L_1 \cap D_0 = (1,0,1,0)(1,0,0,1)(1,0,1,1)(1,1,1,0)(1,1,0,1)(1,1,1,1).$$

Let us consider the set of the triangular prismahedrons of four types E, L, C, D of least - cubed numbers close to the origin from the considered set of the triangular prismahedrons. This set consists of the triangular prismahedrons E_0, L_0, C_0, D_0. By their coordinate expressions we establish that the pairs of the triangular prismahedrons E_0 and C_0, E_0 and L_0, C_0 and D_0, L_0 and D_0 have common trigonal prisms, and the pairs of the triangular prismahedrons E_0 and D_0, C_0 and L_0 have common planar faces of the four coal form

$$E_0 \cap C_0 = (0,1,0,0,)(1,0,0,0)(0,1,0,1)(1,0,0,1)(0,1,1,0)(1,0,1,0),$$

$$E_0 \cap L_0 = (0,0,1,0,)(0,1,1,0)(1,0,1,0)(0,0,0,1)(0,1,0,1)(1,0,0,1),$$

$$C_0 \cap D_0 = (0,1,0,1,)(1,1,0,1)(1,0,0,1)(0,1,1,0)(1,1,1,0)(1,0,1,0),$$

$$L_0 \cap D_0 = (0,1,1,1,)(1,0,1,1)(0,1,0,1)(1,0,0,1)(0,1,1,0)(1,0,1,0),$$

$$E_0 \cap D_0 = (0,1,0,1)(1,0,0,1)(0,1,1,0)(1,0,1,0),$$

$$L_0 \cap C_0 = (0,1,0,1)(1,0,0,1)(0,1,1,0)(1,0,1,0).$$

Thus, the triangular prismahedrons form closed complex of the triangular prismahedrons, which is convex polytope with dimension 4 (see Figure 11).

It is easy to see that the complex of the trigonal prismahedrons E_0, L_0, C_0, D_0 is the fundamental domain of the partition of four - dimensional space by a trigonal prismahedron, since translation of this complex with respect to the coordinates of space leads to the filling of this space without gaps. At the same time, adjacent complexes have common three - dimensional figures (rhombohedrons) included in the complex.

Figure 11. The fundamental domain of the decomposition of a four - dimensional space by a trigonal prismahedron

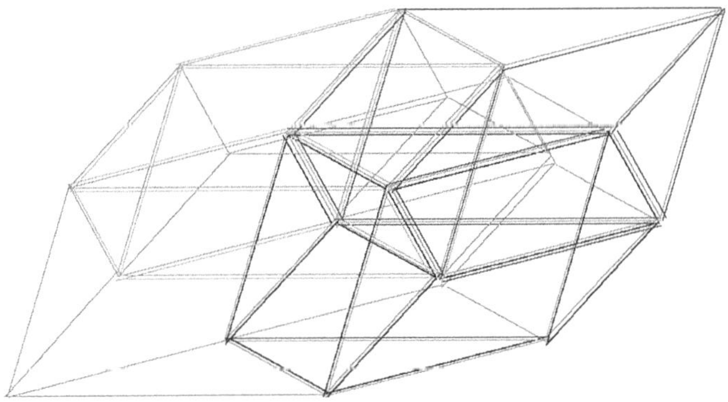

The fundamental area of the partition of a four - dimensional space by a trigonal prismahedron can be conditionally represented as a set of the trigonal prismahedrons $\frac{LD}{EC}$ in which the diagonal elements are joined to each other along two - dimensional quadrangular surfaces and the elements along the vertical and horizontally join each other connected along trigonal prism. The fundamental region with adjacent fundamental regions is bordered by rhombohedrons.

GENERAL PRINCIPLES OF THE PARTITION OF THE *N* - DIMENSIONAL SPACES BY POLYTOPIC PRISMAHEDRONS

The investigations of specific partitions of spaces by the polytopic prismahedrons allow us to formulate some general principles for partitioning spaces by the polytopic prismahedrons.

The polytopic prismahedrons, being a product of geometric figures of different dimensions, necessarily contain in their composition works of facets of figures on one - dimensional segments (edges). Therefore, they always have systems of parallel edges, and therefore parallelogons, in particular, flat faces in the form of quadrilaterals with parallel sides. In space, these parallel edges in the translation of the polytopic prismahedrons along the coordinates of space this gives rise to a system of parallel lines. Thus, between the polytopic prismahedrons in space are formed regions bounded by ribs and flat faces, i.e. again the polytopic prismahedrons are born. Their form is induced by the polytopic prismahedrons during their translation (translated polytopic prismahedrons). It either repeats the form of the translated the polytopic prismahedrons (in the case of a fairly symmetrical form of the latter) or slightly differs from their shape by the principle of "squeezing out" the shape. The number of possible deviations of free regions from translation the polytopic prismahedrons depends on the deviations from the symmetry of the most translation the polytopic prismahedrons. The contact of the polytopic prismahedrons of different shapes is carried out along prism generators (as factors), and on surfaces formed by these generators. In addition, the contact of the polytopic prismahedrons is realized by three - and multidimensional figures connected by these generators. Translated the polytopic prismahedrons form a network in space. If we introduce a coordinate system directed along the edges of some polytopic prismahedron emanating from its arbitrary vertex, taken as the origin, and the length of each edge is considered equal to one, then each vertex of the polytopic prismahedron acquires the notation. This designation consists of a sequence of zeros and ones as the values of the coordinates of the n - dimensional space. The translation of the polytopic prismahedron by the length of the edge along one of the coordinates of the space increases the value of this coordinate at all vertices of the polytopic prismahedron by one. The introduction of integer values of the coordinates of the vertices of a polytopic prismahedron allows analytically describe not only the translated polytopic prismahedron, but also polytopic prismahedrons located between the translated polytopic prismahedrons.

If we denote every translatable polytopic prismahedron by a point in n - dimensional space, then when translating this polytopic prismahedron by one in all coordinates of the n - dimensional space, then the number of these points is equal to $\sum_{i=0}^{n} C_n^i = 2^n$, where C_n^i is the number of combinations of the total number of coordinates n on the number of coordinates equal to unity . These points can be considered vertices of the n - cube, since the number of vertices in the n - cube is equal to 2^n . The number of edges in the n - cube is $n2^{n-1}$ (Zhizhin, 2014a, b, c). In this case, each edge in the n - cube corresponds to translation of the polytopic prismahedron by one from one of the coordinate n – dimension space.

The analytical description of the polytopic prismahedrons allows analytically to describe the partition of an n - dimensional space by the polytopic prismahedrons. In particular, establish exactly the intersection areas of the polytopic prismahedrons during their translation. Translated the polytopic prismahedrons and polytopic prismahedrons located between translated polytopic prismahedrons naturally form the fundamental regions of partition of n - dimensional space by polytopic prismahedrons. An analytic description of the polytopic prismehedrons also allows one analytically describe the fundamental regions of partitions of n - dimensional spaces by the polytopic prismahedrons.

Due to the nature of the polytopic prismahedrons, the fundamental regions of partition of the n -dimensional spaces by the polytopic prismahedrons can contact each other along geometric elements of different dimensions: edges, flat surfaces, three - dimensional and multidimensional figures.

The presented principles of partition of n - dimensional spaces by the polytopic prismahedrons, based on the study of specific geometric constructions, radically differ from the simplified speculative representations of the theory of the stereohedrons of Delone (Delone, 1961).

THE CONNECTION BETWEEN HIERARCHICAL AND TRANSLATIONAL FILLING OF SPACE

The discovery of quasicrystals was associated with the detection of the absence in the diffraction patterns of intermetallic compounds of translational symmetry (Shechtman et al., 1984). This led to large the number of works on the study of the hierarchical and chaotic filling of the plane and space with various geometric figures, considering these ways of filling space as an opposition to the well - known method of translational space filling (Penrose, 1974; Mackay, 2001; Lord, 1991; Lord et al., 2006; Audier & Duneau, 2000; Zhizhin, 2012; Zhizhin, 2014d). However, it was later discovered that the diffraction patterns of intermetallic compounds considered as projections of structures from spaces of higher dimension have a latent periodicity (Shevchenko, Zhizhin, Mackay, 2013a, b; Zhizhin, 2014 c). In the book of Zhizhin (2019) has already dealt with hierarchical filling of spaces. In this section, one show that there is a direct geometric relationship between the two types of filling the hierarchical and translational spaces. Let's return to the hierarchical filling of the two - dimensional plane with the correct pentagon. Following the method outlined in this chapter, continue the sides of the red pentagon to the intersection (Figure 12).

Figure 12. Hierarchical and translational filling of the plane with a regular pentagon

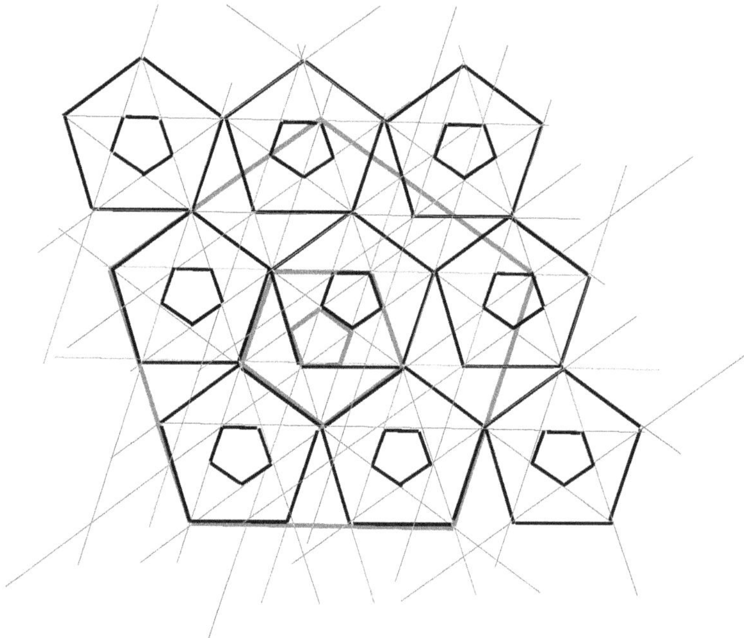

These intersections are the vertices of another regular red pentagon of a larger size. Continuing the sides of this pentagon to the intersection, can to get the vertices of an even larger pentagon. This is how the plane is filled with a regular pentagon. Now we continue the sides of the regular red pentagons into the whole foreseeable part of the plane (within the picture). One also draw part of the lines parallel to these sides. This leads to the formation of a series of intersection points that can be considered as vertices of other regular black pentagons (see Figure 12). It is found that these regular black pentagons translationally fill the plane. Thus, in Figure 12 there is also a hierarchical and translational filling of the plane by the same figures: regular pentagons.

Consider the second example.

Let there be a polytopic prismahedron the 5 - simplex - prism (Zhizhin, 2019). This prismahedron has dimension 5. Can to introduce the coordinate system of the five - dimensional space in one of the vertices of the 5 - simplex - prism.

Figure 13. Hierarchical and translational filling of the five - dimensional space by the 5 - simplex - prism

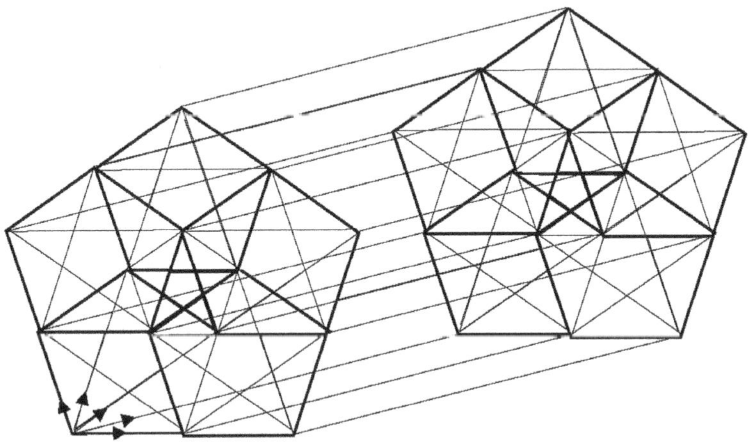

Assuming the length of all edges of the 5 - simplex - prism per unit, one will translate the 5 -simplex prism by the length of the edge along each coordinate direction within the figure. There are four the 5 - simplex - prisms, which together with the original 5 - simplex - prism form a 5 -simplex - prism of a larger size. This process can be continued by combining the translation of a 5 - simplex - prism with growth its size. This is hierarchically and translationally filling the five - dimensional space with a 5 - simplex - prism on the same system of parallel lines. Thus, there is no contradiction between hierarchical and translational filling of spaces.

MODERN CONCEPT OF DESCRIBING THE STRUCTURE OF MATTER

Can to recall briefly the main concepts and results of the classical theory of discrete systems. An example of the realization of constructing a space from congruent polytopes (18 Hilbert's problem) to be the crystalline structures that are widespread in nature (Hilbert, 1901). A crystalline structure is called correct if the groups in all those space motions, like a rigid whole that combine this structure with itself, is discrete and has a fundamental domain. A group G of motions is said to be discrete, if there exists a point A and a positive number r such, that every point different from A and equivalent to A is relative to G (that is, the one into which the point A passes by the motion from G) lies no closer to A then on distance r. The fundamental domain of the group G is a set of points of space such that:

1. All its points are not equivalent to each other with respect G.
2. Any point of the space is equivalent to an equivalent point of this domain with respect G.

Such groups G are called crystallographic. These classical definitions (Delone, Padurov & Alexandrov, 1934) implicitly assume that the considered spaces are infinite, although all crystal structures in nature have a boundary and a finite volume. It was first established by Fedorov (Fedorov, 1890) and somewhat later by Schönfles (1891) that there are 17 on the two-dimensional plane, and 230 crystallographic groups in the three - dimensional space. If one add to the conditions of discreteness and finiteness of the fundamental domain the requirement that in the group G there exist an n - dimensional subgroup T of parallel transfers(if G is n -dimensional), then the proof of the finiteness of the number of such groups follows directly from the Frobenius argument (Frobenius, 1911). Evidence for the existence of such n - dimensional subgroups T was obtained in a number of papers (Schönfles, 1891; Bieberbach, 1910, 1911, 1912). Delone (Delone, 1937) introduced the concept about system of the distribution of discrete matter, based on two conditions:

- The existence of a finite smallest radius of the ball inside which there are no points of the system (the concept of an empty ball).
- The existence of a radius of the ball inside which there is necessarily at least one point of the system.

The question arises as to what are the convex polytopes that play the role of fundamental domains in n - dimensional Euclidean space. Based on the introduced notion of a discrete system Delone proved (Delone, 1961) that a normal partition of a space (polytopes are adjacent to each other along entire (n - 1) - dimensional faces) for any n there are only a finite number of topologically different types of partition. In the definition of normal partitions is assumed that for every (n - 1) face of polytope P in the normal decomposition there is one and only one other polytope S having this same face. If we take the group G and repeat a point A, then we obtain a regular system of points. The Dirichlet domains of its points form some regular Dirichlet decomposition connected with the group G. It is normal. Delone and Sandakova proved (Delone & Sandakova, 1961) that the stereohedrons of these partitions for a fixed n can only be of a finite number of topologically different types. The edge grids of the stereohedrons do not stretch to each other. If one do not require the normality of partitions into convex fundamental domains, then there can be infinitely many such topologically distinct partitions (Zamorzayev, 1965).

It is clear absolutely that the concept of discrete systems introduced by Delone clearly contradicts the existence of a scaling process, which has been proved experimentally when studying phase transitions of the second kind in condensed substances. The concept of an empty ball does not allow a continuous reduction in scale. In this connection, it is necessary to harmonize the theory of describing the structure of matter with scaling processes. We consider an arbitrary n - dimensional system of points in n - dimensional space. The system is discrete, i.e. between points there is always some distance, which can only tend asymptotically to zero, not reaching an exact equality to zero. The system has in the neighborhood of each point the hierarchical distribution of points (Mackay, 2001). The latter leads to the existence of a periodic distribution of points in space.

The introduction of such a method of describing the structure of matter corresponds to the diffraction patterns of the nanostructures and to the scaling process discovered in recent years. It is essential that the lattice of points of this system allows for the separation in it of all Platonic solids, Bravais and Delone cells. Thus, this discrete system is universal. Based on this new conception of describing the structure of matter, it was possible to obtain and systematize (Zhizhin, 2019) the polytopic prismahedrons of the higher dimension, which, as shown in this Chapter, allow filling an n - dimensional space without gaps. The direct construction of the polytopic prismahedrons showed that the obligatory condition accepted in the theory of the Delone stereohedrons (Tarasov, 1997), does not hold, the separation of vertices of neighboring stereohedrons along the n - 1 plane.

The discrete Delone system has so far not been able to obtain a single example of a stereohedron higher dimensional (Galiulin, 2003). Until no enumeration of the stereohedrons of Delone with dimension 3 of normal partitions has been obtained. There are only isolated examples of these stereohedrons (Shtogrin, 1973; Peter, 1981). Moreover, the use of discrete Delone systems can lead to incorrect results. For example, in the work of Ryzhkov, Shushbaev (Ryzhkov & Shushbaev, 1981) is asserted on the basis of ideas about these systems, that with the help of a 4 - cross - polytope one can obtain the correct partition of the space 4*D*. However, a direct construction by the methods developed in this book can show that this is not so (Zhizhin, 2019).

The most complete variants of normal monohedral, dihedral and polyhedral partitions of the two - dimensional Euclidean plane was considered by Zhizhin (Zhizhin, 1993, 2010).

The scaling process is inextricably linked with the hierarchical filling of space. In nature it corresponds, in particular, to the formation of clusters. In the hierarchical filling of space a natural restriction of the geometric analysis follows the restriction of the volume of the space occupied by such a filling. This is observed in clusters, which always have a limited volume. It is important that hierarchical filling of space can be coordinated with the translational filling of space. In this case, a model of a uniformly expanding universe appears.

Hierarchical filling of space leads to the existence of two types of clusters. In one of these types the entire volume of the cluster it is occupied by the polytope boundary complex. This corresponds to the notion that the cluster surface occupies all of its volume. In the cluster of second type boundary complex does not occupy the entire volume of the polytope, it has internal points. This corresponds to the fact that there is free space inside the cluster.

Thus, a modern description of the structure of matter leads to a new content of Hilbert's 18th problem.

CONCLUSION

It is proved that clusters in the form of the polytopic prismahedrons have the necessary properties for partitioning the n - dimensional spaces of a face into a face, that is, they satisfy the conditions for solving the eighteenth Hilbert problem of the construction of n - dimensional spaces from congruent figures. Moreover, they create extended nanomaterials, in principle, of any size. General principles and an analytical method for constructing n - dimensional spaces with the help of the polytopic prismahedrons are developed. On the example of specific types of the polytopic prismahedrons (tetrahedral prism, triangular prismahedron), the possibility of such constructions is analytically proved. It was found that neighboring the polytopic prismahedrons in these constructions can have common geometric elements of any dimension less than n or do not have common elements. This refutes the speculatively accepted condition in the theory of the Delone stereohedrons that neighboring stereohedrons it is necessary to have common facets (that is, faces of dimension n - 1) in or not to have common elements. It is shown that the set of lines arising in the hierarchical filling of spaces allows the existence of translational symmetry.

Therefore, there are no fundamental contradictions between the hierarchical and translational filling of space, which the researchers so loved to speak of in the past. Based on the results of the research, a modern concept of describing the structure of matter is formulated, free from previously accepted conditions in the model of a discrete world. The new concept is consistent with a number of experimental data obtained in recent years on the structure of quasicrystals, phase transitions of the second kind, and so on. This leads to saturation of the eighteenth problem of Hilbert about the construction of spaces by geometric figures with a new content.

REFERENCES

Audier, M., & Duneau, M. (2000). Icosahedral quasiperiodic pacing of fibers parallel to fivefold and threefold axes. *Acta Crystallographica. Section A, Foundations of Crystallography*, *56*(1), 49–51. doi:10.1107/S0108767399010533 PMID:10874416

Bieberbach, L. (1910). Uber die Bewegungsruppen des *n* - dimensionalen Euklidischen Räumes mit einem endlichen fundamentalbereich. *Gött. Nachr.*, 75 - 84.

Bieberbach, L. (1911). Uber die Bewegungsruppen der Euklidischen Räume. I. *Mathematische Annalen*, *70*(3), 207–336. doi:10.1007/BF01564500

Bieberbach, L. (1912). Uber die Bewegungsruppen des Euklidischen Räume. II. *Mathematische Annalen*, *72*, 400–412. doi:10.1007/BF01456724

Delone, B. (1937). Geometry of positive quadratic forms. *Uspekhi Matematicheskhi Nauk*, *3*, 16–62.

Delone, B. (1961). Proof of the main theorem of the theory of stereohedrons. *Reports of the Academy of Sciences of the USSR*, *138*(6), 1270–1272.

Delone, B. (1969). To the eighteenth problem of Hilbert. In P. Aleksandrov (Ed.), *Problems of Hilbert* (pp. 200–203). Moscow: Science.

Delone, B., Padurov, N., & Aleksandrov, A. (1934). *Mathematical foundations of the structural analysis of crystals*. Leningrad: ONTI.

Delone, B., & Sandakova, N. (1961). Theory of stereohedrons. *Proceedings of the Mathematical Institute. V.A. Steklov, 64,* 28 - 51.

Fedorov, E. S. (1885). *The beginning of the doctrine of figures*. Saint Petersburg: Academy of Sciences.

Frobenius, G. (1911). Ueber die unzerlegbaren diskreten Bewegungsgruppen. *Sitzb. Preuss. Akad. Wiss., Phys.- Math. K, 1*, 654–655.

Galiulin, R. (2003). Systems B.N. Delone as the mathematical foundation of a discrete world. *J. Computed. Mat. and Math. Physics, 43*(6), 790–801.

Hilbert, D. (1901). Article. *Gesammelte Abhandlungen, 3*(1), 44 - 63, 213 - 237.

Lord, E. A. (1991). Quasicrystals and Penrose patterns. *Current Science, 61*, 313–319.

Lord, E. A. (2006). *New Geometry for New Materials*. Cambridge: Cambridge University Press.

Mackay, A. (2001). On Complexity. *Crystallography Reports, 46*(4), 524–526. doi:10.1134/1.1387117

Penrose, R. (1974). The role of aesthetics in pure and applied mathematical research. *Bulletin - Institute of Mathematics and Its Applications, 10*, 266–271.

Peter, E. (1986). *Geometric Crystallography*. Dordrecht: Springer.

Reinhardt, K. (1928). Zur Zerlegung der Euklidischen Raume in Kongruente Polytope. *Sitzb. Preuss. Akad. Wiss.*, 150 - 155.

Ryzhkov, S. S., & Shusbaev, S. Sh. (1981). The structure of the L - partition for the second perfect lattice. *Mat. Collection, 146*(2), 218–231.

Schönfless. (1891). *Kristallsysteme und Kristallstruktur*. Liepzig.

Shechtman, D., Blech, I., Gratias, D., & Cahn, J. W. (1984). Metallic phase with longerange orientational order and no translational symmetry. *Physical Review Letters, 53*(20), 1951–1953. doi:10.1103/PhysRevLett.53.1951

Shevchenko, V., Zhizhin, G., & Mackay, A. (2013a). On the structure of the quasicrystals in the high dimension space. In M. V. Diudea (Ed.), *Diamond and Related Nanostructures*. Dordrecht: Springer. doi:10.1007/978-94-007-6371-5_17

Shevchenko, V.Y., Zhizhin, G.V. & Mackay, A.L. (2013b). On the structure of quasicrystals in a space of higher dimension. *Izvestiya RAS. Chemical Series, 2*, 269 - 274.

Stogrin, M. I. (1973). Correct Dirichlet -Voronoi decompositions for the second triclinic group. *Trudy Mat. Institute of Steklov*, *123*, 128.

Tarasov, A.S. (1997). Complexity of convex stereohedrons. *Mathematic Notes, 61*(5), 797 - 800.

Zamorzayev, A. M. (1965). On abnormal regular partitions of Euclidean space. *Reports of the Academy of Sciences of the USSR*, *161*(1), 30–31.

Zhizhin, G. V. (1993). *Parquetages of equal triangles, adjoining on whole sides*. St. Petersburg: Polytechnic.

Zhizhin, G. V. (2010). *Geometrical bases of the dissipative structures*. St. Petersburg: Polytechnic.

Zhizhin, G. V. (2012, October). *Hierarchical filling of spaces with polytopes.* Paper presented at "St. Petersburg Scientific Forum: Science and Human Progress". 7th St.- Petersburg meeting of Nobel Prize laureates, St. Petersburg, Russia.

Zhizhin, G. V. (2014a). The fractal nature of disproportionate phases. In *10th all - Russian Scientific School "Mathematical research in the natural sciences"*. Apatity: Geological institute KSC RAS.

Zhizhin, G. V. (2014b). Incommensurable and fluctuating structures in the terrestrial space. *Biosphere*, *3*, 211–221.

Zhizhin, G. V. (2014c). *World - 4D*. St. Petersburg: Polytechnic Service.

Zhizhin, G. V. (2014d). On the higher dimension in nature. *Biosphere*, *6*(4), 313–318.

Zhizhin, G. V. (2015, November). *Polytopic prismahedrons - fundamental regions of the n -dimension nanostructures.* Paper presented at The International conference "Nanoscience in Chemistry, Physics, Biology and Mathematics", Cluj-Napoca, Romania.

Zhizhin, G. V. (2018). *Chemical Compound Structures and Higher Dimension of Molecules: Emerging Research and Opportunities*. Hershey, PA: IGI Global. doi:10.4018/978-1-5225-4108-0

Zhizhin, G. V. (2019). *The Geometry of Higher - Dimensional Polytopes*. Hershey, PA: IGI Global. doi:10.4018/978-1-5225-6968-8

Zhizhin, G. V., & Diudea, M. V. (2016). Space of Nanoworld. In M. V. Putz & M. C. Mirica (Eds.), *Sustainable Nanosystems, Development, Properties, and Applications* (pp. 214–236). Hershey, PA: IGI Global.

Zhizhin, G. V., Khalaj, Z., & Diudea, M. V. (2016). Geometrical and topology dimensions of the diamond. In A. R. Ashrafi & M. V. Diudea (Eds.), *Distance, symmetry and topology in carbon nanomaterials* (pp. 167–188). New York: Springer. doi:10.1007/978-3-319-31584-3_12

KEY TERMS AND DEFINITIONS

5-Simplex-Prism: The product of a 4-simplex by a segment.

Congruent Polyhedrons: Polyhedrons that are compatible with motion.

New Paradigm of Discrete *n*-Dimension World: The elementary cells of the translational filling of the *n*-dimensional space are the polytopic prismahedrons-the stereohedrons, of which Delone spoke, but he did not give a single concrete example of stereohedron. Polytopic prismahedrons, filling the *n*-dimensional space, as shown by direct construction, can have common elements in the entire range of dimensions up to the dimension of the facets, or do not have any common elements. In addition to the translational filling of the *n*-dimensional space, there is a hierarchical filling of the space, which is inextricably linked with the scaling process, that is, discrete scale change of the figure. Translational filling of space can be combined with hierarchical filling of space. In this case, in principle, in each point of space there is an asymptotic decrease or increase in the scale of the figure (as an expansion of the Universe from each of its points). Delone's provisions on an "empty" ball, the finite minimum and maximum distances between the points of a discrete system, are not used. Thus, Hilbert's problem acquires a completely new content.

Polytopic Prismahedron: The product of polytopes.

Tetrahedral Prism: The product of a tetrahedron by a segment.

Triangular Prismahedron: The product of a triangle by a triangle.

Related Readings

To continue IGI Global's long-standing tradition of advancing innovation through emerging research, please find below a compiled list of recommended IGI Global book chapters and journal articles in the areas of nanotechnology, nanomaterials, and materials science. These related readings will provide additional information and guidance to further enrich your knowledge and assist you with your own research.

Agboola, O., Popoola, P. A., Sadiku, R., Sanni, S. E., Babatunde, D. E., Abatan, O. G., ... Fasiku, O. V. (2019). Polymer Nanocomposites for Advanced Automobile Applications. In N. Ramdani (Ed.), *Polymer Nanocomposites for Advanced Engineering and Military Applications* (pp. 96–130). Hershey, PA: IGI Global. doi:10.4018/978-1-5225-7838-3.ch004

Ashrafiyan, O., Saremi, M., Pakseresht, A., & Ghasali, E. (2018). Oxidation-Protective Coatings for Carbon-Carbon Composites. In A. Pakseresht (Ed.), *Production, Properties, and Applications of High Temperature Coatings* (pp. 429–446). Hershey, PA: IGI Global. doi:10.4018/978-1-5225-4194-3.ch016

Aslam, S., & Ahmad, M. (2019). Nanoparticles for Degradation of Organic Pollutants. In R. Nazir (Ed.), *Nanotechnology Applications in Environmental Engineering* (pp. 241–267). Hershey, PA: IGI Global. doi:10.4018/978-1-5225-5745-6.ch011

Azadi, M. (2018). Deposition of TiN, TiC, and DLC Coatings by PACVD. In A. Pakseresht (Ed.), *Production, Properties, and Applications of High Temperature Coatings* (pp. 381–402). Hershey, PA: IGI Global. doi:10.4018/978-1-5225-4194-3.ch014

Babatunde, D. E., Denwigwe, I. H., Babatunde, O. M., Agboola, O., & Akinsipe, G. D. (2019). Relevance of Chemically Functionalized Nano-Fillers and Modified Nanocomposite in Energy Systems. In N. Ramdani (Ed.), *Polymer Nanocomposites for Advanced Engineering and Military Applications* (pp. 10–65). Hershey, PA: IGI Global. doi:10.4018/978-1-5225-7838-3.ch002

Bhattacharyya, S., & Majumdar, P. S. (2018). An Overview on Peak Shape Method for Thermoluminescence Glow Curve Analysis: Application on Tremolite and Actinolite Glow Peaks. In R. Tiwari, V. Dubey, & S. Dhoble (Eds.), *Emerging Synthesis Techniques for Luminescent Materials* (pp. 26–52). Hershey, PA: IGI Global. doi:10.4018/978-1-5225-5170-6.ch002

Bolleddu, V., Racherla, V., & Bandyopadhyay, P. P. (2018). Microstructural and Tribological Characteristics of Air Plasma Sprayed Alumina-Titania Coatings. In A. Pakseresht (Ed.), *Production, Properties, and Applications of High Temperature Coatings* (pp. 268–298). Hershey, PA: IGI Global. doi:10.4018/978-1-5225-4194-3.ch011

Borah, R., & Kumar, A. (2019). Enhanced Cellular Activity on Conducting Polymer: Chitosan Nanocomposite Scaffolds for Tissue Engineering Application. In N. Ramdani (Ed.), *Polymer Nanocomposites for Advanced Engineering and Military Applications* (pp. 150–189). Hershey, PA: IGI Global. doi:10.4018/978-1-5225-7838-3.ch006

Chawla, R., Singhal, P., & Garg, A. K. (2018). Investigation of Structural, Morphological and Electro-Optical Behavior of GO/rGO. *International Journal of Surface Engineering and Interdisciplinary Materials Science*, *6*(1), 32–43. doi:10.4018/IJSEIMS.2018010103

Chopra, S., Sreya, S., Babhulkar, R. V., Halde, S. P., Deshmukh, K. A., & Peshwe, D. R. (2019). Cryogenic Treatment of Polymer/MWCNT Nano-Composites for Mechanical and Tribological Applications. In N. Ramdani (Ed.), *Nanotechnology in Aerospace and Structural Mechanics* (pp. 103–161). Hershey, PA: IGI Global. doi:10.4018/978-1-5225-7921-2.ch004

Darwish, M. (2019). Lightweight Nanocomposites Polymers for Shielding Application. In N. Ramdani (Ed.), *Nanotechnology in Aerospace and Structural Mechanics* (pp. 206–233). Hershey, PA: IGI Global. doi:10.4018/978-1-5225-7921-2.ch006

Di Girolamo, G. (2018). Current Challenges and Future Perspectives in the Field of Thermal Barrier Coatings. In A. Pakseresht (Ed.), *Production, Properties, and Applications of High Temperature Coatings* (pp. 25–59). Hershey, PA: IGI Global. doi:10.4018/978-1-5225-4194-3.ch002

Dubey, V., Dubey, N., Shrivastava, R., KaurSaluja, J., Som, S., Dutta, S., & Chandrakar, R. (2018). Upconversion Luminescence Behaviour of Er3+/Yb3+ Doped MY2O4 (M=Ba, Ca, Sr) Phosphors: Upconversion Study of MY2O4:Er3+/Yb3+. In R. Tiwari, V. Dubey, & S. Dhoble (Eds.), Emerging Synthesis Techniques for Luminescent Materials (pp. 117-148). Hershey, PA: IGI Global. doi:10.4018/978-1-5225-5170-6.ch005

Dutta, M. M. (2019). Characterization of Nanomaterials. In R. Nazir (Ed.), *Nanotechnology Applications in Environmental Engineering* (pp. 44–80). Hershey, PA: IGI Global. doi:10.4018/978-1-5225-5745-6.ch003

Elehinafe, F. B., & Ayeni, A. O. (2019). Processing of Polymer-Based Nanocomposites in Advanced Engineering and Military Application. In N. Ramdani (Ed.), *Polymer Nanocomposites for Advanced Engineering and Military Applications* (pp. 1–9). Hershey, PA: IGI Global. doi:10.4018/978-1-5225-7838-3.ch001

Figueira, R. B. (2018). Hybrid Sol-Gel Coatings: Erosion-Corrosion Protection. In A. Pakseresht (Ed.), *Production, Properties, and Applications of High Temperature Coatings* (pp. 334–380). Hershey, PA: IGI Global. doi:10.4018/978-1-5225-4194-3.ch013

Gebai, S. S., Hallal, A. M., & Hammoud, M. S. (2018). *Mechanical Properties of Natural Fiber Reinforced Polymers: Emerging Research and Opportunities* (pp. 1–228). Hershey, PA: IGI Global. doi:10.4018/978-1-5225-4837-9

Hassanzadeh, M., Saremi, M., Valefi, Z., & Pakseresht, A. H. (2018). Investigation on Improved-Durability Thermal Barrier Coatings: An Overview of Nanostructured, Multilayered, and Self-Healing TBCs. In A. Pakseresht (Ed.), *Production, Properties, and Applications of High Temperature Coatings* (pp. 60–78). Hershey, PA: IGI Global. doi:10.4018/978-1-5225-4194-3.ch003

Henao, J., & Sotelo, O. (2018). Surface Engineering at High Temperature: Thermal Cycling and Corrosion Resistance. In A. Pakseresht (Ed.), *Production, Properties, and Applications of High Temperature Coatings* (pp. 131–159). Hershey, PA: IGI Global. doi:10.4018/978-1-5225-4194-3.ch006

Horst, D. J., & Junior, P. P. (2019). 3D-Printed Conductive Filaments Based on Carbon Nanostructures Embedded in a Polymer Matrix: A Review. *International Journal of Applied Nanotechnology Research*, *4*(1), 26–40. doi:10.4018/IJANR.2019010103

Irzhak, T. F., & Irzhak, V. I. (2019). Synthesis of Epoxy Nanocomposites. In N. Ramdani (Ed.), *Nanotechnology in Aerospace and Structural Mechanics* (pp. 34–79). Hershey, PA: IGI Global. doi:10.4018/978-1-5225-7921-2.ch002

Ismar, E., & Sarac, A. (2019). Carbon Nanomaterials: Carbon Nanotubes, Graphene, and Carbon Nanofibers. In N. Ramdani (Ed.), *Nanotechnology in Aerospace and Structural Mechanics* (pp. 1–33). Hershey, PA: IGI Global. doi:10.4018/978-1-5225-7921-2.ch001

Jucha, S. O., & Moskal, G. J. (2018). Thermal Barrier Coatings on the Base of Samarium Zirconates. In A. Pakseresht (Ed.), *Production, Properties, and Applications of High Temperature Coatings* (pp. 79–106). Hershey, PA: IGI Global. doi:10.4018/978-1-5225-4194-3.ch004

Katrenipadu, S. P., & Gurugubelli, S. N. (2018). Regression Modeling and Experimental Investigations on Ageing Behavior of Nano-Fly Ash Reinforced Al-10wt%Mg Alloy Matrix Composites. *International Journal of Surface Engineering and Interdisciplinary Materials Science*, *6*(2), 36–49. doi:10.4018/IJSEIMS.2018070103

Kausar, A. (2019). Aeronautical Impact of Epoxy/Carbon Nanotube Nanocomposite. In N. Ramdani (Ed.), *Nanotechnology in Aerospace and Structural Mechanics* (pp. 80–102). Hershey, PA: IGI Global. doi:10.4018/978-1-5225-7921-2.ch003

Kausar, A. (2019). Scope of Polymer/Graphene Nanocomposite in Defense Relevance: Defense Application of Polymer/Graphene. In N. Ramdani (Ed.), *Polymer Nanocomposites for Advanced Engineering and Military Applications* (pp. 296–315). Hershey, PA: IGI Global. doi:10.4018/978-1-5225-7838-3.ch010

Keshavamurthy, R., Naveena, B. E., & Sekhar, N. (2018). Thermal Spray Coatings for Erosion-Corrosion Protection. In A. Pakseresht (Ed.), *Production, Properties, and Applications of High Temperature Coatings* (pp. 246–267). Hershey, PA: IGI Global. doi:10.4018/978-1-5225-4194-3.ch010

Khelfi, A. (2019). Toxicocinetic and Mechanisms of Action of Nanoparticles. In R. Nazir (Ed.), *Nanotechnology Applications in Environmental Engineering* (pp. 344–368). Hershey, PA: IGI Global. doi:10.4018/978-1-5225-5745-6.ch014

Kosti, S. (2019). Nanomaterials and Nanocomposites Thermal and Mechanical Properties Modelling. In N. Ramdani (Ed.), *Nanotechnology in Aerospace and Structural Mechanics* (pp. 234–256). Hershey, PA: IGI Global. doi:10.4018/978-1-5225-7921-2.ch007

Kumar, I. (2019). Simulation and Modeling of Nanotechnology Aircraft Using MATLAB. In N. Ramdani (Ed.), *Nanotechnology in Aerospace and Structural Mechanics* (pp. 257–290). Hershey, PA: IGI Global. doi:10.4018/978-1-5225-7921-2.ch008

Kumar, S., Krishnan, S., Samal, S. K., Mohanty, S., & Nayak, S. K. (2019). Polymer Nanocomposites Coating for Anticorrosion Application. In N. Ramdani (Ed.), *Polymer Nanocomposites for Advanced Engineering and Military Applications* (pp. 254–294). Hershey, PA: IGI Global. doi:10.4018/978-1-5225-7838-3.ch009

Kuz'min, V. E., Ognichenko, L. N., Sizochenko, N., Chapkin, V. A., Stelmakh, S. I., Shyrykalova, A. O., & Leszczynski, J. (2019). Combining Features of Metal Oxide Nanoparticles: Nano-QSAR for Cytotoxicity. *International Journal of Quantitative Structure-Property Relationships*, *4*(1), 28–40. doi:10.4018/IJQSPR.2019010103

Lahariya, V., & Ramrakhiani, M. (2018). Electroluminescence Principle and Novel Electroluminescent Materials. In R. Tiwari, V. Dubey, & S. Dhoble (Eds.), *Emerging Synthesis Techniques for Luminescent Materials* (pp. 415–438). Hershey, PA: IGI Global. doi:10.4018/978-1-5225-5170-6.ch011

Lateef, A., & Nazir, R. (2019). Application of Magnetic Nanomaterials for Water Treatment. In R. Nazir (Ed.), *Nanotechnology Applications in Environmental Engineering* (pp. 201–219). Hershey, PA: IGI Global. doi:10.4018/978-1-5225-5745-6.ch009

Majumdar, D. (2019). Polyaniline as Proficient Electrode Material for Supercapacitor Applications: PANI Nanocomposites for Supercapacitor Applications. In N. Ramdani (Ed.), *Polymer Nanocomposites for Advanced Engineering and Military Applications* (pp. 190–219). Hershey, PA: IGI Global. doi:10.4018/978-1-5225-7838-3.ch007

Majumdar, D. (2019). Polyaniline Nanocomposites: Innovative Materials for Supercapacitor Applications – PANI Nanocomposites for Supercapacitor Applications. In N. Ramdani (Ed.), *Polymer Nanocomposites for Advanced Engineering and Military Applications* (pp. 220–253). Hershey, PA: IGI Global. doi:10.4018/978-1-5225-7838-3.ch008

Mubarak, S. M. (2019). Nanotechnology for Air Remediation. In R. Nazir (Ed.), *Nanotechnology Applications in Environmental Engineering* (pp. 121–142). Hershey, PA: IGI Global. doi:10.4018/978-1-5225-5745-6.ch006

Naik, R., Revathi, V., Prashantha, S. C., & Nagabhushana, H. (2018). Synthesis of Magnesium Based Nanophosphors and Nanocomposites by Different Techniques: Silicates and Ferrites. In R. Tiwari, V. Dubey, & S. Dhoble (Eds.), *Emerging Synthesis Techniques for Luminescent Materials* (pp. 251–276). Hershey, PA: IGI Global. doi:10.4018/978-1-5225-5170-6.ch007

Nasreen, S., & Rafique, U. (2019). Nanotechnology for Water Environmental Application: Functionalized Silica Hybrids as Nano-Sorbents. In R. Nazir (Ed.), *Nanotechnology Applications in Environmental Engineering* (pp. 171–200). Hershey, PA: IGI Global. doi:10.4018/978-1-5225-5745-6.ch008

Nazir, R. (2019). Introduction to Environmental Nanotechnology: E-Nano. In R. Nazir (Ed.), *Nanotechnology Applications in Environmental Engineering* (pp. 1–27). Hershey, PA: IGI Global. doi:10.4018/978-1-5225-5745-6.ch001

Noman, D., & Owens, G. (2019). Wastewater Treatment and Role of Green Synthesized Metal Oxide Nanocomposites. In R. Nazir (Ed.), *Nanotechnology Applications in Environmental Engineering* (pp. 268–307). Hershey, PA: IGI Global. doi:10.4018/978-1-5225-5745-6.ch012

Nosheen, S. (2019). Nanomembrane Applications in Environmental Engineering. In R. Nazir (Ed.), *Nanotechnology Applications in Environmental Engineering* (pp. 103–120). Hershey, PA: IGI Global. doi:10.4018/978-1-5225-5745-6.ch005

Odetola, P. I., Popoola, P. A., & Oladijo, P. (2018). Thin Coating Deposition by Magnetron Sputtering. In A. Pakseresht (Ed.), *Production, Properties, and Applications of High Temperature Coatings* (pp. 403–428). Hershey, PA: IGI Global. doi:10.4018/978-1-5225-4194-3.ch015

Pandey, A., Tiwari, S. P., Dutta, V., & Kumar, V. (2018). Photon Upconversion in Lanthanide-Activated Inorganic Luminescent Materials. In R. Tiwari, V. Dubey, & S. Dhoble (Eds.), *Emerging Synthesis Techniques for Luminescent Materials* (pp. 86–116). Hershey, PA: IGI Global. doi:10.4018/978-1-5225-5170-6.ch004

Pandey, J. (2019). Recent Progresses in Membranes for Proton Exchange Membrane Fuel Cell (PEMFC) for Clean and Environmentally Friendly Applications. In R. Nazir (Ed.), *Nanotechnology Applications in Environmental Engineering* (pp. 308–343). Hershey, PA: IGI Global. doi:10.4018/978-1-5225-5745-6.ch013

Pesetskii, S. S., Bogdanovich, S. P., & Aderikha, V. N. (2019). Polymer/Clay Nanocomposites Produced by Dispersing Layered Silicates in Thermoplastic Melts. In N. Ramdani (Ed.), *Polymer Nanocomposites for Advanced Engineering and Military Applications* (pp. 66–94). Hershey, PA: IGI Global. doi:10.4018/978-1-5225-7838-3.ch003

Purohit, P., & Vagge, S. T. (2018). Oxidation Behavior of LTA/YSZ Intermixed Layer Coating. In A. Pakseresht (Ed.), *Production, Properties, and Applications of High Temperature Coatings* (pp. 107–130). Hershey, PA: IGI Global. doi:10.4018/978-1-5225-4194-3.ch005

Sadegh, H., Ali, G. A., Nia, H. J., & Mahmoodi, Z. (2019). Nanomaterial Surface Modifications for Enhancement of the Pollutant Adsorption From Wastewater: Adsorption of Nanomaterials. In R. Nazir (Ed.), *Nanotechnology Applications in Environmental Engineering* (pp. 143–170). Hershey, PA: IGI Global. doi:10.4018/978-1-5225-5745-6.ch007

Sadiku, E. R., Agboola, O., Mochane, M. J., Fasiku, V. O., Owonubi, S. J., Ibrahim, I. D., ... Ndamase, A. S. (2019). The Use of Polymer Nanocomposites in the Aerospace and the Military/Defence Industries. In N. Ramdani (Ed.), *Polymer Nanocomposites for Advanced Engineering and Military Applications* (pp. 316–349). Hershey, PA: IGI Global. doi:10.4018/978-1-5225-7838-3.ch011

Sadri, E., & Ashrafizadeh, F. (2018). High Temperature Nanocomposite Coatings by Plasma Spraying for Friction and Wear Applications. In A. Pakseresht (Ed.), *Production, Properties, and Applications of High Temperature Coatings* (pp. 216–245). Hershey, PA: IGI Global. doi:10.4018/978-1-5225-4194-3.ch009

Saeedi, F., Montazeri, A., Bahari, Y., Pishvaee, M., & Ranjbar, M. (2018). Synthesis and Characterization of Chitosan-Poly Vinyl Alcohol-Graphene Oxide Nanocomposites. *International Journal of Chemoinformatics and Chemical Engineering*, *7*(1), 1–12. doi:10.4018/IJCCE.2018010101

Shahien, M. (2018). Reactive Plasma Spray: A Method for Nitride Coatings Deposition in Thermal Spray. In A. Pakseresht (Ed.), *Production, Properties, and Applications of High Temperature Coatings* (pp. 299–332). Hershey, PA: IGI Global. doi:10.4018/978-1-5225-4194-3.ch012

Shahzadi, P. (2019). Properties of Nanomaterials and Environment. In R. Nazir (Ed.), *Nanotechnology Applications in Environmental Engineering* (pp. 28–43). Hershey, PA: IGI Global. doi:10.4018/978-1-5225-5745-6.ch002

Shamim, A., Ahmad, S., Khitab, A., Anwar, W., Khushnood, R. A., & Usman, M. (2018). Applications of Nano Technology in Civil Engineering: A Review. *International Journal of Strategic Engineering*, *1*(1), 48–64. doi:10.4018/IJoSE.2018010104

Shikuku, V. O., & Sylvain, T. (2019). Application of Geopolymer Composites in Wastewater Treatment: Trends, Opportunities, and Challenges. In N. Ramdani (Ed.), *Polymer Nanocomposites for Advanced Engineering and Military Applications* (pp. 131–149). Hershey, PA: IGI Global. doi:10.4018/978-1-5225-7838-3.ch005

Singh, J., Bhadane, M. S., Dubey, V., Dhole, S. D., Manam, J., & Dahiwale, S. S. (2018). Phosphors for Various Dosimetry Applications Derived by Different Synthesis Routes. In R. Tiwari, V. Dubey, & S. Dhoble (Eds.), *Emerging Synthesis Techniques for Luminescent Materials* (pp. 53–84). Hershey, PA: IGI Global. doi:10.4018/978-1-5225-5170-6.ch003

Som, S., Dutta, S., Kumar, V., & Swart, H. C. (2018). Swift Heavy Ion Synthesis and Modifications of Nanophosphors for Dosimetric Application: Effect of Swift Heavy Ion Irradiation. In R. Tiwari, V. Dubey, & S. Dhoble (Eds.), *Emerging Synthesis Techniques for Luminescent Materials* (pp. 1–25). Hershey, PA: IGI Global. doi:10.4018/978-1-5225-5170-6.ch001

Suresh, S. S., George, K., Mohanty, S., & Nayak, S. K. (2019). Architect of Polymer Nanocomposites for Aerospace Applications. In N. Ramdani (Ed.), *Nanotechnology in Aerospace and Structural Mechanics* (pp. 163–205). Hershey, PA: IGI Global. doi:10.4018/978-1-5225-7921-2.ch005

Tahir, L. (2019). Green Approaches to Environmental Sustainability. In R. Nazir (Ed.), *Nanotechnology Applications in Environmental Engineering* (pp. 81–101). Hershey, PA: IGI Global. doi:10.4018/978-1-5225-5745-6.ch004

Tripathi, S., Mishra, M. K., Jain, V. K., Tiwari, R., & Dubey, N. (2018). Thermo, Photo and Mechanoluminescence Studies of Eu3+ Doped Y4Al2O9 Phosphors: TL, PL, and ML Analysis of YAM: Eu3. In R. Tiwari, V. Dubey, & S. Dhoble (Eds.), *Emerging Synthesis Techniques for Luminescent Materials* (pp. 389–414). Hershey, PA: IGI Global. doi:10.4018/978-1-5225-5170-6.ch010

Ugemuge, N. S., Gayner, C., Natrajan, V., & Dhoble, S. J. (2018). Perspectives of Hydrothermal Synthesis of Fluorides for Luminescence Applications: Fluorides Phosphors for Luminescence. In R. Tiwari, V. Dubey, & S. Dhoble (Eds.), *Emerging Synthesis Techniques for Luminescent Materials* (pp. 277–303). Hershey, PA: IGI Global. doi:10.4018/978-1-5225-5170-6.ch008

Umm-i-Kalsoom Ali, N., Begum, N., Hussnain, A., & Ahmad, R. (2018). Hardness of Duplex Treated Stainless Steel: Influence of the Architecture and Composition of Films. In A. Pakseresht (Ed.), Production, Properties, and Applications of High Temperature Coatings (pp. 447-466). Hershey, PA: IGI Global. doi:10.4018/978-1-5225-4194-3.ch017

V., S., R., M., S., K., & S., S. K. (2018). White Light Emitting Phosphors for Solid State Lighting. In R. Tiwari, V. Dubey, & S. Dhoble (Eds.), *Emerging Synthesis Techniques for Luminescent Materials* (pp. 150-250). Hershey, PA: IGI Global. doi:10.4018/978-1-5225-5170-6.ch006

Vaidyanathan, S., & Tagare, J. (2018). Phenanthroimidazole-Based Organic Fluorophores for Organic Light-Emitting Diodes: An Overview. In R. Tiwari, V. Dubey, & S. Dhoble (Eds.), *Emerging Synthesis Techniques for Luminescent Materials* (pp. 305–388). Hershey, PA: IGI Global. doi:10.4018/978-1-5225-5170-6.ch009

Van Miert, S., Creylman, J., & Verheyen, G. R. (2019). Mining a Nanoparticle Dataset, Compiled Within the MODENA-COST Action. *International Journal of Quantitative Structure-Property Relationships*, *4*(1), 1–17. doi:10.4018/IJQSPR.2019010101

Verma, R., Suri, N. M., & Kant, S. (2018). Slurry Sprayed Mullite Coatings and Their Corrosion Performances. In A. Pakseresht (Ed.), *Production, Properties, and Applications of High Temperature Coatings* (pp. 187–214). Hershey, PA: IGI Global. doi:10.4018/978-1-5225-4194-3.ch008

Xie, W. (2018). Thermal Spray Methods and Splat Formation. In A. Pakseresht (Ed.), *Production, Properties, and Applications of High Temperature Coatings* (pp. 1–24). Hershey, PA: IGI Global. doi:10.4018/978-1-5225-4194-3.ch001

Yaseen, T., & Yaseen, A. (2019). Wastewater Treatment in Removal of Heavy Metals: Nanotechnology Applications in Environmental Engineering. In R. Nazir (Ed.), *Nanotechnology Applications in Environmental Engineering* (pp. 220–240). Hershey, PA: IGI Global. doi:10.4018/978-1-5225-5745-6.ch010

Zhizhin, G. V. (2019). Higher Dimensions of Clusters of Intermetallic Compounds: Dimensions of Metallic Nanoclusters. *International Journal of Applied Nanotechnology Research*, *4*(1), 8–25. doi:10.4018/IJANR.2019010102

Zhu, S., & Wang, F. (2018). Nanocrystalline, Enamel and Composite Coatings for Superalloys. In A. Pakseresht (Ed.), *Production, Properties, and Applications of High Temperature Coatings* (pp. 160–186). Hershey, PA: IGI Global. doi:10.4018/978-1-5225-4194-3.ch007

Index

Q

T

U

www.ingramcontent.com/pod-product-compliance
Lightning Source LLC
LaVergne TN
LVHW081331021026
849161LV00008BA/1929